WORKS ISSUED BY
THE HAKLUYT SOCIETY

SEARCHING FOR FRANKLIN:
THE LAND ARCTIC SEARCHING EXPEDITION

THIRD SERIES
NO. 1

Telephone: 01986 788359 Fax: 01986 788181 E-mail: haksoc@paston.co.uk

Postal address only:
Hakluyt Society, c/o The Map Library, The British Library, 96 Euston Road,
London NW1 2DB

Website: www.hakluyt.com

Registered Charity No. 313168 VAT No. GB 233 4481 77

INTERNATIONAL REPRESENTATIVES

Australia: Ms Maura O'Connor, Curator of Maps, National Library of Australia, Canberra, ACT 2601

Canada: Dr Joyce Lorimer, Department of History, Wilfred Laurier University, Waterloo, Ontario, N2L 3C5

Germany: Thomas Tack, Ziegelbergstr. 21, D-63739 Aschaffenburg

Japan: Dr Derek Massarella, Faculty of Economics, Chuo University, Higashinakano 742–1, Hachioji-shi, Tokyo 192–03

New Zealand: J. E. Traue, Department of Librarianship, Victoria University of Wellington, PO Box 600, Wellington

Portugal: Dr Manuel Ramos, Av. Elias Garcia 187, 3Dt, 1050 Lisbon

Russia: Professor Alexei V. Postnikov, Institute of the History of Science and Technology, Russian Academy of Sciences, 1/5 Staropanskii per., Moscow 103012

South Africa: Dr F. R. Bradlow, 28/29 Porter House, Belmont Road, Rondebosch, Cape 7700

USA: Dr Norman Fiering, The John Carter Brown Library, Box 1894, Providence, Rhode Island 02912 *and* Professor Norman Thrower, Department of Geography, UCLA, 405 Hilgard Avenue, Los Angeles, California 90024–1698

Western Europe: Paul Putz, 54 rue Albert 1, 1117 Luxembourg, Luxembourg

PLATE I. One of the surviving chimneys of Fort Reliance, built by Alexander McLeod for Captain George Back's expedition, 1833 and refurbished by James Lockhart for the Anderson/Stewart expedition, 1855
(Photo taken 1981; W. Barr collection)

Searching for Franklin: The Land Arctic Searching Expedition

James Anderson's and James Stewart's Expedition via the Back River

1855

Edited by

WILLIAM BARR

THE HAKLUYT SOCIETY
LONDON
1999

Published by the Hakluyt Society
c/o The Map Library
British Library, 96 Euston Road,
London NW1 2DB

ISBN 0 904180 61 1
ISSN 0072 9396

British Library Cataloguing-in-Publication Data
A catalogue record for this book is
available from the British Library

Typeset by Waveney Typesetters, Wymondham, Norfolk
Printed in Great Britain at
the University Press, Cambridge

CONTENTS

List of Plates ix
Preface xi
Abbreviations xv

I Genesis of the expedition
Introduction 1
Documents 18
II The Company accepts the challenge
Introduction 25
Documents 28
III The wheels in motion
Introduction 37
Documents 46
IV Getting to the rendez-vous
Introduction 67
Documents 72
V Journey to the Arctic Ocean
Introduction 105
Documents 107
VI James Lockhart's support expedition to Fort Reliance
Introduction 149
Documents 151
VII The expedition disperses
Introduction 161
Documents 161
VIII Reports of the expedition's achievements
Introduction 166
Documents 167
IX Contemporary reactions to the results of the expedition
Introduction 196
Documents 198
X Complaints against Stewart
Introduction 211
Documents 213
XI Accounting, awards and relics
Introduction 219
Documents 220

LIST OF PLATES

PLATE I. One of the surviving chimneys of Fort Reliance, built by Alexander McLeod for Captain George Back's expedition, 1833 and refurbished by James Lockhart for the Anderson/Stewart expedition, 1855 (Photo taken 1981; W. Barr collection) *Frontispiece*

PLATE II. Sir George Simpson, *c.* 1856–60. Copy of a daguerrotype (Hudson's Bay Company Archives) 34

PLATE III. James Anderson, wearing the Arctic Medal he received for his leadership of the Land Arctic Expedition (C. S. Mackinnon collection) 39

PLATE IV. Central Rupert's Land, showing how the various components of the expedition converged on Fort Resolution 68

PLATE V. Anderson's draft of his map of the southern portion of the 'Mountain Portage'. South is at the top of the map (Hudson's Bay Company Archives) 104

PLATE VI. Great Slave Lake and area 110

PLATE VII. The 'Mountain Portage', the expedition's route north from Great Slave Lake to Aylmer Lake 114

PLATE VIII. Aylmer Lake and the headwaters of the Back River 117

PLATE IX. General map of the expedition's route down the Great Fish (Back) River 120

PLATE X. The lakes on the middle section of the Great Fish (Back) River 127

PLATE XI. Chantrey Inlet, showing the farthest point (Maconochie Island) reached by the expedition 132

PLATE XII. The Lockhart River, Pike's Portage, Artillery Lake and the site of Fort Reliance 146

PLATE XIII. James Lockhart (Hudson's Bay Company Archives) 150

PREFACE

On 12 April 1902, while sledging north across the tundra from the Thelon River to the arctic coast with some Inuit companions, the British explorer David Hanbury crossed the Back River at the head of Pelly Lake. Here he met an Inuk named Itkelek with his family.

> Questioned about Back's journey,[1] he said that his father had told him that, long ago, two large boats [Back had only one] and about twenty white men [Back had about half that number] had come down Back's River and returned again the same summer. He was a small boy at the time and did not remember it. Back descended and ascended the river in 1834. This man could not be more than forty years of age, and his statement could not possibly be correct. I asked him if he was sure it was not his father who was the small boy when Back passed, but he adhered to his original statement.[2]

In fact it was Hanbury who was misinformed, not Itkelek. His father and family had encountered not Back's party in 1834, but the expedition led by James Anderson and James Stewart, travelling down and up the Back River in two canoes in 1855. That an explorer who should have been knowledgeable about prior expeditions to the area was not aware of the Anderson/Stewart expedition provides some measure of how little it was known in the early years of this century. Despite the fact that Anderson's journal has twice been published since then (see below) the achievements of Anderson, Stewart and their men are still little known. While they may not have made a very substantial contribution to solving the mystery of the disappearance of Sir John Franklin's expedition (the primary objective of the expedition) those achievements deserve to be better known.

In the summer of 1845 Sir John Franklin's ships HMS *Erebus* and *Terror* had sailed for the Arctic in search of the Northwest Passage – and disappeared. Over the next nine years some fifteen expeditions, naval and private, British and American, travelling by sea and overland, and involving at least twenty-four ships, scoured the Arctic – and discovered only the site at Beechey Island (south-western Devon Island) where Franklin's expedition spent its first arctic winter (1845–6). No clue was discovered as to what had happened to the two ships (and their crews totalling 126 men) thereafter.

Then in the spring of 1854, while employed by the Hudson's Bay Company on

[1] The pioneer trip by Lieutenant George Back, RN, down the Back River to the arctic coast and back, in a boat with nine companions, in the summer of 1834. See George Back, *Narrative of the Arctic Land Expedition to the Mouth of the Great Fish River and along the Shores of the Arctic Ocean, in the Years 1833, 1834, and 1835*, London, 1836.

[2] David Hanbury, *Sport and Travel in the Northland of Canada*, London, 1904, p. 123.

geographical exploration and survey work totally unrelated to the search for the missing expedition, Dr John Rae elicited from the Inuit at Pelly Bay and later at Repulse Bay, stories of white men having been seen trudging south hauling sledges and of dead bodies having later been found somewhere west of the mouth of a large river. He also purchased from the Inuit a selection of silverware and other items incontestably belonging to members of Franklin's expedition, which the Inuit had found with the bodies. Abandoning his survey work Rae travelled south to York Factory and hurried back to England, where the news of his finds broke in the pages of *The Times* on the morning of 23 October 1854.

Embroiled in the Crimean War, and no doubt very conscious that five ships had been abandoned in the Arctic during its latest series of search expeditions, the Admiralty was less than enthusiastic about sending out yet another naval search expedition. Instead it turned to the Hudson's Bay Company and requested that, using its manpower, its expertise in river travel and sledge travel, and its network of trading posts, the Company should rapidly mount a search expedition by way of the Back River. This the Company proceeded to do.

The expedition, led by two experienced northern travellers, Chief Factor James Anderson and Chief Trader James Green Stewart, was plagued by ill luck, a late breakup on the lakes and rivers, and by bad weather, not to mention some friction between the two leaders. Nonetheless they reached the arctic coast, confirmed the stories elicited by Rae from the Inuit, recovered an array of artefacts derived from the Franklin expedition, and significantly narrowed the search area for the subsequent, entirely successful search expedition mounted by Captain Francis Leopold McClintock in the steam yacht *Fox* in 1857–9. The verdict of history thus far has inclined to the view that the expedition was ineffectual, as a result of the incompetence or stupidity of the expedition leaders and/or of the shortcomings of the Company's organization. This reassessment of the expedition will demonstrate that the Company and its officers and men tackled a well-nigh impossible task expeditiously and competently, and that its limited success was due entirely to bad luck, bad weather and an unusually late season which greatly delayed canoe travel and ultimately forced the party to turn back only a short distance from the area where the main concentrations of artefacts and skeletal remains from the Franklin expedition were later found.

Versions of the journal of Chief Factor James Anderson have twice previously been published, but in each case, depending on one's viewpoint, in relatively obscure places. On the first occasion it appeared in the *Transactions of the Women's Canadian Historical Society of Toronto*, in 1919–20.[1] This version is accompanied by a brief introduction and is almost totally free of annotation.

A version of the journal was again published in 1940–41, in instalments in the *Canadian Field Naturalist*.[2] This was published from a typed, bound transcription of the journal (i.e. a copy of the full journal held in the Hudson's Bay Company

[1] Mickle, S., ed., 'The Hudson Bay Expedition in Search of Sir John Franklin', *Transactions of the Women's Canadian Historical Society of Toronto*, 20, 1919–20, pp. 11–45.

[2] Anderson, J. 'Chief Factor James Anderson's Back River Journal of 1855', *Canadian Field Naturalist*, 54, 1940, pp. 63–7, 84–9, 107–9, 125–6, 134–6; 55, 1941, pp. 9–11, 21–6, 38–44.

Archives),[1] held in the library of the Lands, Parks and Forests Branch of the Department of Mines and Resources in Ottawa. The then editor of the *Canadian Field Naturalist* added an introduction, quite extensive annotation, and copies of a limited number of Anderson's letters. These ancillary materials provide little or no information as to the genesis of the expedition, or its aftermath.

I wish to thank various archives for permission to publish documents and maps in their possession: the Hudson's Bay Company Archives, Provincial Archives of Manitoba; the Provincial Archives of Alberta; the British Columbia Provincial Archives and Records Service; and the Archives of the Scott Polar Research Institute, Cambridge, England. I am particularly indebted to Ms Judith Hudson Beattie, Keeper of the Archives, Hudson's Bay Company Archives and to members of her staff, especially Ms Ann Morton. I am also grateful to Dr C. S. McKinnon of Edmonton for his advice and suggestions and for a photograph of Chief Factor James Anderson, and to Mr Warren Baker of Montreal for information on Anderson's career and for a further photograph of Anderson. Mr Michael Payne of Edmonton, generously gave me access to his detailed notes on the Oman family of Churchill. And I owe a great debt of gratitude to Mr Keith Bigelow of the Department of Geography, University of Saskatchewan, for drafting the maps and for copying photographs. Finally I wish to thank Jim and Loreen Gardner for their unfailing hospitality on my frequent visits to Winnipeg to pursue this and other projects in the Hudson's Bay Company Archives.

My travel to the various archives in which the relevant documents are distributed, was facilitated by a grant from the President's Social Science and Humanities Research Fund, University of Saskatchewan.

[1] HBC E.37/3.

ABBREVIATIONS

BCA	British Columbia Archives, Victoria
HBC	Hudson's Bay Company Archives (Provincial Archives of Manitoba), Winnipeg
PAA	Provincial Archives of Alberta, Edmonton
SPRI	Scott Polar Research Institute, Cambridge

CHAPTER I

GENESIS OF THE EXPEDITION

INTRODUCTION

Carrying probably the best-manned and best-equipped expedition ever to sail in search of the Northwest Passage, two barque-rigged bomb vessels, HMS *Erebus* and *Terror* (370 and 340 tons, respectively) put to sea from Greenhithe on the River Thames on the morning of 10 May 1845. Their combined complements amounted to 134 officers and men, commanded by Captain Sir John Franklin. Their orders were to sail through the Northwest Passage from Baffin Bay to Bering Strait, then proceed to the Sandwich Islands (Hawaii) and return to Britain via Cape Horn.[1]

These orders, in hindsight, may appear to have been dangerously, if not fatally over-optimistic; but from the vantage point of London in 1845 the task which Franklin was being ordered to tackle appeared to be quite feasible, even pedestrian. In 1819–20, in HMS *Hecla* and *Griper*, Captain William Edward Parry had encountered relatively few impediments from the ice and had managed to penetrate as far west through Parry Channel as the south-west corner of Melville Island, and had wintered at Winter Harbour.[2] In 1829–33 Sir John Ross had led a private expedition aboard *Victory*; penetrating south through Prince Regent Inlet and the Gulf of Boothia he had reached Felix Harbour on the east side of the Isthmus of Boothia; however, he had had to abandon his ship and he and his men had made their way north by sledge and boat to be rescued by a whaling vessel. But during the expedition, in the spring of 1830 Ross's nephew, James Clark Ross, had sledged west across the isthmus to Spence Bay and had explored the north coast of King William Island as far west as Victory Point.[3]

Less than a decade later another expedition would almost reach the same point from the west. In the summer of 1839 a boat-borne expedition, mounted by the Hudson's Bay Company and led by Peter Dease and Thomas Simpson, travelled

[1] Richard Cyriax, *Sir John Franklin's Last Expedition; A Chapter in the History of the Royal Navy*, London, 1939.

[2] William Edward Parry, *Journal of a Voyage for the Discovery of a North-West Passage from the Atlantic to the Pacific; performed in the years 1819–20, in His Majesty's Ships Hecla and Griper*, London, 1821.

[3] John Ross, *Narrative of a Second Voyage in Search of a North-west Passage, and of a Residence in the Arctic Regions during the Years 1829, 1830, 1831, 1832, 1833*, London, 1835.

coastwise from the Coppermine, towards the east.[1] Discovering Simpson Strait they explored the south coast of King William Island, visited Montreal Island, and turned back at Castor and Pollux Bay on the west coast of Boothia Peninsula. At Cape John Herschel, where they erected a large cairn, they were within 100 km of Victory Point, on the north-west coast of King William Island, the most westerly point reached by James Ross; hence a gap of only 100 km remained to be filled to complete the picture of the northern limits of the continent. But travelling across sea ice and tundra on foot, man-hauling sledges, as Ross's party had done, or boating along the coast in the ice-free waters of summer, as Dease and Simpson had done, are totally different kettles of fish from tackling those same ice-infested waters, totally uncharted and often foul with shoals and reefs, in relatively deep-drafted sailing ships.

Erebus and *Terror* could scarcely have been improved upon, in terms of the technology of the day. As bomb vessels they had been designed as floating platforms for large-calibre mortars capable of hurling enormous shells on a high trajectory into walled coastal towns or entrenched positions under siege. The mortars sat on massive wooden beds located amidships, and since there was no built-in mechanism for absorbing the recoil of these formidable weapons, the ships' timbers and hulls were extraordinarily strongly built and hence ideally suited for withstanding the crushing pressures which they might encounter in arctic sea ice. If caught in a nip between two large floes colliding under the impetus of wind or current, they stood a much better chance of surviving than an ordinary wooden vessel.

Neither ship was a stranger to ice-infested polar waters. Under the command of Captain George Back, in 1836 *Terror* had attempted to reach Wager Bay, which had been selected as the starting point for an overland attempt at exploring the midsection of the Northwest Passage; but she was caught in the ice off the north-east coast of Southampton Island and had drifted south, beset in the ice, through Foxe Channel and east into Hudson Strait. She fought clear of the ice, seriously damaged and leaking, in July of 1837 and limped back to Britain.[2]

In 1839, along with *Erebus*, *Terror* headed for the Antarctic, under the command of James Clark Ross.[3] They circumnavigated Antarctica, entered the Ross Sea, and discovered the Ross Ice Shelf and the active volcano of Mount Erebus. Both ships came close to being crushed and suffered serious damage during this impressive voyage.

For their attempt on the Northwest Passage, both ships were thoroughly overhauled. Additional ice-sheathing was added to their hulls, and their already massive bow timbers were reinforced to the point where they consisted of a mass of timber almost 2.5 m thick. The bows were then armoured with iron plates to protect them from the abrasion of the ice. For the first time both ships were provided with auxiliary steam power to assist them in ice or during calms. Minimally modified railway engines were mounted athwartships in the holds, with a drive shaft extending aft

[1] Thomas Simpson, *Narrative of the Discoveries on the North Coast of America; effected by the Officers of the Hudson's Bay Company during the years 1836–39*, London, 1843.

[2] George Back, *Narrative of an Expedition in H.M.S. Terror, Undertaken with a View to Geographical Discovery on the Arctic Shores, in the Years 1836–7*, London, 1838.

[3] James Clark Ross, *A Voyage of Discovery and Research in the Southern and Antarctic Regions during the Years 1839–43*, London, 1847.

from one of the drive wheels and driving a two-bladed propeller.[1] The engines were of only 20 hp and gave the ships a speed of only four knots even under the most advantageous conditions. But in that the maximum speed of these bluff-bowed vessels under sail was only about eight knots, the auxiliary power probably proved very useful. If the propellers were threatened by ice they could be hoisted out of harm's way via a vertical well. Further innovations boasted by Franklin's ships were condensers attached to the galley stoves to produce drinking water from sea water, and hot-water heating systems throughout the living quarters.

Provisions were calculated to last three years. They included some 70 tonnes of flour, 48 tonnes of meat (one-third canned and two-thirds salt), 3,684 gallons of liquor and 3.5 tonnes of tobacco. Antiscorbutics included 4.5 tonnes of lemon juice and 170 gallons of cranberries.

The expedition leader, Sir John Franklin, was fifty-eight years old; he had joined the Navy at the age of fourteen and had participated in two of the Royal Navy's bloodiest battles of the Napoleonic Wars; he had served aboard HMS *Polyphemus* at the Battle of Copenhagen, and as signal midshipman aboard HMS *Bellerophon* at the Battle of Trafalgar.[2] He had first gone to the Arctic in command of HMS *Trent* in 1818; in company with HMS *Dorothea* (Captain David Buchan) he had tried unsuccessfully to reach the North Pole via a route between Svalbard and Greenland.[3] This venture, twenty-seven years earlier, represented Sir John's only experience of shipborne arctic exploration. On the other hand few could match him for experience of overland travel and longshore travel in small boats and canoes in the Arctic. In 1819–22 , travelling by canoe, Franklin had explored the coast from the mouth of the Coppermine to Point Turnagain on Kent Peninsula; unfortunately eleven of his companions died of starvation and exposure during the desperate overland trek back to their base at Fort Enterprise.[4] On a second expedition in 1825–7, he led a party which explored westwards by boat from the mouth of the Mackenzie to Return Reef (west of Prudhoe Bay).[5] Sir John had not been in the Arctic since then, and had spent most of the intervening period as an administrator, serving as Governor of Tasmania from 1837 until 1843.

HMS *Terror* was under the command of Captain Francis Rawdon Moira Crozier, aged forty-eight; he had by far the greatest experience of shipborne polar exploration of any of the members of the expedition. During Captain William Edward Parry's second expedition of 1821–3, which had wintered first at Winter Island in Frozen Strait, and then at Igloolik, he had served as midshipman on board HMS

[1] Cyriax, *Sir John Franklin's Last Expedition*.

[2] Ibid.

[3] Frederick William Beechey, *A Voyage of Discovery Towards the North Pole in His Majesty's Ships 'Dorothea' and 'Trent', under the Command of Captain David Buchan, R.N., 1818*, London, 1843.

[4] John Franklin, *Narrative of a Journey to the Shores of the Polar Sea, in the Years 1819, 20, 21 and 22*, London, 1823. See also C. Stuart Houston, ed., *To the Arctic by Canoe, 1819–1821: The Journal and Paintings of Robert Hood, Midshipman with Franklin*, Montreal and London, 1974; *Arctic Ordeal: The Journal of John Richardson, Surgeon-Naturalist with Franklin, 1820–1822*, Kingston and Montreal, 1984; *Arctic Artist: The Journal and Paintings of George Back, Midshipman with Franklin, 1819–1822*, Montreal and Kingston, 1994.

[5] John Franklin, *Narrative of a Second Expedition to the Shores of the Polar Sea in the Years 1825, 1826, and 1827*, London, 1828.

Fury.[1] At both locations the expedition had established close and generally friendly relations with the local Inuit, and it is likely that Crozier had acquired at least some knowledge of Inuktitut. Crozier had also served as midshipman in *Hecla* on Parry's third expedition, and thus had wintered at Port Bowen in Prince Regent Inlet, and had seen HMS *Fury* driven ashore by the ice at Fury Beach.[2] Finally, as captain of HMS *Terror* he had participated in James Ross's epic antarctic voyage of 1839–43, and hence had vast experience of handling a ship in ice.[3]

On the other hand Commander James Fitzjames, in command of *Erebus* on the new expedition, had no previous polar experience, although he had explored the Euphrates River by steamer in 1835 and had been in charge of a rocket brigade during the hostilities in China in 1841. His lack of experience was to some degree offset by the selection of his First Lieutenant, Graham Gore; in 1836–7 he had served as mate under Captain George Back aboard HMS *Terror* and hence was only too familiar with the dangers of a wintering spent adrift in the pack. A few of the other officers had previous arctic experience: Mr Charles Osmer, purser and paymaster in *Erebus*, had been to the Bering Strait area with Beechey in HMS *Blossom*.[4] *Terror*'s assistant surgeon, Alexander McDonald, had previously been to the Arctic in whaling ships. And finally, each ship carried an ice-master to advise on ice navigation. James Reid served in this capacity aboard *Erebus* and Thomas Blanky aboard *Terror*; both were experienced whaling captains.

Most of the men were recruited especially for the expedition from the north of England and many had probably served previously aboard whalers sailing out of ports such as Hull and Whitby. A few, however, were veterans of the Royal Navy. As an inducement to volunteer, everyone, both officers and men, was to receive double pay for the duration of the expedition, in keeping with Admiralty practice as regards arctic expeditions. Of the total complement of 134 men who sailed from the Thames, five would be invalided home from Greenland.

Franklin's orders specified quite narrowly the route he was to attempt through the labyrinth of what is now the Canadian Arctic Archipelago. Heading west through Lancaster Sound and Barrow Strait along the parallel of approximately 74°15′N as far as the meridian of Cape Walker (the north-east tip of Russell Island, off the north coast of Prince of Wales Island at about 98°W), he was to steer south and west from there, towards Bering Strait, keeping as straight a course as ice and/or unknown land would permit. But if ice prohibited progress in this direction, he was to try northwards via Wellington Channel, between Cornwallis and Devon islands.[5]

[1] William Edward Parry, *Journal of a Second Voyage for the Discovery of a North West Passage from the Atlantic to the Pacific; Performed in the years 1821–22–23, in His Majesty's Ships Fury and Hecla, under the Orders of Captain William Edward Parry*, London, 1824.

[2] William Edward Parry, *Journal of a Third Voyage for the Discovery of a North-west Passage from the Atlantic to the Pacific; Performed in the Years 1824–25, in His Majesty's Ships Hecla and Fury, under the Orders of Captain William Edward Parry*, London, 1826.

[3] See J. C. Ross, *A Voyage of Discovery and Research.*

[4] Frederick William Beechey, *Narrative of a Voyage to the Pacific and Beering's Strait, to Co-operate with the Polar Expeditions; Performed in His Majesty's Ship Blossom, under the Command of Captain F .W. Beechey … in the years 1825, 26, 27, 28*, London, 1831.

[5] See Cyriax, *Sir John Franklin's Last Expedition.*

Having called at the Whalefish Islands in West Greenland, *Erebus* and *Terror* pushed north into the ice of Melville Bay in north-eastern Baffin Bay. There, on 25 July 1845, they encountered the whalemen *Enterprise* (Captain Robert Martin) of Peterhead and *Prince of Wales* (Captain Dunnett) of Hull. The four ships kept company among the ice for several days, with men and officers visiting to and fro between the ships. When last seen by Captain Martin the expedition vessels were moored to an iceberg. He and his crew were the last Europeans to see any member of Franklin's expedition alive.[1]

Not suprisingly, no further word from the expedition reached Britain in 1846, but no particular concern was felt. But by 1847, with still no news, the level of anxiety began to rise a little and plans for search expeditions were laid. In 1848 the Admiralty mounted searches from three different directions. Sir John Richardson (who had participated in both of Franklin's overland expeditions, and had led a boat party which explored the coast from the Mackenzie Delta to the mouth of the Coppermine in 1826),[2] and Dr John Rae, an employee of the Hudson's Bay Company renowned for his extended overland trips in the Arctic, descended the Mackenzie and searched the coast eastward to the mouth of the Coppermine by boat.[3] Rae then wintered at Fort Confidence on Great Bear Lake, then resumed his search in 1849; he planned to search the coasts of Victoria Island but ice prevented him from crossing Dolphin and Union Strait.[4] As the second prong of this search campaign the Admiralty dispatched HMS *Plover* (Captain Thomas Moore) to Bering Strait via the Pacific; she reached her destination in October 1848 and wintered in Emma Bay, near the present town of Provideniya, on the Siberian side of Bering Strait.[5] To reinforce this component of the search HMS *Herald* (Captain Henry Kellett), already in the Pacific, engaged in surveying the coasts of Central America, was ordered northward from Panama in the summer of 1848 to rendezvous with *Plover* in Kotzebue Sound;[6] on failing to make contact she returned south.

Meanwhile a third expedition had been dispatched to Baffin Bay to attempt to follow the same route as outlined in Franklin's orders. HMS *Enterprise* and HMS *Investigator* (Captain Sir James Ross and Captain Edward Bird, respectively) sailed from the Thames in May 1848. Prevented by ice from pushing farther west through Barrow Strait or farther south down Prince Regent Inlet, they wintered at Port Leopold on north-eastern Somerset Island; in the spring sledging parties searched long sections of the coasts of Somerset Island and also crossed Prince Regent Inlet

[1] A. G. E. Jones, 'Captain Robert Martin: a Peterhead Whaling Master in the Nineteenth Century', *Scottish Geographical Magazine*, 85, 3, 1969, pp. 196–202.

[2] See Franklin, *Narrative of a Second Expedition.*

[3] John Richardson, *Arctic Searching Expedition: a Journal of a Boat-Voyage through Rupert's Land and the Arctic Sea, in Search of the Discovery Ships under Command of Sir John Franklin*, London, 1851.

[4] E. E. Rich and A. M. Johnson, eds, *John Rae's Correspondence with the Hudson's Bay Company on Arctic Exploration 1844–1855*, London, 1953.

[5] William H. Hooper, *Ten Months among the Tents of the Tuski, with Incidents of an Arctic Boat Expedition in Search of Sir John Franklin, as far as the Mackenzie River and Cape Bathurst*, London, 1853.

[6] Berthold Seemann, *Narrative of the Voyage of H.M.S. Herald during the Years 1845–51, under the Command of Captain Henry Kellett, R.N., C.B.; being a Circumnavigation of the Globe, and Three Cruises to the Arctic Regions in Search of Sir John Franklin*, London, 1853.

and Lancaster Sound to visit points on the coasts of Baffin Island and Devon Island.[1] No trace of the missing expedition was found and the ships returned to England in 1849.

That year, too, Captain Henry Kellett took *Herald* north again through Bering Strait, made rendezvous with *Plover* in Kotzebue Sound, then pushed north to Wainwright Inlet, before being blocked by ice. Swinging west across the Chukchi Sea, Kellett discovered Ostrov Geral'da (Herald Island) before retreating back south through Bering Strait to winter in Mexican waters.[2] In the meantime a party from HMS *Plover*, led by Lieutenants William Pullen and William Hooper, travelling in two boats, searched the coast eastwards from Point Barrow to the Mackenzie Delta,[3] but without success.

Meanwhile, also in 1849, Captain James Saunders took HMS *North Star* north through Baffin Bay with instructions first to resupply James Ross's expedition, and then to search Smith and Jones sounds. Difficult ice conditions severely delayed *North Star*'s progress across Melville Bay, and she was forced to winter in Wolstenholme Sound, Northwest Greenland. Getting under way again on 1 August 1850 Saunders pushed west through Lancaster Sound but due to ice was unable to leave depots at Port Leopold or Port Bowen as intended and landed a depot in Navy Board Inlet instead.[4]

In 1850 the tempo of the Franklin search picked up markedly; it involved both private and naval search expeditions and even included an American component. The Admiralty dispatched four ships to the Eastern Arctic, *Resolute* (Captain Horatio Austin), *Assistance* (Captain Erasmus Ommanney), and the steam tenders *Intrepid* (Captain Bertie Cator) and *Pioneer* (Captain Sherard Osborn); this squadron sailed from England in May.[5] Lady Jane Franklin organized and equipped a search initiative of her own: Captain Charles Forsyth sailed from England in *Prince Albert.*[6] The arctic veteran, Sir John Ross, led an expedition financed by the Hudson's Bay Company, aboard *Felix*, and with *Mary*, a 12-ton yacht, as tender.[7] Yet another expedition sailed from Britain that summer; the Aberdeen whaling captain William Penny, with instructions from the Admiralty and financed by the British Government, led an expedition consisting of the ships *Lady Franklin* and *Sophia*.[8] Finally the US

[1] James D. Gilpin, 'Outline of the Voyage of H.M.S. Enterprize and Investigator to Barrow Strait in Search of Sir John Franklin,' *Nautical Magazine*, 19, 1850, pp. 8–9, 82–90, 160–70, 230.

[2] Seemann, *Narrative of the Voyage of H.M.S. Herald.*

[3] H. F. Pullen, ed., *The Pullen Expedition in Search of Sir John Franklin*, Toronto, 1979.

[4] Cyriax, *Sir John Franklin's Last Arctic Expedition.*

[5] Sherard Osborn, *Stray Leaves from an Arctic Journal; or, Eighteen Months in the Polar Regions, in Search of Sir John Franklin's Expedition, in the Years 1850–51*, London, 1852.

[6] William Parker Snow, *Voyage of the Prince Albert in Search of Sir John Franklin: a Narrative of Every-Day Life in the Arctic Seas*, London, 1851.

[7] Ernest S. Dodge, *The Polar Rosses: John and James Clark Ross and their explorations*, London, 1973; M. J. Ross, *Polar Pioneers: John Ross and James Clark Ross*, Montreal and Kingston, 1995; Malcolm Wilson, 'Sir John Ross's Last Expedition, in search of Sir John Franklin,' *The Musk-Ox*, 13, 1973, pp. 3–11.

[8] Peter C. Sutherland, *Journal of a Voyage in Baffin's Bay and Barrow Straits, in the Years 1850–1851, Performed by H.M. Ships, 'Lady Franklin' and 'Sophia', under the Command of Mr. William Penny, in Search of the Missing Crews of H.M. Ships Erebus and Terror*, 2 vols, London, 1852.

Navy's contribution, consisting of the ships *Advance* (Captain E. J. De Haven) and *Rescue* (Captain Samuel Griffin), sailed north from New York in May 1850.[1]

All five expeditions pushed north through Baffin Bay, entered Lancaster Sound, and congregated at Beechey Island at the south-western tip of Devon Island in late August 1850. It was here that the first traces of the missing expedition were found. On Cape Riley, across Erebus Bay from Beechey Island, Captain Erasmus Ommanney and some of his officers found traces of an encampment on 23 August 'and collected the remains of materials which evidently prove that some party belonging to Her Majesty's ships have been detained on this spot'.[2] The remains included rope, canvas, wood and animal bones; experts such as Captain William Edward Parry, Dr John Richardson and Colonel Sabine who later examined these remains and read Ommanney's report, deduced that a boat party from *Erebus* and *Terror* had camped here, probably to carry out magnetic observations.

The next traces of the missing expedition were found three days later by Captain William Penny; leading a boat party from *Lady Franklin* he found traces of a camp 10 km north of Cape Spencer on the shores of Wellington Channel. Discarded soup cans, barrel fragments, charred pieces of wood, rope, mittens, rags and a scrap of newspaper dated 1844 were scattered around the remains of a circular stone hut with a neatly paved floor.[3] Sledge tracks were also found in the gravel nearby.

On the 27th *Lady Franklin*, *Sophia*, *Felix*, *Advance* and *Rescue* congregated in Union Bay, on the north-west side of Beechey Island. When boat parties went ashore to investigate the shingle isthmus which joins Beechey Island to the mainland, they quickly realized that this was where the Franklin expedition had spent the winter of 1845–6.[4] The definitive evidence was provided by the headboards of three graves, two of the occupants being from *Erebus* and one from *Terror*; all had died in the early months of 1846.[5] The site of a smith's forge was clearly identifiable, while a general litter of tin cans, wood chips, fragments of canvas, rope, etc. showed that there had been a substantial amount of activity on shore. Captain Osborn of *Pioneer*, which arrived on the 28th, even found a pair of gloves laid out to dry, weighted down with pebbles in their palms to prevent them from blowing away.[6] Despite a thorough search, however, no records were found, and, perhaps strangest of all, no message to indicate the expedition's intentions. Where Franklin's ships had gone from Beechey Island, presumably in the summer of 1846, remained as deep a mystery as ever.

The Royal Navy vessels (*Resolute*, *Assistance*, *Pioneer* and *Intrepid*) spent the winter

[1] Elisha Kent Kane, *The U.S. Grinnell Expedition in Search of Sir John Franklin: a Personal Narrative*, London, 1854.

[2] Great Britain. Parliament, *Return to an Address of the Honourable The House of Commons, dated 7 February 1851; – for, 'Copy or Extracts from any Correspondence or Proceedings of the Board of Admiralty…', 'Copies of any Instructions from the Admiralty to any Officers in Her Majesty's Service, Engaged in Arctic Expeditions…', and, 'Copy or Extracts from any Correspondence or Communications from the Government of the United States … in Relation to any Search to be Made on the Part of the United States…'*, London (House of Commons, Sessional Papers, Accounts and Papers, 1851, 33, 97), p. 70.

[3] Sutherland, *Journal of a Voyage in Baffin's Bay.*

[4] Kane, *The U.S. Grinnell Expedition*; Sutherland, *Journal of a Voyage into Baffin's Bay.*

[5] Owen Beattie and John Geiger, *Frozen in Time: the Fate of the Franklin Expedition*, London, 1987.

[6] Osborn, *Stray Leaves.*

of 1850–51 off Griffith Island, off the south coast of Cornwallis Island, while Penny's ships *Lady Franklin* and *Sophia* shared winter quarters with John Ross's *Felix* in Assistance Harbour on the south coast of Cornwallis Island.[1] The two American ships were not so fortunate; beset in the ice, *Advance* and *Rescue* drifted north down Wellington Channel as far as 75°24′N, then back south to Barrow Strait. Throughout the winter their drift continued east down Lancaster Sound and south to Davis Strait where they were released in June 1851.[2]

In the spring of 1851 sledge parties from the wintering ships at Griffith Island and Assistance Harbour fanned out to search an amazingly wide area of the Arctic Archipelago. Lieutenant McClintock pushed west to Winter Harbour and searched the coasts of Dundas Peninsula, south-western Melville Island; other parties from Austin's squadron crossed Barrow Strait and searched much of the coastline of Prince of Wales Island and parts of Somerset Island. Penny, meanwhile focused his attention northwards; his sledge parties searched most of the shores of Wellington Channel;[3] a party led by Commander Charles Phillips, Ross's second-in-command, even trekked right across the barren interior of Cornwallis Island.[4] But none of these parties found any further signs of the Franklin expedition.

Farther west Dr John Rae had resumed his search; dispatched by the Hudson's Bay Company to renew his efforts to search the coasts of Victoria Island, he wintered at Fort Confidence on Great Bear Lake.[5] In the spring of 1851 he sledged across to the mouth of the Coppermine and searched a considerable stretch of the south-west coast of Victoria Island. After breakup he switched to boat-travel and searched the south-east coasts of the island from Cambridge Bay to Pelly Point, the eastern tip of Collinson Peninsula.

Forsyth's *Prince Albert* had returned to Britain in October 1850, bringing the news of the finds at Beechey Island.[6] Inspired by these finds, a further expedition, sponsored by Lady Franklin, sailed from Aberdeen aboard *Prince Albert* in May 1851.[7] The leader, William Kennedy, was a former Hudson's Bay Company man, of Métis origin; second-in-command was a French naval officer, Joseph-René Bellot, who had volunteered his services. Kennedy tried, unsuccessfully, to get into Port Leopold, but instead was forced to winter at Batty Bay, on the east coast of Somerset Island. From there, in the spring of 1852, they sledged south to Bellot Strait and westwards through it. Their route then took them right across Prince of Wales Island and back, then back to their winter quarters via the north coast of Somerset Island. No traces of the Franklin expedition were found.

[1] M. J. Ross, *Polar Pioneers*; Sutherland, *Journal of a Voyage in Baffin's Bay*.

[2] Kane, *The U.S. Grinnell Expedition*.

[3] Sutherland, *Journal of a Voyage in Baffin's Bay*.

[4] Wilson, 'Sir John Ross's Last expedition'.

[5] John Rae 'Journey from Great Bear Lake to Wollaston Land', *Royal Geographical Society Journal*, 22, 1852, pp. 73–82; John Rae, 'Recent Explorations along the South and East Coast of Victoria Land', *Royal Geographical Society Journal*, 22, 1852, pp. 82–96.

[6] Snow, *Voyage of the 'Prince Albert'*.

[7] Joseph-René Bellot, *Memoirs of Lieutenant Joseph René Bellot ... with his Journal of a Voyage in the Polar Seas, in Search of Sir John Franklin*, 2 vols, London, 1855; William Kennedy, *A Short Narrative of the Second Voyage of the Prince Albert in Search of Sir John Franklin*, London, 1853.

While *Prince Albert* was getting under way to start for home, the next, and most ambitious, of the Royal Navy's search expeditions was heading into the Arctic. Commanded by Captain Sir Edward Belcher, it consisted of five vessels: *Resolute* (Captain Henry Kellett) and *Assistance*, the steam tenders *Pioneer* (Captain Sherard Osborn) and *Intrepid* (Captain F. Leopold McClintock), and the depot ship *North Star* (Captain William Pullen).[1] This squadron reached Beechey Island in August 1852, and then split into three components. Kellett took *Resolute* and *Intrepid* west along Parry Channel and on being blocked by ice, found secure winter quarters at Dealy Island in Bridport Inlet on the south coast of Melville Island. Meanwhile Belcher headed north with *Assistance* and *Pioneer* through Wellington Channel and wintered in Northumberland Sound on the north-west coast of Devon Island. *North Star* (Captain William Pullen) remained at Beechey Island to supply a rendezvous on which the other two components could fall back in case of emergency.

From Dealy Island in the spring of 1853 Kellett dispatched sledge parties which explored all the western, north-western and northern coasts of Melville Island, as well as those of Prince Patrick and Eglinton Islands. One party, led by Lieutenant Bedford Pim even crossed to Mercy Bay on Banks Island (of which more anon). From *Assistance* and *Pioneer* in Northumberland Sound further sledge parties explored and searched Grinnell Peninsula, Bathurst Island, and the north-east coasts of Melville Island.

In the summer of 1853 Kellett's ships began to retreat eastwards but got only as far as Cape Cockburn, the south-west tip of Bathurst Island, before being forced by the ice to winter again. Belcher meanwhile was able to make even less progress in his retreat south towards Beechey Island: *Assistance* and *Pioneer* were forced to winter north of Cape Osborn on the west coast of Devon Island.

Anticipating that neither pair of ships would be able to get free in the summer of 1854, Belcher ordered the abandonment of all four vessels, their crews retreating by sledge to *North Star* at Beechey Island in the spring of that year.[2] In August *North Star* broke free of the ice in Erebus Bay; shortly afterwards the supply vessels *Phoenix* and *Talbot* arrived and helped to evacuate the crews of the abandoned vessels back to Britain.

Simultaneously with the dispatch of Austin's squadron to the eastern Arctic in the spring of 1850, the Admiralty had also sent a two-ship squadron to the Pacific with instructions to pass through Bering Strait and to search the western limits of the

[1] Edward Belcher, *The Last of the Arctic Voyages; being a Narrative of the Expedition in H.M.S. Assistance, under the Command of Captain Sir Edward Belcher, C.B., in Search of Sir John Franklin, during the Years 1852–53–54*, 2 vols, London, 1855; Emile F. de Bray, *A Frenchman in Search of Franklin: De Bray's Arctic journal 1852–1854*, W. Barr, trans. and ed., Toronto, 1992; Robert McCormick, *Voyages of Discovery in the Arctic and Antarctic Seas, and Round the World; being Personal Narratives of Attempts to Reach the North and South Poles; and of an Open-Boat Expedition up the Wellington Channel in the year 1852, under the command of R. McCormck, R.N., F.R.C.S., in H.M.B. "Forlorn Hope", in search of Sir John Franklin*, 2 vols, London, 1884; George F. McDougall, *The Eventful Voyage of H.M. Discovery Ship "Resolute" to the Arctic Regions in Search of Sir John Franklin and the Missing Crews of H.M. Discovery Ships "Erebus" and "Terror", 1852, 1853, 1854*, London, 1857.

[2] Belcher, *The Last of the Arctic Voyages*; de Bray, *A Frenchman in Search of Franklin*; McDougall, *The Eventful Voyage*.

Archipelago. The ships concerned were HMS *Enterprise* (Captain Richard Collinson) and HMS *Investigator* (Captain Robert M'Clure).

The two ships sailed via the Straits of Magellan, but soon afterwards were separated and never again made contact. M'Clure[1] pushed through Bering Strait, rounded Point Barrow by squeezing between the fast ice and the pack, ran east into Amundsen Gulf and all the way north through Prince of Wales Strait before being blocked by heavy ice in M'Clure Strait. M'Clure then retreated south to the Princess Royal Islands where he wintered. Sledge parties searched considerable stretches of the south-west, west and north coasts of Victoria Island in the spring of 1851. After breakup *Investigator* ran back south, west around Nelson Head, the southern tip of Banks Island, and north along the west coast of that island. Rounding Cape Prince Alfred into M'Clure Strait, the ship was finally forced by ice to winter in Mercy Bay.

The ice of Mercy Bay did not break up the following summer (1852) and early in the spring of 1853 M'Clure was on the point of implementing a desperate plan for most of his crew to try to escape overland, some east to Port Leopold, the others south to Fort McPherson in the Mackenzie Delta, when Lieutenant Pim arrived by sledge from Dealy Island to inform M'Clure that Kellett's ships were wintering there. The decision was then taken to abandon *Investigator* and her crew sledged across M'Clure Strait to the relative security of Kellett's ships. After experiencing a further wintering in the ice (their fourth) off Cape Cockburn, on the abandonment of *Resolute* and *Intrepid*, with their hosts they sledged to *North Star* at Beechey Island and were evacuated to England in the summer of 1854.

Meanwhile Captain Richard Collinson in HMS *Enterprise*, after losing contact with M'Clure in *Investigator*, despite a slower start, had also been active. Reaching Point Barrow some weeks after M'Clure in the summer of 1850, Collinson had been unable to push past that cape between the pack ice and the shore;[2] turning back he had run south to Sitka, then, after a call at the Sandwich Islands (Hawaii) he had wintered at Hong Kong.

Running north through Bering Strait again in the summer of 1851, he followed almost exactly in M'Clure's track of the previous year. Heading east across the Beaufort Sea and Amundsen Gulf, he swung north from Cape Parry to Nelson Head and pushed north into Prince of Wales Strait (where he narrowly missed *Investigator*). Like M'Clure before him he was blocked by ice in the northern part of Prince of Wales Strait; turning back, he followed M'Clure round the southern tip of Banks Island, but decided to turn back because of ice conditions considerably farther south than Cape Prince Alfred. Swinging south around Nelson Head, he chose Walker Bay, on the south-west coast of Victoria Island as his wintering site.

During the winter of 1851–2, sledge parties explored parts of the south-west, west

[1] Alexander Armstrong, *A Personal Narrative of the Discovery of the North-west Passage; with Numerous Incidents of Travel and Adventure during nearly Five Years' Continuous Service in the Arctic Regions while in Search of the Expedition under Sir John Franklin*, London, 1857; Sherard Osborn, ed., *The Discovery of the North-west Passage by H.M.S. "Investigator", Capt. R. M'Clure, 1850, 1851, 1852, 1853, 1854 ... from the Logs and Journals of Capt. Robert M. M'Clure*, London, 1856.

[2] Richard Collinson, *Journal of H.M.S. Enterprise, on the Expedition in Search of Sir John Franklin's Ships by Behring Strait, 1850–55*, London, 1889.

and north coasts of Victoria Island, discovering, to their frustration, that they had been forestalled by parties from *Investigator* almost everywhere they went. In the summer of 1852 Collinson took his ship south, then east through Dolphin and Union Strait, Coronation Gulf and Dease Strait before finding winter quarters at Cambridge Bay on southern Victoria Island. In the spring of 1853 Collinson led a sledge party which searched the south-east and east coast of Victoria Island as far as Gateshead Island. Here the level of frustration must have risen even higher, since they found several messages from Dr John Rae, left during his search of this same coast (by boat from Great Bear Lake) in the summer of 1851. Collinson made no attempt to search the mainland coast of Queen Maud Gulf or the west coast of King William Island. At one stage during his spring sledge trip he was within 100 km of where HMS *Erebus* and *Terror* had been abandoned!

When the ice of Cambridge Bay broke up in the summer of 1853 Collinson started back westwards; he safely negotiated the hazards of Dease Strait, Coronation Gulf and Dolphin and Union Strait, but was blocked by ice and forced to winter (for the third time) at Camden Bay on the North Slope of Alaska. Getting free again in the summer of 1854, he pushed west to Point Barrow and south through Bering Strait before heading for home via Hong Kong and the Cape of Good Hope.

Even before HMS *Enterprise* was released from the ice of Camden Bay, however, the first solid news of the ultimate fate of the missing Franklin expedition had been garnered – by Dr John Rae,[1] during an expedition which was not even aimed at

[1] John Rae was born on 30 September 1813 at Hall of Clestrain near Stromness in the Orkneys, sixth son of John and Margaret Rae. His father was factor to the Laird of Armadale and sometime local agent for the Hudson's Bay Company. As a boy John Rae learned to shoot, sail, fish and hike and at an early age developed remarkable endurance as a walker and a capacity for tolerating discomfort. At the age of 16 he was enrolled at the University of Edinburgh and in 1833 qualified as a licentiate of the Royal Scottish College of Surgeons.

His first professional appointment was as surgeon on board the Hudson's Bay Company's supply ship, *Prince of Wales*, on her annual voyage to the Bay in 1833. After calling at Moose Factory the ship was prevented from leaving the Bay by ice in Hudson Strait and was forced to winter at Charlton Island in James Bay. Here, late in the winter, scurvy broke out among the crew but Rae was able to control it with cranberries which he found beneath the snow. Before heading for home, in the spring of 1834 the ship again called at Moose Factory, where Rae signed on as surgeon. He stayed for ten years, as surgeon and clerk; it was during this period that he perfected the technique of walking on snowshoes, learning from the local Indians.

In 1844 he was chosen by Sir George Simpson to survey the last remaining gap (left by Dease and Simpson at the end of their surveys of 1837–9) in the arctic coastline, namely from Chantrey Inlet to Foxe Basin. In preparation Rae travelled to Toronto to learn surveying from Lieutenant John Henry Lefroy; en route he snowshoed from Fort Garry to Sault Ste Marie (a distance of 1920 km) in two months. On his first arctic expedition in 1846 Rae travelled by boat from York Factory to Repulse, where he wintered. Then in 1847 he surveyed Committee Bay, Simpson Peninsula, Pelly Bay and the west coast of Melville Peninsula. On his return south he found he had been promoted Chief Trader.

His next two expeditions were aimed at searching for the missing Franklin expedition. In 1848, travelling with Sir John Richardson by boat, he searched the coast from the Mackenzie Delta to the mouth of the Coppermine; after wintering at Fort Confidence on Great Bear Lake, they attempted to cross Dolphin and Union Strait to search the coasts of Victoria Island, but were foiled by ice.

In the fall of 1849, now made Chief Factor, Rae took charge of the Mackenzie River district with his base at Fort Simpson. But in 1850, this time operating on his own, he resumed his search for Franklin. He wintered at Fort Confidence again, then in the spring of 1851 he searched the south-west coasts of Victoria Island by dogteam, and in the summer the south and east coast, as far as Collinson Peninsula, by boat. He was prevented by ice from crossing to King William Island.

searching for the Franklin expedition. Rae had been sent north from York Factory with a small group of men by the Hudson's Bay Company to survey the relatively short uncharted gap in the mainland coast between Dease and Simpson's farthest east (the Castor and Pollux River) and Bellot Strait.[1]

Rae's plans for his expedition were as follows:

> ...The party is to consist of one officer and twelve men, including two Esquimaux interpreters, in two boats, the one boat light and small for convenient transport over land and for river navigation, the other large, strong and well fitted for encountering rough weather in an open sea, but without any deck or other covering except tarpaulins. Our stock of provisions will be sufficient for three months, which, with an ample supply of ammunition, nets and articles to barter with, and for presents to the Esquimaux, will be enough for every purpose.
>
> ...As the navigation on the great American lakes does not open until April, I shall not leave Canada for the north until the latter part of that month. After calling at Lachine, to receive the final instructions of Sir George Simpson, Governor-in-Chief of the Company's territories, I shall proceed by steamboat as far as Sault St Mary's, and thence northward in a large bark canoe, manned by Iroquois and Canadians, by Lakes Superior, Rainy, and Winnipeg, to York factory, where I hope to arrive about the 13th of June. Here, or at Norway house, I expect to find my men waiting my arrival, and should the sea-ice be broken up, the party will immediately embark in the boats provided for the service, and push northwards along the west shores of Hudson's Bay.
>
> Having reached Chesterfield inlet, we shall advance to its western extremity, and there leave the large boat under charge of three men: while the remainder of the party, dragging the smaller boat, are to take a direct course over land for the nearest point of the Back or Great Fish river, the distance to which is estimated at about 90 miles. Having reached the river, three of the men will be sent back to the inlet to aid those left there in laying up a supply of fish, venison, and musk-ox meat, to guard against contingencies. The small boat, with a crew of seven persons, will descend the Back and push its way northwards, following closely the windings of the west coast of Boothia as far as lat. 72°N, which is now supposed to be the extreme north point of the American continent. From this point we shall commence our return by the same route as that by which we came, unless the state of the ice may permit us to cross Victoria Channel; in that case I shall visit the east shore of Victoria Land, and trace its coast southward from the spot where Kennedy touched upon it in his winter journey to my furthest north (lat. 70 deg. 30 min.) in the summer of 1851.
>
> Having reascended the Back as far as requisite, we shall leave the boat and cross over

---

It was on his fourth arctic expedition (see pp. 13–15) that he stumbled over the first clues as to the fate of the Franklin expedition.

Rae retired from the Hudson's Bay Company in 1856. He died in London on 22 July 1893 and was buried in the churchyard of St Magnus Cathedral in Kirkwall in the Orkneys. Few explorers, if any, have mapped more of the Arctic as economically as Rae: during his four expeditions he mapped 2,800 km of previously unmapped territory, including 2,460 km of northern coastline. On the four expeditions he walked a total of 10,400 km and travelled another 10,600 km by small boat. Throughout his expeditions he lived off the land to a large degree. For further details see: Robert L. Richards, 'John Rae', *Dictionary of Canadian Biography*, XII (1891–1900), Toronto, 1990, pp. 876–8; Robert L. Richards, *Dr John Rae*, Whitby, 1985; Stuart C. Houston, 'John Rae (1813–93)', in *Lobsticks and Stone Cairns*, Richard C. Davis, ed., Calgary, 1996, pp. 87–9.

[1] Rich and Johnson, *John Rae's Correspondence.*

> to Chesterfield inlet on foot, embark the whole party in the large boat, and start for York factory, where I hope to arrive on or about the 30th of September.[1]

As far as the mouth of Chesterfield Inlet, Rae's journey went exactly according to plan, the party arriving there on 17 July 1853.[2] But three days later fate took a hand; while Rae was asleep, in foggy conditions the steersman turned into the mouth of a large river entering Chesterfield Inlet from the north; when Rae woke, he realized the error, but seeing that they were in a large river flowing from the north-west, he felt it should lead them by a relatively easy portage to the Back, and decided to push up it, leaving three men and the larger boat near Chesterfield Inlet to hunt. He named the river after the River Quoich in Glen Garry, Invernessshire, where Rae had spent three weeks stalking and shooting grouse and ptarmigan as a guest of Mr Edward Ellice the previous autumn.[3] But to Rae's disappointment, after some distance the river swung towards the north-east, and at the point where it became unnavigable he estimated that they were still seventy miles from the Back River.

Running back down the Quoich Rae decided to head for Repulse Bay, since it was now too late in the season to attempt his original plan of portaging from the head of Chesterfield Inlet to the Back. Sending half his men back south from the mouth of Chesterfield Inlet, he ran north up Roe's Welcome Sound in one boat with seven men. They reached Repulse Bay on 15 August.[4] Rae was a little disturbed to find no Inuit there, but decided to winter at Repulse Bay, assuming that the prospects for hunting looked favourable by the start of freeze-up. In fact, over the following six weeks Rae and his men were able to kill 109 caribou, 1 muskox, 106 ptarmigan and 1 seal, while their nets produced 54 salmon. By 28 October there was sufficient wind-packed snow to allow them to build snow houses, in which they spent a cosy and comfortable winter.

On 31 March, leaving three men to look after the boat and stores, Rae set off northwards with the remainder, intending to fulfil his aim of charting the west coast of Boothia.[5] On 20 April, to the west of the head of Pelly Bay, they encountered a group of Inuit, some of whom Rae knew from his sojourn at Repulse Bay in 1847. Next day, an Inuk who had been away seal hunting the previous day, caught up with them, and from him Rae learned the first news of the fate of the Franklin expedition. He reported that he had never seen white men before, but he had heard of a number of Whites dying a long distance to the west, pointing in that direction, beyond a large river. Round his head he wore a gold cap band, which he said had been found where the Whites had died. He himself had never been to the spot; he did not know where it was; and was unable to guide Rae there.[6] Rae bought the cap band from him, and, although understandably intrigued by this information,

[1] *The Times*, 27 November 1852, p. 6.

[2] HBC E.15/9, letters from John Rae to George Simpson and Archibald Barclay, 9 August 1853; Rich and Johnson, *John Rae's Correspondence*, pp. 260, 262.

[3] Rich and Johnson, *John Rae's Correspondence*, p. 234.

[4] Ibid., p. 266.

[5] Ibid., p. 269.

[6] Ibid., p. 274; D. Murray Smith, *Arctic Expeditions from British and Foreign Shores from the Earliest Times to the Expedition of 1875–76*, Edinburgh, 1877, p. 644.

decided that it was too vague to permit him to pursue it. Instead he continued with his planned survey.

He reached the west coast of Boothia Peninsula on the evening of the 27th and shortly afterwards the mouth of a small river which Rae was satisfied was the Castor and Pollux River, i.e. the farthest east point reached by Dease and Simpson in 1839. He located a cairn which he was sure was the one left by Dease and Simpson, despite the fact that the upper part had collapsed.[1]

Over the next few days Rae and his party worked their way north along the west coast of Boothia. At Point de la Guiche, just north of Cape Colville, which they reached on 6 May, Rae was satisfied that he had reached the area mapped by James Ross. One of his men, Thomas Mistegan, walked a further six miles north and could see the coast still trending north for a further five miles. Realizing that, due to various delays, there was not enough time to complete the survey of the remaining unmapped section of coast from the Isthmus of Boothia north to Bellot Strait, Rae contented himself with the knowledge that there was no strait across to Pelly Bay south of this point, and that hence Boothia Peninsula was indeed a peninsula and not an island. Having built a cairn at Point de la Guiche, he started back for Repulse Bay.

He reached Pelly Bay again in the early hours of 17 May, where he again encountered some Inuit, who had in their possession further articles which had clearly belonged to members of the Franklin expedition:

> From these people I bought a silver spoon and fork. The initials F.R.M.C. [Francis Rawdon Moira Crozier], not engraved, but scratched with a sharp instrument, on the spoon, puzzled me much, as I knew not at the time, the Christian names of the Officers of Sir John Franklin's Expedition; and thought possibly that the letters above named might be the initials of Captn McClure, the small c between the M C being omitted.[2]

Rae and his party were back at Committee Bay on 21 May and at Repulse Bay on the 26th, to find the three men who had been left there in good health. Numbers of Inuit, some of whom Rae had seen earlier at Pelly Bay, were encamped nearby.

Rae had to wait for over two months for the ice to break up to allow him to start south by boat. It was during this period that, through repeated questioning with the help of his Inuktitut interpreter, William Ouligbuck, Rae was able to piece together a more detailed picture of what the Inuit knew of the fate of the Franklin expedition. In brief they reported that in the spring of 1850, while hunting seals near the north coast of King William's Island, some Inuit families had seen about forty white men travelling south on the sea ice, dragging a boat and sledges. None of them could speak Inuktitut well enough to make themselves understood, but they mimed that their ship or ships had been crushed by the ice, and that they were heading south in the hope of finding caribou. The men hauling the sledges looked thin and they bought a small seal or a piece of seal meat from the Inuit. The officer was described as tall, stout and middle-aged. The entire party camped in tents at night.

Later that same year, but before breakup, the Inuit found some thirty corpses and

[1] Rich and Johnson, *John Rae's Correspondence*, p. 279.

[2] Ibid., p. 283.

some graves on the mainland, and a further five corpses on a nearby island 'about a long day's journey to the north west of the mouth of a large stream',[1] which Rae presumed to be the Back River. Some of the bodies were in a tent or tents, some under an overturned boat, and some lay scattered around. One of the corpses was assumed to be that of an officer as he had a telescope slung over his shoulders and a double-barrelled gun lay under him.

The time of year, and the fact that the Inuit had narrowly missed seeing some of the survivors alive again is indicated by Rae's statement:

> A few of the unfortunate Men must have survived until the arrival of the wild fowl, (say until the end of May), as shots were heard, and fresh bones and feathers of geese were noticed near the scene of the sad event.[2]

Rae included a further detail in his report, one which would arouse the ire of a substantial portion of Victorian England and bring down considerable calumny on his head:

> From the mutilated state of many of the bodies and the contents of the kettles, it is evident that our wretched Countrymen had been driven to the last dread alternative, as a means of sustaining life.[3]

Rae purchased from the Inuit quite a substantial number of silver spoons and forks (many with the initials or crests of various members of the Franklin expedition), a small silver plate engraved with 'Sir John Franklin K.O.H' and an assortment of other small items which had belonged to expedition members. However they stressed that none of them had seen the white men, dead or alive, that they had never been to the places where the corpses were found, and that the details of the corpses, and the various relics had been obtained from other Inuit who had seen them.

Breakup was very late, and even when Rae and his party managed to start south on 4 August, they had great difficulty due to ice in Roe's Welcome Sound; it was not until they reached Cape Fullerton that they reached open water. They reached Churchill on 28 August 1854 and York Factory on the 31st.[4]

On 20 September Rae sailed for England on board the Company's ship *Prince of Wales*, and on 20 October, as the ship was working her way up the Channel, wrote a letter to the editor of *The Times*, in which he gave a brief outline of what he had discovered as to the fate of the Franklin expedition, diplomatically avoiding any reference to cannibalism. He landed at Deal on the 22nd, and by that evening was at his usual London hotel, the Tavistock, in Covent Garden. He immediately went to the Admiralty and submitted the report (focusing mainly on what he had found concerning the Franklin expedition) which he had written on 29 July, shortly before leaving Repulse Bay. Also on the 22nd he delivered to the Hudson's Bay Company a letter which he had written earlier that day, giving a very brief account of the results of his expedition, apologizing for its brevity, and promising a more complete report later. This report was forwarded by the Company to the Admiralty the next

[1] Ibid., p. 275.
[2] Ibid., p. 276.
[3] Ibid., p. 276.
[4] Ibid., p. 285.

day.[1] The Admiralty immediately sent Rae's letter of 29 July (complete with the references to cannibalism) to *The Times*, where it was published on 23 October, together with a few additional details, and along with Rae's own letter of the 20th. Rae's detailed report to the Hudson's Bay Company, which he had begun at York Factory on 1 September, was not finished till after he returned to England. It was read to the Committee of the Hudson's Bay Company on 13 November 1854, was passed to the Admiralty on 1 December, but was not published until later.[2] This, from Rae's point of view, was extremely unfortunate, in that publication of the detailed report would have averted one of the most severe criticisms which were levelled at him, namely that he should immediately have investigated the Inuit reports himself, rather than bringing the unsubstantiated reports back to Britain.

But this was not his worst crime, as far as Victorian England was concerned. Even worse was the fact that he should have relayed the Inuit reports that officers and men of the Royal Navy had indulged in cannibalism. Typical of those criticisms is that contained in an anonymous letter (in fact from the Reverend Edward Hornby, of Bury, whose brother Frederick was mate on board HMS *Terror*), published in *The Times* :

> It appears to me that Dr Rae has been deeply reprehensible either in not verifying the report which he received from the Esquimaux, or, if that was absolutely out of the question, in publishing the details of that report, resting as they do on grounds most weak and unsatisfactory. He had far better have kept silence altogether than have given us a story which, while it pains the feelings of many, must be very insufficient for all.[3]

In a very restrained response, published in *The Times* the following day, Rae justified his actions in coming back to England with the news, rather than attempting to verify the reports of the Inuit on the ground:

> To have verified the report which I brought home would, I believe, have been no difficult matter, but it could not possibly be done by my party in any other way than by passing another winter at Repulse Bay, and making another journey over the ice and snow in the spring of 1855. This I could not easily have done without exposure to more privations than persons accustomed to the Hudson's Bay Company's service are in the habit of enduring; but I had a deeper motive in returning to England with the information I had obtained than the mere selfish feeling of regard for my own comfort, which is a point I have never much studied. My reason for returning from Repulse Bay without having effected the survey I had contemplated was to prevent the risk of more valuable lives being sacrificed in a useless search in portions of the Arctic seas hundreds of miles distant from the sad scene where the lives of so many of the long-lost party terminated; and I am happy to say that my conduct has been approved by all those whose opinion I value, and with whom I have conversed on the subject.[4]

The officers and men of Captain Sir Edward Belcher's squadron (HMS *Assistance*, *Pioneer*, *Resolute*, *Intrepid* and *North Star*) and of Captain Robert M'Clure's *Investigator*, had returned to England on board HMS *North Star*, *Phoenix*, and *Talbot*, in early

[1] Ibid., p. lxxxii.
[2] Ibid., p. 265, footnote.
[3] *The Times*, 30 October 1854, p. 10.
[4] *The Times*, 31 October 1854, p. 8.

October,[1] and the court-martials of Belcher and the other captains for the abandonment of their ships had seized the attention of the public for several days (18–20 October), only ten days earlier. This, however, Rae could not have foreseen the previous April. Even in October, Captain Richard Collinson's *Enterprise* which had pushed north through Bering Strait in the summer of 1851, had not been heard of for three years, and hence Rae's fear that further major ship-borne expeditions might be dispatched was quite a reasonable one.

Undoubtedly part of the problem derived from the misunderstanding under which the Reverend Hornby (and the public in general) were labouring, as to the timing and sequence of Rae's obtaining the information as to the Inuit's encounters with the Franklin survivors and their remains. His own rough notes on his initial learning of the disaster from the Inuk near Pelly Bay in April (written on 20 and 21 April), read as follows:

> Met a very communicative and apparently imaginative Esquimaux; had never met whites before, but said that a number of Kabloonans, at least 35 or 40, had starved to death west of a large river a long distance off. Perhaps about 10 or 12 days' journey? Could not tell the distance, never had been there, and could not accompany so far. Dead bodies seen beyond two large rivers; did not know the place. Could not or would not explain it on the chart.[2]

To this, Rae had added the comment: 'This information too vague to act upon, particularly at this season, when everything is covered with snow'.

Unfortunately, in his report to the Admiralty, dated 29 July, and in his initial, preliminary report to the Hudson's Bay Company (both published in *The Times* on 23 October), it is far from clear that Rae did not learn that the Inuit reports referred to King William Island, and to the area west of the mouth of the Back River, until he got back to Repulse Bay on 26 May. By that time it was much too late for him to attempt a trip to verify the stories himself.

But in the public mind, Rae's real sin was to have repeated the Inuit stories of cannibalism among the survivors of the Franklin expedition. Rae himself had made it clear in a letter to Reverend Hornby that the details which caused so much distress were included in private reports to the Admiralty and the Hudson's Bay Company; in fact he had made no mention of them in his own letter to *The Times*.

> As to my having mentioned the Esquimaux's report of the distressing details connected with the last days of the unfortunate people, the latest survivors, I believe, of Sir John Franklin's party, and of which you complain, my duty in the matter was perfectly clear. It was to report to the authorities of the Admiralty, or of the Hudson's Bay Company, the substance of what I had heard from the Esquimaux, as fully, but, at the same time, as concisely as possible, leaving it for them to publish as much, or as little, of such report, as they considered right and proper.[3]

[1] De Bray, *A Frenchman in search of Franklin*, pp. 180–81.

[2] Great Britain. Parliament, *Further Papers Relative to the Recent Arctic Expeditions in Search of Sir J. Franklin and the Crews of Her Majesty's Ships "Erebus" and "Terror" etc., Presented to the House of Commons*, London, 1856.

[3] *The Times*, 3 November 1854, p. 7.

But this disclaimer did little to appease the horrified Victorian public. Even Charles Dickens was provoked into a lengthy discussion of the topic in his magazine *Household Words*, in which he promised his readers that he would demonstrate that 'there is no reason whatever to believe, that any of its members prolonged their existence by the dreadful expedient of eating the bodies of their dead companions' and 'that it is in the highest degree improbable that such men as the officers and crews of the two lost ships would, or could, in any extremity of hunger, alleviate the pains of starvation by this horrible means'.[1] To give him his due, however, Dickens went to great lengths to acquit Rae of any blame:

> He has himself openly explained, that his duty demanded that he should make a faithful report, to the Hudson's Bay Company or the Admiralty, of every circumstance stated to him; that he did so, as he was bound to do, without any reservation; and that his report was made public by the Admiralty: not by him. It is quite clear that if it were an ill-considered proceeding to disseminate this painful idea on the worst of evidence, Dr Rae is not responsible for it. It is not material to the question that Dr Rae believes in the alleged cannibalism; he does so, merely 'on the substance of information obtained at various times and various sources', which is before us all.[2]

Subsequently various other investigators garnered reports from the Inuit that the survivors of the Franklin expedition had indulged in cannibalism. But it is only in recent decades that scientific analysis of skeletal remains has confirmed these stories. In 1981 Beattie discovered cut marks on a right femur from a Franklin site on south-eastern King William Island.[3] Even more convincing proof was uncovered on a small island in Erebus Bay in 1992. Of a total of some four hundred bones or bone fragments derived from a minimum of eleven individuals, incontestably from the Franklin expedition, ninety-two bones revealed cut marks, over 55% of them showing multiple cuts.[4] Locations of the cut marks were consistent with the removal of muscle tissue. Thus Rae and his Inuit informants have been fully vindicated.

DOCUMENTS

1. John Rae's letter to *The Times*, on board *Prince of Wales*, English Channel, 20 October 1854.[5]

As any information, however meagre, regarding Sir John Franklin and his party must be of deep interest to every one, I take the earliest opportunity of communicating the following particulars:-

During my journey from Repulse Bay this spring over the ice, with the view of

[1] 'The Lost Arctic Voyagers', *Household Words*, 245, (2 December 1854), p. 361.

[2] Ibid.

[3] Owen Beattie, 'A Report on Newly Discovered Human Skeletal Remains from the Last Sir John Franklin Expedition', *The Musk-Ox* 33, 1983, pp. 68–77; Owen Beattie and James Savelle 'Discovery of Human Remains from Sir John Franklin's Last Expedition', *Historical Archaeology*, 17, 1983, pp. 100–105.

[4] A. Keenleyside, M. Bertulli and H. Fricke, 'The Final Days of the Franklin Expedition: New Skeletal Evidence', *Arctic* 50(1), 1997, pp. 36–46.

[5] *The Times*, 23 October 1854, p. 7.

completing the survey of the west coast of Boothia, I then and subsequently obtained information, and purchased articles of the natives, which prove beyond a doubt that a portion (if not all) of the then survivors of Sir John Franklin's long-lost and ill-fated party perished of starvation in the spring of 1850, on the coast of America, a short distance west of a large stream, which, by the description given of it, can be no other than Back's Fish River.

Among the articles purchased (all of which are now in my possession) which the Esquimaux found where the corpses of the 'white men' were discovered, are a small silver plate with 'Sir John Franklin, K.C.B.', engraved upon it, several silver spoons and forks, with initials of the following officers, viz. Captain Crozier, Lieutenant G. Gore, Assistant-Surgeon A. M'Donald, Assistant-Surgeon J. S. Peddie and Second-Master G. A. M'Bean.

Further particulars on this melancholy subject will appear in my report to the Hon. Hudson's Bay Company.

I may add that my small party wintered in snow houses comfortably enough at Repulse Bay, after some very hard work in the autumn laying up a supply of venison and fuel. We returned to York Factory all well on the 30th of August, but without having completed the contemplated survey.

2. John Rae's letter to the Admiralty, 29 July 1854.[1]

I have the honour to mention, for the information of my Lords Commissioners of the Admiralty, that during my journey over the ice and snows this spring, with the view of completing the survey of the west shore of Boothia, I met with Esquimaux in Pelly Bay, from one of whom I learnt that a party of 'white men' (Kabloonans) had perished from want of food some distance to the westward, and not far beyond a large river containing many falls and rapids. Subsequently, further particulars were received and a number of articles purchased, which places the fate of a portion, if not of all, of the then survivors of Sir John Franklin's long-lost party beyond a doubt – a fate as terrible as the imagination can conceive.

The substance of the information obtained at various times and from various sources was as follows:

In the spring, four winters past (spring, 1850), a party of 'white men', amounting to about forty, were seen travelling over the ice and dragging a boat with them by some Esquimaux, who were killing seals near the north shore of King William's Land, which is a large island. None of the party could speak the Esquimaux language intelligibly, but by signs the natives were made to understand that their ship, or ships, had been crushed by ice, and that they were now going to where they expected to find deer to shoot. From the appearance of the men, all of whom with the exception of one officer, looked thin, they were then supposed to be getting short of provisions, and they purchased a small seal from the natives. At a later date the same season, but previously to the breaking up of the ice, the bodies of some thirty persons were discovered on the continent, and five on an island near it, about a long day's journey to the NW of a large stream, which can be no other than Back's

[1] *The Times*, 23 October 1854, p. 7.

Great Fish River (named by the Esquimaux Oot-ko-hi-ca-lik), as its description and that of the low shore in the neighbourhood of Point Ogle and Montreal Island agree exactly with that of Sir George Back. Some of the bodies had been buried (probably those of the first victims of famine); some were in a tent or tents; others under the boat, which had been turned over to form a shelter, and several lay scattered about in different directions. Of those found on the island, one was supposed to have been an officer, as he had a telescope strapped over his shoulders and his double-barrelled gun lay underneath him.

From the mutilated state of many of the corpses, and the contents of the kettles, it is evident that our wretched countrymen had been driven to the last resource – cannibalism – as a means of prolonging existence.

There appeared to have been an abundant stock of ammunition, as the powder was emptied in a heap on the ground by the natives out of the kegs or cases containing it; and a quantity of ball and shot was found below high water mark, having probably been left on the ice close to the beach. There must have been a number of watches, compasses, telescopes, guns (several double-barrelled) etc., all of which appear to have been broken up, as I saw pieces of these different articles with the Esquimaux, and, together with some silver spoons and forks, purchased as many as I could get. A list of the most important of these I enclose, with a rough sketch of the crests and initials on the forks and spoons. The articles themselves shall be handed over to the Secretary of the Hon. Hudson's Bay Company on my arrival in London.

None of the Esquimaux with whom I conversed had seen the 'whites', nor had they ever been at the place where the bodies were found, but had their information from those who had been there and who had seen the party when travelling.

I offer no apology for taking the liberty of addressing you, as I do so from a belief that their Lordships would be desirous of being put in possession at as early a date as possible of any tidings, however meagre, and unexpectedly obtained, regarding this painfully interesting subject.

I may add that, by means of our guns and nets, we obtained an ample supply of provisions last autumn, and my small party passed the winter in snow houses in comparative comfort, the skins of the deer shot affording abundant warm clothing and bedding. My spring journey was a failure in consequence of an accumulation of obstacles, several of which my former experience in Arctic travelling had not taught me to expect.

*List of Articles purchased from the Esquimaux, said to have been obtained at the place where the bodies of the persons reported to have died of famine were found, viz:-

1 silver table fork – crest, an animal's head with wings, extended above; 3 silver table forks – crest: a bird with wings extended; 1 silver table spoon – crest, with initials F.R.M.C. (Captain Crozier, Terror); 1 silver table spoon and 1 fork – crest, bird with laurel branch in mouth, motto *Spero meliora*; 1 silver table spoon, 1 tea spoon and 1 dessert fork – crest, a fish's head looking upwards, with laurel branches on either side; 1 silver table fork – initials A.M'D. (Alexander M'Donald, assistant surgeon, Erebus); 1 silver table fork – initials G.A.M. (Gillies A. Macbean, second

master, Terror); 1 silver table fork, initials J.T.; 1 silver dessert spoon – initials J.S.P. (John S. Peddie, surgeon, Erebus); 1 round silver plate, engraved 'Sir John Franklin, K.C.B.'; a star or order, with motto *Nec aspera terrent*, G.R. III, MDCCCXV'.

Also a number of other articles with no marks by which they could be recognized, but which will be handed over with those above-named to the Secretary of the Hon. Hudson's Bay Company.

3. Letter from John Rae to Archibald Barclay, 1 September 1854, York Factory.[1]
I have the honour to report for the information of the Governor, Deputy Governor and Committee, that I arrived here yesterday with my party all in good health, but from causes which will be explained in their proper place, without having effected the object of the Expedition. At the same time information has been obtained, and articles purchased from the Natives which prove, beyond a doubt, that a portion, if not all, of the then survivors of the long lost and unfortunate party under Sir John Franklin, had met with a fate as melancholy and dreadful as it is possible to imagine...

The morning of the 21st [April 1854] was extremely fine and at 3 a m we started across land towards a very conspicuous hill, bearing west of us [in the Pelly Bay area]...

...We were now joined by another of the Natives who had been absent Seal hunting yesterday, but being anxious to see us, had visited our snow house early this morning, and then followed up our track. This man was very communicative, and on putting to him the usual questions as to his having seen 'white men' before, or any ships or boats – he replied in the negative; but said, that a party of 'Kabloonans' had died of starvation, a long distance to the west of where we then were, and beyond a large River. He stated that, he did not know the exact place; that he had never been there; and that he could not accompany us so far.

The substance of the information then and subsequently obtained from various sources, was to the following effect:

In the Spring, four winters past, (1850) whilst some Esquimaux families were killing Seals near the north shore of a large Island named in Arrowsmith's Charts, King William's Land, about forty white men were seen travelling in company southward over the ice, and dragging a boat and sledges with them. They were passing along the west shore of the above named Island. None of the party could speak the Esquimaux language so well as to be understood, but by signs the Natives were led to believe that the Ship or Ships had been crushed by ice, and that they were then going to where they expected to find deer to shoot. From the appearance of the Men (all of whom with the exception of one Officer, were hauling on the drag ropes of the sledge and were looking thin) they were then supposed to be getting short of provisions, and they purchased a small Seal or piece of Seal from the natives. The Officer was described as being a tall, stout, middle aged man. When their days journey terminated, they pitched Tents to rest in.

[1] HBC E.15/9. Published in Great Britain, 1855:835–44 and in *Journal of the Royal Geographical Society*, 25 (1855), pp. 246–56; and Rich and Johnson, *John Rae's Correspondence*, pp. 265–87.

At a later date the same Season but previous to the disruption of the ice, the corpses of some thirty persons and some Graves were discovered on the Continent, and five dead bodies on an Island near it, about a long day's journey to the north west of the mouth of a large stream, which can be no other than Back's Great Fish River, (named by the Esquimaux Oot-koo-i-hi-ca-lik,) as its description, and that of the low shore in the neighbourhood of Point Ogle and Montreal Island agree exactly with that of Sir George Back. Some of the bodies were in a tent or tents; others were under the boat which had been turned over to form shelter, and some lay scattered about in different directions. Of those seen on the Island, it was supposed that one was that of an Officer, (chief) as he had a telescope strapped over his shoulders, and his double barrelled gun lay underneath him.

From the mutilated state of many of the bodies and the contents of the kettles, it is evident that our wretched Countrymen had been driven to the last dread alternative, as a means of sustaining life. A few of the unfortunate Men must have survived until the arrival of the wild fowl, (say until the end of May,) as shots were heard, and fresh bones and feathers of geese were noticed near the scene of the sad event.

There appears to have been an abundant store of ammunition, as the Gunpowder was emptied by the Natives in a heap on the ground out of the kegs or cases containing it and a quantity of shot and ball was found below high water mark, having probably been left on the ice close to the beach before the spring thaw commenced. There must have been a number of telescopes, guns, (several of them double-barrelled,) watches, compasses, etc. all of which seem to have been broken up, as I saw pieces of these different articles with the Natives, and I purchased as many as possible, together with some silver spoons and forks, an order of merit in the form of a Star, and a small silver plate engraved Sir John Franklin K.O.H.

Inclosed is a list of the principal articles bought, with a note of the initials and a rough pen and ink sketch of the crests on the forks and spoons. The articles themselves I shall have the honour of handing over to you, on my arrival in London.

None of the Esquimaux with whom I had communication saw the 'white men' either when living or after death, nor had they ever been at the place where the Corpses were found, but had their information from Natives who had been there, and who had seen the party when travelling over the ice. From what I could learn, there is no reason to suspect that any violence had been offered to the sufferers by the Natives....

... We reached Pelly Bay at 1 a m on the 17th [May], and built a snow house about 2½ miles south, and the same distance west, of my observations of the 20th April.

Observing traces of Esquimaux, two men were sent after supper to look for them. After eight hours absence they returned with ten or twelve native men, women, and children. From these people I bought a silver spoon and fork. The initials F.R.M.C. not engraved, but scratched with a sharp instrument, on the spoon, puzzled me much, as I knew not at the time, the Christian names of the Officers of Sir John Franklin's Expedition; and thought possibly that the letters above named might be the initials of Captn. McClure, the small c between the M [and] C being omitted.

On the morning of the 1st [May] we arrived at Committee Bay, from thence our route to Repulse Bay was almost the same as before, and I shall not therefore advert to it further than to mention that we arrived at our winter home at 5 a m on the 26th May...

It was from this time until August, that I had opportunities of questioning the Esquimaux regarding the information which I had already obtained of the party of whites who had perished of starvation, and of eliciting the particulars connected with that sad event, the substance of which I have already stated.

[Enclosure]

List of Articles purchased at Repulse and Pelly Bays said to have been found with the party of men that starved to the West of Back's River in the Spring of 1850

1 Silver table Fork		Crest No. 1
4 " " Do.		" " 2
	Motto	
1 " " Do.	*Spero Meliora*	" " 4
2 " " Do.		" " 5
1 Silver dessert Fork		Crest No. 5
1 " table Do.	with initials	H.D.S.G.
1 " " Do.	"	A.Mc.D.
1 " " Do.	"	G.A.M.
1 " " Do.	"	J.T.
1 " " Spoon		Crest No. 3
	Motto	
1 " " Do.	*Spero Meliora*	" 4
1 " " Do.		" 5
1 " tea Do.		" 5
1 " " Do.	initials	J.S.P.
1 " dessert Do.	"	J.S.P.
1 " " Do.	"	G.G.
1 round silver Plate	Sir John Franklin K.O.H.	

1 Star or Order
2 Pieces Gold Watch Case
1 Case silver gilt pocket Chronometer & dial
7 pieces Cases of Silver Watches
1 small silver pencil Case
1 piece of Silver Tube
1 piece of an Optical Instrument
1 old cold [gold] Cap Band
2 pieces (about 2 inches) gold Watch Chain
2 Sovereigns
1 Half crown
4 Shillings
2 Leaves of the Students Manual
1 surgeons Knife

1 Scalpel	
2 Knives	
1 Do. Womens' or Shoemaker's	
1 pocket Compass Box	
1 Ivory Handle of a table Knife	Marked 'Hickey'
1 narrow tin case	Marked Fowler
1 " " Do. no cover	W.M.

Sundry other articles of little consequence

CHAPTER II

THE COMPANY ACCEPTS THE CHALLENGE

INTRODUCTION

While the furore was raging about Rae's reports of cannibalism among the Franklin expedition members, more practical steps were being taken to verify the Inuit reports, and at the same time to provide assistance, if necessary, to Captain Richard Collinson's ship, *Enterprise*, about which nothing had been heard since the summer of 1851. There was a very understandable reluctance on the part of their Lords Commissioners of the Admiralty to dispatch another major seaborne expedition to the Arctic. Over the period 18–20 October the court-martials of Captains Sir Edward Belcher, Henry Kellett, Robert M'Clure and Commander George Richards had been taking place at Sheerness for having abandoned five ships in the Arctic, namely HMS *Investigator*, *Resolute*, *Intrepid*, *Assistance*, and *Pioneer*;[1] undoubtedly the fact that these ships had discovered not the slightest clue as to the fate of the Franklin expedition may have helped to sway their Lordships' opinions against a further similar shipborne expedition. Their Lordships must have welcomed the support of the editor of *The Times* as expressed in remarks such as:

> The Arctic ice is already studded, as it were, with great ships which have been despatched in search of our lost friends, and it is matter of notoriety that but little benefit has been derived from all the energy and endurance of their captains and crews, and from all the expense which has been lavished upon them.[2]

The reluctance to become embroiled in further expensive arctic searches was undoubtedly also influenced by the outbreak of the Crimean War in March 1854. The Royal Navy had been engaged in operations in the Baltic, the White Sea and in the Pacific since the summer, and the Allied force had landed in the Crimea and commenced the siege of Sevastopol in September. In short, the Admiralty had more pressing claims on its ships and manpower.

For all these reasons the Admiralty turned to the Hudson's Bay Company to mount a search of the area which the Inuit had generally indicated as being the site where the Franklin expedition had come to grief. Very sensibly, the Admiralty's first step was to seek Rae's expert advice, via Sir Francis Beaufort, the Hydrographer. In

[1] *The Times*, 18 October 1854, p. 8.
[2] *The Times*, 2 November 1854, p. 6.

a letter dated 26 October,[1] Rae recommended that an expedition be dispatched by canoe down the Back River from a base on Great Slave Lake for the purpose of verifying the reports he had obtained from the Inuit, and if possible of recovering documents. Two canoes would be required, each with a crew of six or seven men, under the command of two officers. As a safety precaution, a small party with a boat should be stationed at the east end of Great Slave Lake, in case the river party were forced to fall back there. Since communication with the Inuit at the mouth of the Back River would be vital in achieving the expedition's aims, Rae recommended that William Ouligbuck, the Inuk interpreter who had been with him on his expedition to Repulse Bay and beyond, should be hired to accompany the new expedition. Finally Rae stressed that haste was critical, in that a great deal of preparations would be necessary if the search expedition were to be mounted in the summer of 1855.

These excellent suggestions of Rae's were forwarded by Captain T. B. Collinson, Captain Richard Collinson's brother, to *The Times* on 26 October. In his covering letter Collinson noted that:

> ... I must state, what will be heard with regret by all interested in this subject, that the enterprising and experienced Dr. Rae had determined previous to his present return, in consequence of the effect of his numerous Arctic travels upon him, not to return again to those latitudes.[2]

Simultaneously with these recommendations, Rae was also preparing a slightly more detailed memorandum for the Hudson's Bay Company; in it he addressed such topics as his recommendations as to the composition of the expedition, the survey and navigation instruments required, and trade goods for the Inuit. By references to the competence of Mr Bernard Ross and Chief Trader James Anderson in handling survey and navigation instruments, Rae also obliquely recommended these individuals as possible expedition leaders.[3]

The Admiralty, equally sensibly, sought the opinion of Sir George Back, who had led the pioneer expedition down the river named after him, in the summer of 1834. This had been a privately sponsored expedition to search for Sir John Ross's expedition which had disappeared into the Arctic in search of the Northwest Passage in 1829 and had not been heard of since. Accompanied by Dr Richard King, two carpenters brought from England, four men from the Royal Artillery, recruited in Montreal, and a party of voyageurs, Back travelled by the standard fur-trade route from Montreal to Fort Resolution on Great Slave Lake in the summer of 1833. Heading east down Great Slave Lake, with some of his party he struck north via the Hoarfrost River to Lake Walmsley, then east to Artillery Lake and north via Clinton-Colden Lake to Aylmer Lake.[4] From there he reconnoitred the portage over the

[1] *The Times*, 27 October 1854, p. 8.

[2] Ibid.

[3] HBC A.8/17, f. 194.

[4] George Back, *Narrative of an Expedition in H.M.S. Terror, Undertaken with a View to Geographical Discovery on the Arctic Shores, in the Years 1836–7*, London, 1838; Richard King, *Narrative of a Journey to the Shores of the Arctic Ocean in 1833, 1834 and 1835, under the Command of Captain Back, R.N.*, 2 vols, London, 1836.

divide to Sussex Lake, one of the sources of the Back River. After this reconnaissance Back travelled back south to Fort Reliance at the head of the East Arm of Great Slave Lake, where the remainder of the party had been erecting winter quarters.

In April 1834 news arrived from England that Ross and his party had returned safely from the Arctic. Nonetheless Back decided to proceed with his plans, since a secondary objective had been to determine where the Back River entered the sea, and to survey as much of the arctic coastline as possible. In the spring of 1834 Back and his party travelled north to the south end of Artillery Lake, where an advance party had built a 30-foot boat especially for the trip down the Back River. They started north up Artillery Lake on 10 June, hauling the boat over the ice on runners. They crossed the divide to Sussex Lake and started down the Back River on 28 June; following the river through Pelly, Garry and Franklin lakes, they reached the sea on 29 July and explored the coast as far north as Point Ogle. They started back upriver on 19 August and had returned safely to Fort Reliance by 27 September. After a further wintering at Fort Reliance Back started for home on 21 March 1835 and reached Liverpool on 8 September.[1]

In his reply to Sir James Graham,[2] Back, like Rae, recommended the use of two canoes, rather than a boat or boats, and that the expedition should start down the Back River not later than the end of June. Like Rae, he recommended that a depot of provisions should be cached at the east end of Great Slave Lake, in case the expedition had to fall back there.[3]

On 27 October Mr W. A. B. Hamilton, Secretary at the Admiralty, wrote to Mr Alexander Barclay at the Hudson's Bay Company on behalf of their Lords Commissioners of the Admiralty 'to express their earnest anxiety that the Company, through Sir George Simpson, their Resident Agent, should take immediate steps for organizing in the most effective manner two Expeditions on an adequate scale to be sent into the Arctic Regions, so soon in next spring as the weather will permit'.[4] One expedition was to proceed down the Mackenzie River, in an attempt to locate Collinson's expedition; the other was to proceed to the mouth of the Back River to verify the Inuit reports brought back to England by Rae, ascertain whether there were any survivors and locate, if possible, remains or records which might throw light on the fate of the Franklin expedition. The selection of the expedition members would be left to the Company, and the British Government undertook to cover the entire expense of both expeditions; Hamilton specifically noted that 'it is the wish of the British Government, and its declared intention to pay liberally for the service to be rendered, and to reward specially any acts of signal daring and distinguished merit'.[5] Hamilton enclosed a copy of Sir George Back's letter of the 27th to Sir James Graham with his recommendations as to the Back River expedition.

[1] Back, *Narrative of the Arctic Land Expedition*; King, *Narrative of a Journey to the Shores*.

[2] First Lord of the Admiralty.

[3] HBC A.8/17, f. 194.

[4] HBC A.8/17, f. 191; Great Britain. Parliament, *Further Papers relative to the Recent Arctic Expeditions in Search of Sir John Franklin and the Crews of H.M.S. "Erebus" and "Terror"; Presented to both Houses of Parliament by Command of Her Majesty*. London, 1855, p. 846.

[5] HBC A.8/17, f. 191; Great Britain, *Further Papers*, 1855, p. 857.

DOCUMENTS

1. Rae's recommendations, forwarded to Sir Francis Beaufort, 26 October 1854.[1] The following statement is drawn up at the request of Sir F. Beaufort, R.N., Hydrographer to the Admiralty, and explains what I consider the best mode of carrying into effect two overland expeditions, the one for the purpose of searching for and aiding, if requisite, Captain Collinson's ship the Enterprise, and the other to make further inquiry into the fate of, and securing, if possible, some documents from Sir John Franklin's party, the traces of which I discovered this year. I hasten with as little delay as possible to comply with this request. The route of both these expeditions would be as far as Slave Lake the same; here they would separate...

[For the Mackenzie expedition] I can recommend two excellent men as steersmen, who have had some experience, and who will engage probably; and these are John Fidler, who was with me in 1851, and Thomas Mistigan, who was last year with me at Repulse Bay, and is an excellent deerhunter besides...

With regard to the other expedition, which, I think, should descend the Back River, although, in general, no advocate for employing canoes on Arctic expeditions, yet in the present case these light craft should be used instead of boats, as there is a large amount of land portage, and perhaps ice portage, and the greater part of the route is river-way. These canoes – two in number – can be prepared either at Athabasca or at Fort Resolution, on Slave Lake, or perhaps one at each place. They ought to be rather shorter, but about equally broad with the usual north canoe; and a crew of six, or at most seven men, would be enough for each. Two officers should likewise accompany this party; one of whom might be, as in the other case, an experienced active clerk in the Hudson's Bay Company's service.

In like manner as for the Mackenzie expedition, a boat and small party should be stationed at the east end of Great Slave Lake in the autumn, with provisions and ammunition and hunters, as a precaution for the safety of the coast party.

The Esquimaux interpreter, William Ouligbuck, who accompanied me, should form one of this expedition, and he can be sent from Churchill, where he is at present stationed, to Athabasca before the breaking up of the ice, provided immediate information is forwarded to Sir G. Simpson at Lachine regarding these expeditions; the communication to him, I need scarcely say, should pass through the hands of the Governor and Committee of the Hudson's Bay Company in London.

At the same time that Ouligbuck travels from Churchill by York Factory to Athabasca a very excellent Halkett's boat,[2] which I left at York Factory, can be forwarded.

[1] *The Times*, 27 October 1854, p. 8.

[2] The Halkett boat was an inflatable boat, the invention of Lt Peter Halkett, RN. It was made of rubber with a canvas cover and measured about 9′ by 4′ (2.7 x 1.2 m) when inflated and was designed to carry two men. When deflated it could be folded up into a bundle easily carried by one man. It came equipped with paddles and bellows for inflating it. For further details see: *The Beaver*, 1955.

Dr John Rae had used a Halkett boat on the first three of his expeditions (to Repulse Bay and Melville Peninsula in 1846–7; to Great Bear Lake and Coronation Gulf with Dr John Richardson in 1847–9; and to Great Bear Lake and Victoria Island in 1850–51). With reference to its use on the first expedition he wrote:

I know of no [other] Esquimaux interpreter in the Hudson's Bay Company territories within reach at such short notice; for the party on the Mackenzie River, perhaps one can be obtained here.

I do not know at what date it will be requisite that the officers to command these expeditions should leave this country. They will have to travel 1,500 miles on snowshoes, aided, if they desire it, by a sledge or carriole drawn by dogs. The best route appears to me to be through the United States to St. Paul's, and thence across the plains to Red River Colony, where the men for the party would be engaged, and dogs for sledge-hauling could be purchased to continue the route. This arrangement had better be left to Sir G. Simpson, whose knowledge of the Indian country, its resources, and the best route through it, is superior to that of any other man living.

I believe that neither provisions (unless it may be a little tea and sugar) nor clothing will be required to be sent from England or Canada, and the tea and sugar may be had at Red River if required. No spirits are allowed by the Company to enter the Mackenzie River.

Three boatloads of provisions, amounting to nearly eight or nine tons, were sent to Mackenzie River district last summer on account of Government.[1] These will of course be available.

Permit me to impress upon you the necessity of haste in setting the expedition in train. If much time is lost at this late season, it is impossible to make the arrangements in a complete manner.

2. Rae's recommendations, solicited by Archibald Barclay and written 27 October 1854.[2]

There were three boats loads of provisions sent to McKenzie River District last summer on account of Her Majesty's Government, which, if provisions are scarce in the district should be available for these Expeditions. These stores can be

---

> During the whole of our spring [1847] fishing [in Repulse Bay] Halkett's air-boat was used for setting and examining the nets, and was preferred by the fishermen to the large canvas canoe, as it was much lighter, and passed over and round the nets with more facility. Notwithstanding its continued use on a rocky shore, it never required the slightest repair. It is altogether a most useful little vessel, and ... ought to form part of the equipment of all surveying parties, whether by land or sea (*Narrative of an Expedition to the Shores of the Arctic Sea in 1846 and 1847*, London, 1850, p. 176).

Rae had planned to use one of these boats on his final expedition (1852–4), but it did not reach York Factory from England in time. This was the boat which Rae was now suggesting for the Back River expedition.

[1] In a letter to the London headquarters of the Company, dated 30 June 1854, Sir George Simpson wrote:

> I have the satisfaction to inform you that the instructions which were given last winter to the officer in charge at Red River, to forward to M'Kenzie River a supply of clothing and provisions for the use of any parties who may seek refuge at the Company's posts in that quarter from exploring vessels now in the Arctic seas, were received in sufficient time to enable the three boats conveying those supplies to take their departure on the 1st instant, under the charge of Mr D. A. Harrison, apprentice clerk. The crews are reported to be efficient; and as the season is favourable I have little doubt the boats will be enabled to proceed to such point on the M'Kenzie River to deliver their cargoes as chief trader Anderson may think most expedient, and to return to this place before the close of the navigation.

Reprinted in Great Britain, *Further Papers Relative*, 1855, p. 846.

[2] HBC A.8/17, ff. 194–5.

replaced from Red River Settlement in summer 1855 if requisite. In the event of there being no spare sails, sheeting or canvas at Athabasca,[1] it might be as well that sails of Russia or Duck Sheeting be made at Red River, and forwarded by sledges to Athabasca. These sails should be lugs – two for each boat – and of a size to suit a boat 22 feet in length, and 7 feet 6 inches beam.

William Ouligbuck[2] should be sent from Churchill for the canoe party by the Back River, and if no other interpreter can be procured, one of the Churchill Esquimaux ought to accompany Ouligbuck, whose services would be useful with the boat party down the McKenzie.

An assortment of beads, files, Knives, daggers, thimbles, fish-hooks, needles, hoop-iron, axes, ice chisels etc., etc. should be carried to the sea by each party, for presents to, and to barter with the Esquimaux.

One or two of the Esquimaux of Back River should be engaged, by promise of large reward, to accompany the canoes to the sea, and point out the scene where terminated the mortal career of so many of our countrymen.

Among the men known to me at Red River and elsewhere whom I would recommend to be engaged on this important service are

Thomas Mistegan*	Steersman of Boat	Norway House
Murdoch McLellan*	Middleman	do
John Fidler*	Steersman of Boat	Red River
James Johnston*	Bowsman & Fisherman	do
John McDonald	Middleman	do
Jacob Beads	Bowsman & Carpenter	do
Charles Kennedy	do	do
Samuel Sinclair	do	do
Henry Fidler*	Middleman	do

* These men took part in the expedition.

Mr. Bernard Ross[3] in McKenzie River has some knowledge of Astronomy, and can take and compute observations for latitude & chronometers.

Chief Trader Anderson, in charge of the same district knows, I believe, how to use the sextant.

The following articles will be required by the Officers who are to command the Expeditions about to be sent in search of Captain Collinson, and of any further traces of Sir John Franklin's ill-fated party:

2 good Pocket Chronometers
2 small Sextants – or as portable as possible
2 Artificial Horizons – do
2 Prismatic or Azimuth Compasses – small size
say 3 to 4 ins. diameter

[1] Fort Chipewyan, headquarters of the Athabasca District.
[2] The Inuk who had served as Rae's interpreter on his last expedition to Repulse Bay and Boothia Peninsula.
[3] In charge of Fort Resolution.

2 Spirit Thermometers, graduated to 60° or 70° below Zero
2 Telescopes, small but good
2 small cases Mathematical Instruments, as light and portable as possible
2 sets of best Charts of the Coast from the McKenzie to Back's Fish River, and of the Arctic Regions generally
1 Copy of Sir George Back's Narrative of his descent of the Fish River
2 Nautical Almanacs for 1855 reduced to the smallest bulk compatible with the requisite details for computing solar astral & lunar observations
1 Copy of Raper's Navigation
2 Esquimaux Vocabularies

3. Recommendations of Sir George Back, forwarded to Sir James Graham, First Lord of the Admiralty, 27 October 1854.[1]

In recommending the best mode of organizing an Expedition to the mouth of the Back River, I may mention that the whole of the details are so well known to Sir George Simpson, with whom I cooperated that it is scarcely necessary to refer to them.

In obedience to your wishes, however, I may briefly state that, instead of a Boat I would suggest on this occasion that two Canoes be chosen, somewhat shorter though equally broad with those in common use.[2]

Outfits and Pemmican, with the usual stock of provision to fall back upon in the Season, will be required to be deposited at the East end of Great Slave Lake, either at Fort Reliance or at a fishing station a little to the westward.

The two canoes complete should be at the Back River not later than the end June 1855, and cachés of Pemmican should be made at different places along the route for the support of the party on its return.

The arrangements which the knowledge and experience of Sir George Simpson will enable him to make, render these remarks almost superfluous, and I have full confidence under his able direction that the thorough examination of the coast at the mouth of Back River wil be satisfactorily completed next year.

4. Letter from W. A. B. Hamilton, Admiralty to Alexander Barclay, Hudson's Bay Company, requesting assistance, 27 October 1854.[3]

With reference to the Report of Dr. Rae transmitted by the Governor and Committee of the Hudson's Bay Company on the 23rd Instant. I am commanded by my Lords Commissioners of the Admiralty to express their earnest anxiety that the Company, through Sir George Simpson, their Resident Agent, should take immediate steps for organizing in the most effective manner two Expeditions[4] on an adequate scale to be sent into the Arctic Regions, so soon in next spring as the weather will permit.

The first Expedition must proceed to the mouth of the Back River, and institute a

[1] HBC A.8/17, f. 194.
[2] The normal 'canot du nord' was about 25 feet (7.5 m) in length.
[3] HBC A.8/17, ff. 191–3.
[4] The other expedition was to try to make contact with HMS *Enterprise* (Captain Richard Collinson), which had headed north through Bering Strait in search of the missing Franklin expedition in the summer of 1851 and had not been heard of since.

diligent search throughout the Islands and space pointed out by the Esquimaux to Dr. Rae as the place where, in 1850, a portion of the crews of the Erebus and Terror were last seen, and where, by the same Report, it is stated that they perished.

Every effort should be made to find some of the Esquimaux who themselves saw the body of Englishmen, and, in 1850, communicated with them. Diligent search should be made for any records which may have been deposited by Sir John Franklin or his Officers in that neighbourhood; every exertion must be exhausted on the spot to find traces of the survivors, if happily they exist, and if not, every portable relic which may serve to throw light on the fate of these gallant men should be brought home.

Any proofs of wreck, which might show where the Erebus and Terror were lost, would be valuable, but the principal object of the Expedition is to ascertain whether there are any survivors of these two ships' Companies – whether the Report made to Dr. Rae by the Esquimaux be true – and, if true, whether any remains can be discovered on the spot, which may further explain the proceedings and events which terminated so fatally.

Their Lordships desire me to enclose you herewith a Memorandum drawn by Sir George Back, which gives an outline of the measures, in his judgment necessary, to render the search confided to this Expedition most effective.

But their Lordships having indicated the objects in view, are disposed to leave the arrangements in detail to the known zeal and direction of Sir George Simpson and of the Hudson's Bay Company's servants. My Lords are confident that a wise selection will be made of a person competent to command the Expedition, who is inured to such hardships and perils, and who is accustomed to communicate with the natives. It will be an honor to be selected for this service, for it is one in which the feelings of the British nation are deeply interested, and the fate of Sir John Franklin and his men has been regarded for years with intense anxiety both in Europe and America....

The British public will cheerfully bear the whole expense of these two Expeditions, of which the Company will be pleased to keep an account, and to the credit of which advances will be made from time to time, if required by the Company.

The Servants of the Company, who may be selected to serve on these two Expeditions may be assured that it is the wish of the British Government, and its declared intention to pay liberally for the service to be rendered, and to reward specially any acts of signal daring and distinguished merit.

5. Letter of instructions from John Shepherd, Deputy Governor of the Hudson's Bay Company, to Sir George Simpson, Lachine, 27 October 1854.[1]

The public journals will inform you of the intense interest which the information brought by Dr. Rae, regarding the melancholy fate of Sir John Franklin and his

[1] HBC A.6/31, f. 142. Sir George Simpson, the Company's Governor-in-chief in North America, based at Lachine near Montreal, was born in Lochbroom parish, i.e. the Ullapool area of the Scottish Highlands, around 1786 or 1787. Born out of wedlock, he was brought up by his aunt, Mary Simpson. Around 1800 he travelled south to London where he was employed by the sugar brokerage firm of Graham and Simpson, in which his uncle Geddes Mackenzie Simpson, was a partner. In 1812 this company merged with Wedderburn and Company. Andrew Wedderburn (later Colvile) was a member of

party has excited in this country, and of the general desire which has been manifested that additional exertions should be made to satisfy the public that every possible means has been taken with the view of rescuing any portion of our Countrymen that may be still alive, and exposed to peril in the Arctic Regions. Deeply sympathizing in this feeling the Lords Commissioners of the Admiralty have applied to us to engage your valuable and zealous services, and of any of our officers who may be available in furtherance of this humane object, and you will easily imagine that the Governor and Committee, warmly participating in the general feeling adverted to, have cordially acceded to their Lordships' wishes.

With the view therefore of fully informing you of their Lordships' views, copies of the several documents received by them will accompany this letter, and, in the absence of the Governor, I have to request, in his name and that of the Committee, that you will spare no expence nor exertion which you may think calculated to accomplish their Lordships' wishes. You will observe that on the subject of expence,

---

the Committee of the Hudson's Bay Company, and by this chain of coincidence George Simpson became involved with that company.

In 1820, at Colvile's urging, George Simpson was appointed Governor-in-chief, *locum tenens*, for the Company in North America; it was thought that his demonstrated business acumen and energies would equip him well for the organization of the combined operations of the Hudson's Bay Company and the North West Company after their merger in 1821. Simpson travelled to North America in 1820 and, after successfully holding his own against the officers of the North West Company during a turbulent winter at Athabasca, took up residence first at York Factory then at Lachine, from where he quickly demonstrated his amazing talents in the running of the Company's far-flung commercial empire.

For almost forty years he made long annual canoe trips to Norway House for the annual meeting of the Council of the Northern Department, or to more remote Company posts. Throughout his life he prided himself on the speed with which he made these trips. In 1841–2 he also made a trip around the world, promoting Company interests and handling Company business at Sitka, Honolulu and Okhotsk, from where he travelled across Russia to St Petersburg in connection with negotiations with the Russian Government and with the Russian American Company.

Throughout his life Simpson maintained an amazing volume of correspondence, both professional and private; in both cases his mastery of detail was astounding, as his correspondence associated with the Anderson/Stewart expedition will reveal. Where Company business was concerned he was extremely hard-nosed, but kind and generous to friends. Both aspects of the man will be revealed in his dealings with Anderson and Stewart.

Simpson married his cousin Frances Simpson in London in 1830 and by her had five children. Prior to his marriage, however, he had had liaisons with several Indian women, *à la façon du pays*, liaisons resulting in several children. Simpson supported these children, but they were kept separate from his 'other' family.

Simpson was deeply interested in, and supportive of arctic exploration. He supported Sir John Franklin's first expedition to some degree, and played a critical role in the organization and support of Franklin's second expedition of 1825–7; he was entirely responsible for the expedition of Peter Dease and Thomas Simpson of 1837–9 which explored an impressive stretch of the arctic coast. It was for this contribution to geographical exploration that Simpson was knighted in 1841. And after the disappearance of the Franklin expedition, he was responsible for various expeditions sent in search of it, namely those led by Rae and Richardson in 1847–9, and by Rae in 1850–51. It was on yet another expedition authorized by Simpson, in 1853–4, that Rae found the first clues to the fate of the Franklin expedition, clues which the Anderson/Stewart expedition was designed to clarify.

Simpson died on 7 September 1860 and was buried in Mount Royal Cemetery in Montreal. The enduring success of the Hudson's Bay Company as a commercial empire was due in large part to George Simpson's remarkable drive and ability. He would have become a business leader of world-wide reputation in any age. For further details see: John S. Galbraith, *The Little Emperor: Governor Simpson of the Hudson's Bay Company*, Toronto, 1976; John S. Galbraith, 'Sir George Simpson', *Dictionary of Canadian Biography*, vol. VIII (1851–60), Toronto, 1985, pp. 812–18.

PLATE II. Sir George Simpson, *c.* 1856–60. Copy of a daguerreotype (Hudson's Bay Company Archives)

it is clearly stated that not only will all the cost actually incurred be defrayed by Her Majesty's Government, but that the Lords of the Admiralty will be prepared to award special remuneration to those who may deserve it for valuable service.

A very important point will be, the selection of the officers to command the two Expeditions, the Lords of the Admiralty having decided that the whole of the arrangements, conduct & management thereof shall be left to the Hudson's Bay Company's officers; their Lordships point out merely the several routes which they desire each party should adopt. We trust that you will be enabled to select Officers to command the two Parties, of experience, ability, and energy, and that their exertions will be ultimately crowned with success.

Inclosed is a Memorandum from Dr. Rae, published in the 'Times' of this morning, which contains his deliberate opinion as to the best mode of proceeding. He has mentioned to us that Mr. Anderson, now on the McKenzie River, would be admirably qualified for the command of one of the parties, but, of course, you will be best able to judge of those whose services may be available at the proper time; – a certain degree of knowledge of astronomy and taking and applying observations with nautical instruments will of course be indispensible in one at least of each party.

This letter and its accompaniments have been very hastily prepared, in order to save time by forwarding them by the present packet, and also to give you all the time possible to organise the details of the two Expeditions. The necessary instruments and charts will be forwarded by next packet.

[Enclosures]
Copy Letter from Capt. Hamilton to A. Barclay Esq. 27th Oct. 1854
Memorandum from Admiralty respecting Capt. Collinson
Printed Records of Captn. Collinson, June 14th to Aug. 27 1854
Copy letter from Sir G. Back to Sir J. Graham, Oct. 27 1854.
Slip from 'Times' newspaper, Octr. 27 1854 containing Dr. Rae's Report – Proposed Arctic relief party.
M.S. Memorandum from Dr. Rae in reference to the contemplated Expeditions.

6. Extract of a letter to Sir George Simpson from W.G. Smith re instruments and charts being forwarded for the search expedition, 3 November 1854.[1]

... A case containing the Instruments, Charts, etc. for the intended Arctic Expeditions alluded to in the Deputy Governor's letter of the 27th Ultimo, has been forwarded to Liverpool for transmission by the packet of tomorrow – it has been addressed to the care of Messrs. Maitland Phelps & Co., who have been instructed to send it on from New York immediately on its arrival.

A small parcel containing two sets of Maps from Mr. Arrowsmith has also been sent to Messrs. Maitland & Co.'s care. All the other articles (a list of which is included) have been furnished by the Lords Commissioners of the Admiralty and, as they have to be returned on the completion of the service, it will be necessary that

[1] HBC A.6/31, f. 148.

you give particular instructions, that every care be taken of them, and that they be forwarded to England as opportunity shall offer...

P.S. The Ratings of the two Chronometers mentioned in the accompanying List having been omitted in the Case, are herewith enclosed.

List of Articles forwarded by the Admiralty to the Hudson's Bay Company for the use of the Officers to be sent in command of the Expeditions about to proceed in search of Captain Collinson, and of any further traces of Sir John Franklin's ill-fated party, November 2nd 1854.

1 Pocket Chronometer, No. 5158, Arnold & Dent.
1 Ditto, No. 516, Murray
1 Artificial Horizon, No. F114, Watkins & Hill
1 Dark Horizon, ditto.
1 Small Sextant, No. D 167, Cary
1 ditto, No. DA 65, ditto.
1 Prismatic Compass, No. H214, Watkins & Hill
1 ditto, Thomas Jones
2 Spirit Thermometers graduated to 60° below zero, No. 6016, Adie & Son
1 ditto, graduated to 70° below zero, No. 483, Pastorelle
1 Telescope, R. Adie, Liverpool
1 ditto, Watkins & Hill
2 Small Cases of Mathematical Instruments
2 pocket Arctic Compasses
1 Copy of Sir George Back's Narrative of his descent of the Fish River
2 Nautical Almanacs for 1855
1 Copy of Raper's Navigation
2 Esquimaux Vocabularies
2 Tables of the Sun's Bearings 1852

Charts
2 Sheets Arctic America (No. 1 Sheet) No. 1711
2 ditto (no. 2) No. 1320
2 ditto Chart of North Polar Sea No. 260
2 ditto, Discoveries in Arctic Sea up to 1852, No. 2118.
2 ditto Arctic Sea, Baffins Bay (No. 1 Sheet) No. 2177
2 Sheets Chart shewing North West passage discovered by Captn. McClure, October 14th 1853
Ratings of the 2 Chronometers mentioned above
Hudson's Bay House, November 2, 1854

CHAPTER III

THE WHEELS IN MOTION

INTRODUCTION

The Company's instructions reached Sir George at Lachine in mid-November and the latter swung into action with his usual energy and efficiency. Since news of Collinson's safe emergence from the Arctic had reached England via San Francisco and the electric telegraph on 7 November,[1] there was no longer any need to dispatch an expedition down the Mackenzie, and Sir George could concentrate his attention on the Back River expedition. His first (and probably most important) decisions involved the selection of the two leaders.

His choice as overall commander of the expedition was James Anderson, then a Chief Trader, in charge of the Mackenzie District and based at Fort Simpson. He was born on 15 January 1800 in Calcutta, the son of Robert Anderson and Eliza Charlotte, née Simpson.[2] After retiring from a military career, his father ran a plantation in India until he and the family returned to England in 1817. Then in 1831 they emigrated to Upper Canada.

James Anderson joined the Hudson's Bay Company in that same year, and was posted to Moose Factory at the southern end of James Bay. In his very first winter with the Company he played a major role in the arrest, trial and execution of the Indians who murdered a white family at Hannah Bay. He was next posted to Nipigon and while there, in 1847 was promoted to Chief Trader. Then in 1850 he was posted to Fort Chipewyan, to take charge of the Athabasca District, and in 1851 to Fort Simpson to run the remote and sprawling Mackenzie District. There he found the accounting system in disarray, but had considerable success in improving it. It was his decision (despite the protestations of Robert Campbell) to cut the Company's losses by the retrenchment in the Yukon which saw the abandonment of the Frances Lake, Pelly and Fort Selkirk posts. His promotion to Chief Factor would occur while he was embroiled in the preparations for the Back River expedition.

On his return to Fort Simpson after the expedition, Anderson resumed his duties of running the Mackenzie River District. At George Simpson's suggestion, in the summer of 1857 he dispatched a small party led by Roderick McFarlane down the

[1] *The Times*, 8 November 1854, p. 6.

[2] C. Stuart Mackinnon, 'James Anderson', *Dictionary of Canadian Biography*, IX (1861–70), Toronto, 1976, p. 5.

Beghula (Anderson) River to investigate the potential for trade with the Inuit of the Liverpool Bay area.[1] On the basis of this reconnaissance McFarlane established Fort Anderson (named after his superior) in the summer of 1861 but it operated for only five years.

Anderson's health had been severely undermined by the rigours of the trip down the Back River and in the autumn of 1857 he left the Mackenzie District. On this occasion William Hardisty, in charge of Fort Yukon, the most remote post in the district, sent Anderson a gift of a pencil case with the accompanying letter:

> My dear Sir,
>
> To mark the high sense I entertain of your amiable character, as a man and a Bourgeois and the numerous acts of kindness I have received from you and Mrs. Anderson, during your stay in McK Rr I take the liberty of presenting you with this slight token of my sincere regard. Trifling as it is, I know you won't value it the less, because it comes from me, not for my sake, but on account of old times from a fellow feeling, caused by a sojourn together in McK. Rr.
>
> I have – in common with the other Officers of this District – to thank you for innumerable acts of disinterested Kindness. You first taught us to know, and to perform our duties aright, and should any of us hereafter be considered worthy of being placed in charge of Districts we shall follow the example you set us, while in charge of Posts – and ever look up to you as our "Head", "Our Bourgeois" —in the sense that we of McK. Rr. apply the term – Our Master in business – Our friend in social intercourse.
>
> May God bless you, and may health, happiness and prosperity attend you and your family wherever you go, is the earnest prayer of
>
> My dear Sir
> Yours most sincerely
> Wm. Lucas Hardisty[2]

After his furlough Anderson was posted to Mingan on the North Shore of the St Lawrence as being a relatively less remote posting. Here, as at Fort Simpson, he demonstrated his skills at making a marginal district profitable; he also entertained the Governor-General and other dignitaries who came for salmon fishing.

Anderson resigned his commission due to ill health in 1863 and retired to Sutton West, Upper Canada. Here he died, in October 1867, of tuberculosis. He was survived by his wife Margaret, daughter of Chief Factor Roderick Mackenzie, six sons and a daughter.[3]

James Green Stewart, selected by Simpson as Anderson's second-in-command, was born in Quebec City on 21 September 1825,[4] son of John Stewart,[5] a Scot by birth, a businessman, Justice of the Peace and a Member of the Executive Council of Lower Canada. On 2 November 1843 John Stewart had written to George Simpson

[1] C. S. Mackinnon. 'James Anderson', p. 5.

[2] Letter from William Hardisty to James Anderson, 10 November 1857. Warren Baker collection.

[3] C. S. Mackinon, 'James Anderson', p. 5.

[4] C. Stuart Mackinnon, 'James Green Stewart', *Dictionary of Canadian Biography*, XI (1881–90), Toronto, 1982, pp. 854–5.

[5] Joanne Burgess, 'John Stewart', *Dictionary of Canadian Biography*, VIII (1851–60), Toronto, 1985, pp. 837–9.

PLATE III. James Anderson, wearing the Arctic Medal he received for his leadership of the Land Arctic Expedition (C. S. Mackinnon collection)

requesting that his son James (aged eighteen) be employed by the Company.[1] Replying on the 21st Sir George pointed out that although

> We are at present overstocked with young gentn. in the service ... I shall, nevertheless, recommend your application in such terms, that I think it is more than probable the Govr. & Com. may authorise me to receive your son in the capacity of App. Clerk (the usual footing young gentn. are admitted into the service) at the opening of the navigation, say the latter end of April next.[2]

In his letter of recommendation to the Governor and Committee[3] Simpson pointed out John Stewart's prominence in Lower Canada and suggested that:

> although we do not particularly require the services of an Apprentice Clerk at present, I beg to recommend that his application be favourably considered, as I think it is highly desirable to have a few young gentlemen respectably connected in this country in the concern.

This recommendation was successful and early in 1844 James Stewart was appointed to the Lake Superior District, spending his first winter (1844–5) at either Fort William or Michipicoten. In the spring of 1845 he was transferred to the Mackenzie District, travelling via Norway House, Portage la Loche and Fort Resolution to Fort Simpson, where he arrived in October. Shortly afterwards he travelled up the Liard to Fort Liard where he was to serve under Adam McBeath, who was in ill health.[4] After his first winter at Liard, he took over the post from McBeath, until the summer of 1847.

During that autumn he travelled on up the Liard in the company of Robert Campbell, who was responsible for developing trade in what is now Northern British Columbia and the Yukon.[5] Stewart took charge of Frances Lake while Campbell continued north to the post at Pelly Banks. The following summer Campbell summoned Stewart to Pelly Banks in order to travel down the Pelly with him to establish a new post at the confluence of the Pelly and the Lewes, the latter being the name by which the Upper Yukon was known at the time. The post was named Fort Selkirk. This was in many ways the most remote post in the Company's territories, or at least the most difficult of access. From Fort Simpson the route lay up the Liard past Fort Liard, then up an endless series of rapids, including those of the notorious Whirlpool Canyon, to the mouth of the Frances River (just west of the present settlement of Watson Lake, Yukon), then up the latter river to Frances Lake. From there an overland trek of over 130 km led to the headwaters of the Pelly.

By August 1849 supplies at Fort Selkirk were running low and Campbell sent Stewart back to Frances Lake to meet the supplies coming from Fort Simpson. But by 1 November they had not arrived, and, scrounging what little he could from the Frances Lake and Pelly Banks posts, Stewart returned to Fort Selkirk, arriving in December 1849. Campbell, Stewart and their men somehow struggled through the

[1] HBC D.5/9, f. 158.
[2] HBC D.4/63.
[3] HBC A.12/2.
[4] HBC D.5/15, ff. 460–61.
[5] Clifford Wilson, *Campbell of the Yukon*, Toronto, 1970, p. 72.

long winter of 1849–50, then in the spring Stewart started on the long trek (some 1760 km) to Fort Simpson; en route he rescued P. C. Pambrun, in charge of the Pelly Banks post, which had been attacked and looted by Chilkat Indians.

Reaching Fort Simpson safely Stewart started back with a half boatload of provisions for the three posts. Dr John Rae, his superior as the Chief Trader in charge of the Mackenzie District, remarked in a letter to Sir George Simpson that Stewart would manage to get the supplies through if anyone could and that he was greatly impressed with him 'for having made so long and difficult a journey in coming down with but very little provisions for himself and companions'.[1] From a traveller of Rae's calibre, this was high praise indeed. In his reply Simpson commented that 'Stewart's exertions in this emergency are beyond all praise and have secured him a character for zeal, activity and perseverance, which will bring him prominently into notice'.[2]

In the summer of 1851 Campbell made a trip down the Lewes and, on reaching the Company's post at Fort Yukon (at the confluence of the Porcupine and Yukon rivers), commanded by William Hardisty, established that the Yukon and Lewes were one and the same river. The following summer (1852) Stewart made the trip down to Fort Yukon to fetch a cow. On his return he found that a band of Chilkat Indians had attacked and looted Fort Selkirk; Campbell had survived but had practically no fuel left. Campbell struck south for Fort Simpson to report this latest disaster, while Stewart salvaged what he could and then started back downriver to Fort Yukon. On hearing of these latest setbacks James Anderson, at Fort Simpson, cut his losses and decided to abandon Frances Lake, Pelly Banks and Fort Selkirk posts.[3]

In November 1852 Hardisty sent Stewart with the express mail to Lapierre's House on the Upper Porcupine; he was back at Fort Yukon by 11 January 1853. He finally left Fort Yukon on 24 March for a winter trip to Fort Good Hope on the Mackenzie. That spring he took charge of Peel's River Post (now Fort McPherson), vacated by the death of Augustus Peers. In the summer he headed up the Mackenzie with the returns for his own post and for Fort Yukon. There Chief Trader James Anderson ordered him to proceed to Fort Rae and Fort Resolution, and then he was to accompany an express canoe to Norway House. While he was en route to Fort Resolution James Anderson wrote to him from Fort Simpson on 25 November 1853, couching his letter in the very warmest terms:

> On your departure from the District I beg to express the high sense I entertain of the Energy, Ability and determination you have displayed in the discharge of your duty under circumstances of no ordinary difficulty. I also beg to offer my thanks for the uniform Zeal you have exhibited in supporting my measures and forwarding my views.
>
> Trusting sincerely that you may have a happy meeting with your relatives, and that I shall have the pleasure of seeing you again in Mackenzie River,
>
> I remain, etc., James Anderson (a), C.T.[4]

He was still at Norway House the following June (1854) when Sir George Simpson

[1] HBC James Stewart search file.
[2] Ibid.
[3] Wilson, *Campbell of the Yukon*, p. 126.
[4] HBC B 200/b/30, ff. 33–4.

held a Council meeting there. Thereafter Stewart headed south to Red River (disobeying orders, according to Mackinnon).[1] There, on 28 June 1854 he married Margaret Mowat 'of the Parish of St. Andrew's' in St Andrew's church.[2] By 15 July he was back at Norway House, probably with his new bride. Then on 8 August 1854 he left Norway House to take up a new post (as Chief Trader) at Fort Carlton.

The main function of this post, located on a low terrace on the east bank of the North Saskatchewan River, in the aspen parkland fringe at the northern edge of the prairies, was to provide pemmican (made from the meat of the buffalo which roamed the nearby prairies) to fuel the Company's brigades of canoes and York boats. Some furs were traded at the post, but this was a minor part of its activities. Stewart can have reached Fort Carlton only around the end of August, and thus had had only a few months in which to acquaint himself with the affairs of the post when Sir George Simpson's letter arrived, appointing him to the Back River expedition.

As we will see later, Stewart went on furlough to Britain after carrying the dispatches from the expedition to Sir George Simpson at Lachine. Thereafter (although with some reservations – see pp. 217–18) Sir George appointed him in charge of the Saskatchewan District, based at Cumberland House. He served there from 1856 until 1862, then at Oxford House (1865 until 1867) and finally at Norway House (1867 until 1871). He was promoted to Chief Factor on 1 June 1869. On retiring from the Company, in 1878 he settled at Clover Bar near Edmonton, where he held the post of Indian Agent for one year. He died of heart disease on 1 September 1881.[3]

Several points should be emphasized in any assessment of Simpson's selection of these two men as leaders of the expedition. Both (and especially Stewart, despite his relatively short period of employment with the Company) had enormous experience of canoe, boat and sledge travel. But neither of them had any experience of the tundra; as we shall see Stewart in particular appears to have been overwhelmed by the desolation of the tundra of the Back River basin. And as we shall also see, he found himself missing his new wife very badly. Stewart, as far as we know, was physically fit, but Anderson suffered very badly from varicose veins in his legs, and even before the expedition began, was seriously concerned as to whether his legs would stand the strain of lengthy portages.

Sir George immediately dispatched instructions by winter express to Anderson and Stewart. On receipt of his instructions Stewart was to leave Fort Carlton and travel by dog sledge to Fort Chipewyan on Lake Athabasca and make arrangements for the construction of two canoes there, and a further two at Fort Resolution on Great Slave Lake.[4] Anderson was instructed that on receipt of Simpson's letter he was to start up the Mackenzie from Fort Simpson and rendezvous with Stewart at Fort Resolution; it was hoped that this would before breakup of Great Slave Lake had occurred. In the event of Anderson not arriving before breakup Stewart was to take command of the expedition and proceed on his own, with the support of whichever of the Company's officers was available. The expedition was to proceed

[1] Mackinnon, *James Green Stewart*, p. 854.
[2] HBC James Green Stewart search file.
[3] C. S. Mackinnon, 'James Green Stewart', p. 855.
[4] BCA A/c/10/H862; Great Britain, *Further Papers Relative*, p. 852; HBC A.8/17, f. 238.

to the head of Great Slave Lake, then northwards, across the divide, and down the Back River to its mouth. There the party was to search for traces of the Franklin expedition, especially any documents; if they found any skeletal remains they were to bury them.

To ensure that the expedition would have some of the best canoeists on the continent Simpson also dispatched three Iroquois canoemen, normally members of Simpson's own crew, to Red River, en route to Fort Resolution. In charge of them (as far as Red River), Simpson sent Mr James Bissett,[1] normally senior clerk at Lachine. The remainder of the expedition members were to be hired at Red River and Norway House and forwarded to the rendezvous. Finally, to ensure that Anderson and Stewart could converse freely with the Inuit whom it was assumed they would meet on the coast, instructions were sent to Churchill that William Ouligbuck and one other Inuktitut interpreter should be sent south to Norway House and then west to Fort Resolution.

In case the main expedition ran into trouble, or it was decided to return to the coast again the following summer, a party under Mr James Lockhart[2] (who was to bring men and supplies from Red River) was to proceed from Fort Resolution to the head of Great Slave Lake and erect winter quarters there at, or near, Back's old winter quarters at Fort Reliance. He was also to forward a boat and a cache of supplies to the head of the Back River, in case the main expedition found itself in difficulties.

To implement this plan, which involved relaying messages, men and supplies across half the continent in a matter of a few months, Simpson also had to forward instructions to a large number of posts. A letter went to William MacTavish[3] at York

[1] James Bissett was born in Montreal, where his father, Alexander Bissett, was Superintendent of Lachine Canal, on which the Hudson's Bay Company's North American headquarters were located. By 1853 James was senior clerk at the Company's Lachine depot, then in 1859 he was made officer-in-charge at Honolulu. In 1860 he was promoted to Chief Trader, and until 1871 was in charge variously of Victoria and Esquimalt. He moved to Montreal as Chief Trader in 1871 and was promoted to Chief Factor in 1872, remaining at Montreal until his retirement in 1880. See HBC, James Bissett Biography.

[2] James Lockhart was born in 1831 at Lachine and joined the Hudson's Bay Company on 26 April 1849. He served as apprentice clerk at successively Lachine, York Factory and Fort Garry where he was promoted clerk in 1854. After participating in the Back River expedition he served as clerk in the Mackenzie River District. From 1857 he was Clerk-in-charge at successively Peel River, Fort Resolution, Fort Yukon and again Fort Resolution. He took a year's furlough in 1866–7, then having been promoted Chief Trader, was in charge of Abitibi until his retirement in 1873, when he retired to Whitby, Ontario. See HBC, James Lockhart Biography.

[3] Born on 29 March 1815 in Edinburgh, William MacTavish was the eldest son of the lawyer Dugald Mactavish, and nephew of Chief Factor John George MacTavish of the Hudson's Bay Company. William joined the Company as an apprentice on 2 January 1833 and was posted to Norway House under Donald Ross. Next year he moved north to York Factory where he worked under the supervision of James Hargrave. He proved very diligent and in 1841 was made general accountant for the Northern Department and second-in-command at York Factory. He was made Chief Trader in 1846 and from 1848 to 1850 was in charge of the post at Sault-Ste-Marie. He returned to York Factory in 1850 where he remained in charge until 1856, being made Chief Factor in 1852. In 1857 he took charge of Upper Fort Garry where, soon afterwards he married Sarah McDermot; they had three children. In 1858 he was appointed Governor of Assiniboia, a very difficult position politically; with George Simpson's death in 1860 MacTavish became Acting Governor of Rupert's Land and in 1864 Governor. Hence he held the titles of Governor of Rupert's Land and of Assiniboia during the Red River Rebellion of 1869–70 and for a time was imprisoned by Louis Riel.

MacTavish had been fighting against tuberculosis for many years, and it became particularly acute

Factory[1] asking him to make contact with William Anderson at Churchill with instructions to send William Ouligbuck and another Inuktitut interpreter south to Norway House, where they were to rendezvous with Thomas Mustegan and Murdoch McLellan, members of Rae's recent expedition, for the trip west to Fort Resolution. Another letter[2] went to George Barnston[3] at Norway House, asking him to engage Mustegan and McLellan, and to supply three or four bales of trade goods (beads, files, fish hooks, axes, needles, etc.) for trading with the Inuit at the mouth of the Back River.

Yet another letter was sent to John Ballenden[4] at Fort Garry;[5] it included a list of seven of the men who had been north with Rae and whom he had recommended for the new expedition. Ballenden was to engage them, if possible, or else appropriate substitutes, and forward them to Fort Carlton, along with the Iroquois canoemen whom Simpson had dispatched from Lachine. Like Barnston at Norway House, Ballenden was asked to supply trade goods for trade with the Inuit. Ballenden was also asked to locate William McKay (thought to be either at Egg Lake[6] or at

from the summer of 1869 onwards. He resigned from the governorship on 15 January 1870 and was released from prison by Riel in February. On 17 May he and his family left for Britain via St Paul and New York. He died on reaching Liverpool, on 23 July 1870. For further details see N. J. Goossen, 'William MacTavish', *Dictionary of Canadian Biography*, IX (1861–70), Toronto, 1976, pp. 529–32.

[1] HBC A.8/17, f. 239.

[2] Ibid., f. 241.

[3] George Barnston, Chief Factor at Norway House, was born in Edinburgh around 1801. He joined the Northwest Company as an apprentice clerk in 1820 and after the coalition of the two companies became a clerk in the York Factory District. After a year at Red River he was posted to the Columbia District in 1826 where he served at Fort Langley and Fort Nez Percés (Walla Walla). He resigned after an argument with George Simpson in 1831 but was rehired in 1832 in case he joined the Americans. After serving at various posts in southern Hudson Bay he was made Chief Trader in 1840 and given charge of Albany from then till 1843. After a furlough in Britain he managed the King's Posts on the north shore of the St Lawrence from 1846 until 1850; while there, in 1847 he was promoted Chief Factor. He was in charge of Norway House from 1851 until 1858, then after a furlough, of Michipicoten from 1859 until 1862. He retired to Montreal on 1 June 1863, and died there on 14 March 1883.

Barnston was a keen amateur naturalist, interested especially in botany and entomology. He donated collections of insects to the British Museum and collections of plants to the Royal Society Museum, the Smithsonian Institution and to McGill University. See R. Harvey Fleming, ed., *Minutes of Council, Northern Department, Rupert Land, 1821–31*, London, 1940 (Hudson's Bay Record Society, III), p. 427; and Jennifer S. H. Brown and Sylvia M. Van Kirk, 'George Barnston', *Dictionary of Canadian Biography*, XI (1881–90), Toronto, 1982, pp. 52–3.

[4] Chief Factor John Ballenden was born in Stromness, Orkney around 1812. He joined the Company as an apprentice clerk under James Hargrave at York Factory in 1829. He next moved to Red River and in 1836 was promoted to accountant at Upper Fort Garry. In December of that year he married Sarah McLeod, daughter of Chief Trader Alexander McLeod. In 1844 he was made clerk-in-charge at Sault Ste-Marie, and in 1844 Chief Trader in charge of the Lake Huron District. Promoted Chief Factor in charge of the Lower Red River District, based at Upper Fort Garry in 1848, he was on his way west to take up this new post when he suffered a stroke with partial paralysis. In 1850 a court case against an individual who had made slanderous charges of adultery against his wife, placed additional stress on Ballenden and he took a year's furlough in the autumn. His next posting was to Fort Vancouver, in charge of the Columbia District, but his poor health continued and after another stroke he took a further furlough in 1853–4. He returned to Red River in 1854 but ill health forced him to return to Scotland in 1855 and he retired officially on 1 June 1856 and died in Edinburgh on 7 December that same year. See Fleming, ed., *Minutes of Council*, pp. 426–7; and Sylvia Van Kirk, 'John Ballenden', *Dictionary of Canadian Biography*, VIII (1851–60), Toronto, 1985, pp. 59–60.

[5] HBC A.8/17, f. 243.

[6] Otherwise known as Nut Lake, located about 28 km north-north-west of the present town of Kelvington, Saskatchewan. William McKay did not in fact join the expedition, his place being taken by James Lockhart.

Touchwood Hills), since he was to be attached to the expedition. And finally, Ballenden was also instructed to make arrangements for a boatload of supplies to be sent to Fort Resolution the following summer, in case Anderson found it necessary to extend his search for a second season.

Since Stewart was to locate himself at Fort Chipewyan until the spring, and to make arrangements for canoes to be built there and Fort Resolution, it was essential that he have the cooperation of the officer in charge at Fort Chipewyan. Hence another of Simpson's letter was dispatched to John Bell;[1] he was asked to render Stewart 'your best assistance'. He was also informed that Stewart could call upon 'the services of any of the officers or servants in Athabasca District'.

To facilitate the travel of both Stewart's party and later that of Mustegan, McLellan, Ouligbuck, etc., Simpson also sent a letter to alert George Deschambault[2] at Ile-à-la-Crosse on the upper Churchill River, one of the more important posts on either the winter route or the summer route.[3] In both cases he was asked 'to render every necessary aid and facility for the expedition's prosecution of their march...'.

And finally, Simpson wrote to William Sinclair,[4] in charge of Fort Edmonton

[1] HBC A.8/17, f. 246; Chief Trader John Bell, of Fort Chipewyan, was born around 1799 in Mull, Scotland. He joined the Northwest Company as an apprentice clerk in 1818, and joined the Hudson's Bay Company at the coalition in 1821. After a brief spell in the Winnipeg River District (1821–4) he was posted to the Mackenzie River District, where he would spend the next twenty-five years. From 1826 until 1840 he was clerk-in-charge at Fort Good Hope. He established Peel River Post (now Fort McPherson) in 1840 and ran it for the next six years. He was made Chief Trader in 1841 and made two important exploring expeditions westwards up the Peel and over the northern section of the Mackenzie Mountains in 1839 and 1842. In 1847–8 he was attached to Dr John Richardson's expedition, based at Fort Confidence on Great Bear Lake, in search of the missing Franklin expedition. After postings to Fort Liard, Fort Simpson, Oxford House, and Cumberland House, in 1853 he was given charge of the Athabasca District, based at Fort Chipewyan. After a year's furlough (1857–8) and a final posting to Sept Iles, he retired on 1 June 1860 to Saugeen, Bruce County, Ontario. He married a daughter of Chief Factor Peter Dease. See Fleming, ed., *Minutes of Council*, pp. 427–8.

[2] Chief Trader George Deschambeault, in charge of the post at Isle-à-la-Crosse, was born at Quebec in 1806, son of Colonel Louis Deschambeault, a friend of Lord Selkirk. He joined the Company as an apprentice clerk at Oxford House in 1819. Thereafter he served at Ile-à-la-Crosse, in the Saskatchewan District, in the English River District, at Great Slave Lake Post, and in the Mackenzie River District, until he was made Chief Trader in 1847. In that capacity he managed Cumberland House (1847–9) and Ile-à-la-Crosse before taking a year's furlough in 1850–51. Thereafter he managed the Athabasca, English River and Cumberland House and Red River districts until 1870, when he died at St Boniface. See Fleming, ed., *Minutes of Council*, p. 436.

Simpson contacted him in connection with the impending expedition since many of the men and much of the supplies would have to be forwarded through his district and his post (the English River District and Ile-à-la-Crosse Post).

[3] HBC A.8/17.

[4] Chief Factor William Sinclair, was born in Hudson Bay, the son of William Sinclair from Orkney, and Nahoway, a Cree woman, around 1794. He joined the Company as an apprentice in 1810. He served at a long list of posts and first took charge of a post (Fort Frances on Rainy Lake) in 1832. He was promoted Chief Trader in 1844 and first took charge of a district (Rainy Lake District) in 1848. Made Chief Factor in 1850, he was in charge of the Saskatchewan District (based at Fort Edmonton) from 1854 to 1857, the Rainy River District again (based at Fort Alexander) from 1857 to 1858, and finally Norway House District (based at Norway House) from 1858 until 1862. After a year's furlough he retired on 1 June 1863 to Brockville, Canada West and died there on 12 October 1868. He married Mary McKay of Norway House on 21 June 1823; they had four sons and four daughters. See Fleming, ed., *Minutes of Council*, pp. 456–8.

Simpson wrote to Sinclair since Fort Carlton was in the Saskatchewan District and hence he was Stewart's superior.

advising him that Stewart would be taking part in the expedition, and that if no replacement were sent from Red River, he (Sinclair) should make other arrangements.

DOCUMENTS

1. Letter of instructions from Sir George Simpson to James Anderson and James Green Stewart, Lachine, 18 November 1854.[1]

The mystery which had so long enveloped the fate of the expedition commanded by Sir John Franklin having been partially solved by the information given by the Esquimaux last winter to Dr. Rae, Her Majesty's Government has decided that an effort shall be made to follow up the clue thus unexpectedly obtained, and at the same time to rescue the survivors, if any, of the party of whites who were seen near the outlet of Back's River, or at least to procure any records they may have deposited at the place where they are reported to have perished.

The execution of this deeply interesting service has been confided by Her Majesty's Government to the Hudson's Bay Company and their officers, and I have now to inform you that you have been appointed to the first and second command, respectively, of the expedition which is to be employed upon it. Her Majesty's Government lay much stress on the selection of the persons who may be honoured with this command, and in nominating you I have had in view your tried zeal, discretion, and perseverance in surmounting difficulties, as well as your experience in dealing with the native tribes, and the important fact that you are inured to the hardships and perils which must necessarily attend a service of this description.

2. Before proceeding to detail the organization of the proposed expedition, I will briefly state its scope and object. By the annexed copy of Dr. Rae's report to the Secretary of the Admiralty, it appears that last spring, while at Pelly Bay, he met some Esquimaux, who informed him that in the spring of 1850, some of their tribe who were hunting at King William's Land, saw a party of forty white men travelling southwards towards the Arctic coast, dragging with them a boat and a sledge; that they reported their ships had been lost in the ice, that they appeared to be starving, and that later in the same spring their bodies were found by the Esquimaux, some on the mainland, and some on an island at a day's journey distant from the mouth of a large river called Oot-ko-hi-ca-lik, which there is little doubt is Back's River. Dr. Rae did not meet with any of the Esquimaux who had seen the white men, but from those who gave the information he purchased various articles of silver plate, etc., which had been in possession of the unfortunate party, bearing the names and initials of officers belonging to the missing expedition. It is under these circumstances the British public and Her Majesty's Government are anxious that the spot indicated as that at which was closed the career of so many of our gallant countrymen

[1] HBC D.4/75, ff. 248–58; HBC A.8/17, f. 232; BCA A/c/10/H862; Great Britain, *Further Papers Relative*, 1855, pp. 850–53.

should be explored, in order to test the accuracy of the information already obtained, and to gather, if possible, further details, by the discovery of any written records which may have been deposited on the spot, and which, possessing no value in the eyes of the natives, possibly remain untouched, bearing at the same time particularly in mind the faint hope that some of the party may have survived who may yet be rescued.

3. It is proposed that the expedition to be employed on this service shall be assembled and organized at Great Slave Lake (Fort Resolution) in June next, from whence it will descend Back's River to the coast, and after exploring the mainland and islands, and communicating with the natives, retrace its steps up the river in sufficient time to reach winter quarters at the east end of Great Slave Lake. As almost the whole navigation is river-way, it is proposed that, instead of boats, the expedition shall make use of canoes, to be constructed at Fort Resolution and Athabasca during the ensuing spring. These canoes Dr. Rae recommends should be rather shorter than the usual north canoe, but of the same breadth of beam, etc. The party is to be composed of two officers, twelve canoe-men, and two Esquimaux interpreters and hunters. You must take your departure from Fort Resolution immediately the navigation of Great Slave Lake opens; and I think you should employ three canoes, bearing the fourth as a reserve in case of accidents. My reason for recommending the use of three canoes, although you have only crews for two, is, that it is very desirable you should take with you as large a supply of provisions as possible, and you will be sufficiently well manned for descending the current. On reaching the coast, one canoe should be deposited in a place of safety, available in case of accident to the other two while exploring the coast and islands; and when ascending the river on your return, your cargoes will be so much reduced, that two canoes will be ample for your conveyance.

4. It will be an important part of your duty to open a communication with the Esquimaux, and particular inquiry should be made for any who may have seen the party of white men in 1850. In the first instance, you must endeavour, by those means which your experience in the Indian country will suggest to you as most effectual, to secure the good will, and afterwards, by the offer of liberal rewards, draw from them all the information they possess. On your way down the river, you should induce two or three of the Esquimaux with whom you may meet to accompany your party to the coast, as their presence would greatly facilitate your intercourse with those you might subsequently fall in with. To secure this point, and, in fact, in all your dealings with the natives, you must treat them with great liberality, allowing no mere consideration of outlay to deprive you of any the most remote chance of furthering the objects of your expedition.

5. On receipt of this despatch, Mr. Anderson will resign the important charge he now holds in McKenzie River to any officer who may be on the spot, to relieve him of it (in which, as in all other respects, the Company's interests are to be made subservient to those of the expedition), and proceed without delay to Slave Lake, where it is to be hoped he may arrive before the navigation opens, otherwise, so essential is it to take advantage of every day of the brief Arctic summer, Mr. Stewart must proceed without him in the chief command of the expedition, selecting any person who may be at hand for his second.

Mr. Stewart, when this reaches him at Fort Carlton, by the hands of a party who will be sent thither from Red River Settlement en route to join the expedition, will likewise resign his charge to his assistant, and accompany the Red River party to Athabasca. It is intended he shall be joined by post-master William McKay, now at Egg Lake; but failing him, Mr. Stewart may select and take with him any properly qualified clerk or post-master within reach, whose duty it will be to assist in the preliminary arrangements at Athabasca and Great Slave Lake, and in the event of either Mr. Anderson or Mr. Stewart being prevented proceeding on the expedition, he will accompany the other as his second. Should he not be wanted in that capacity, he is to proceed in the course of the summer to the east end of Great Slave Lake with a boat laden with provisions, warm clothing, net thread,[1] and ammunition for the use of the expedition on its return from the coast, such supplies to be furnished from those in depôt at the company's posts in McKenzie, and for their own trade or for Her Majesty's Government, from Athabasca or other neighbouring districts, on which point you must make the necessary arrangements before leaving [for] the coast. Another boat's cargo of the same descriptions of supplies will be forwarded from Norway House next summer as far as Fort Resolution, where they will be held as a reserve for the expedition, subject to your orders. In the crew of the boat which is sent to meet you at the east end of Slave Lake, should be two good deer hunters, whose services may be turned to profitable account.

6. As soon as Mr. Stewart reaches Athabasca, he must see two canoes built, and also forward instructions to Fort Resolution to build two more there.

The despatches for Mr. Anderson, which he will convey as far as Athabasca, should be from thence forwarded to that gentleman at Fort Simpson, without the loss of a single day.

7. Instructions have been transmitted to Churchill to forward from thence to Athabasca this winter the Esquimaux interpreter William Ouligbuck, and another Esquimaux, who on an emergency could also act as interpreter. The remainder of your party will consist of men now at Red River and Norway House, who have been with Dr. Rae, and who will no doubt be willing to join in this service; there will also be forwarded from hence three of our most experienced and trustworthy Iroquois voyageurs. Should any of the men on whom we rely to make up the party not be forthcoming, you are at liberty to avail yourself of the services of any of the company's servants in McKenzie River, Athabasca, or elsewhere, who may be qualified for, and willing to enter upon such duty, for which they will be entitled to the same scale of remuneration as those men who are specially engaged.

8. By Ouligbuck's party there will be forwarded from York Factory to Athabasca a Halkett's india-rubber boat, left there by Dr. Rae, which may prove useful; also an assortment of articles suitable for presents to the Esquimaux, principally the finer descriptions of iron works. Similar assortments, as also some tea, chocolate, sugar and tobacco will be forwarded from Norway House and Red River Settlement.

9. By the foregoing scheme of operations, you will observe it is supposed you will accomplish the objects of the expedition in the course of one summer, but in case

[1] Twine for making fishing nets.

you may find that impossible, and that there are sufficient grounds to justify your prolonging the search to a second season, you should be prepared to pass the winter of 1855–6 on the coast, renewing your explorations in spring and summer, and returning to Athabasca in the autumn of 1856. If, in your opinion, it would be incurring too much risk to endeavour to maintain so many people on the coast through the winter, you are at liberty to send back one canoe and some of your men, and it is left discretionary for one of yourselves to return at the same time, as health and other circumstances may seem to render expedient. You will of course take care to be well provided with arms, ammunition, and nets, and in the event of your wintering on the coast, you should follow the example of Dr. Rae in eking out your store of provisions by such additions as the meagre resources of the country may afford.

10. There are forwarded herewith a small assortment of astronomical instruments and some charts, which have been furnished by the Lords of the Admiralty, for the use of the expedition, and which at the conclusion of the service are to be returned. You will of course keep a detailed journal of your proceedings, make observations for latitude and longitude as frequently as possible, and as far as your opportunities admit, collect information respecting the country you may visit likely to be of interest to the scientific world, bearing in mind that such matters are of secondary consideration, and must not be allowed to interfere with the main objects of your expedition. Should you discover any traces of Sir John Franklin's party, you will carefully collect and bring back with you whatever may be portable, more especially manuscripts; such articles will most probably be found in the possession of the natives, from whom they should be purchased at any cost. Should you fall in with the remains of any of the unfortunate men who are reported to have perished on the Arctic coast in the spring of 1850, you will have them decently interred, erecting over them a cairn of stones to mark the spot, in which should be deposited a written memorial of all that is known of their career and melancholy fate.

11. Having gone with sufficient detail into the arrangements of the expedition, I will conclude by stating that your proceedings will be watched with deep interest by the whole civilized world. The Lords Commissioners of the Admiralty, in addressing the Company on the subject, desire that those who may be selected to serve on this expedition may be assured that it is the wish of the British Government, and its declared intention, to pay liberally for the service to be rendered, and to reward specially any acts of signal daring and distinguished merit. As a further incentive, if any be wanting, I may add that the Governor and Committee, already cognizant of your past meritorious conduct, have it in contemplation to promote you both in their service in the course of the present winter; and, from their wonted liberality, I feel assured they will mark in a substantial manner their approbation of any extraordinary zeal manifested on the present occasion. Most of the men to be employed on the expedition have previously served under Dr. Rae, whose admirable tact in the command of his people is proverbial, and was highly conducive to the success of his various expeditions. You cannot do better than follow his example on this point, treating the men with kindness and consideration, and maintaining subordination more by your influence over them than by a resort to strict discipline. I have so

much confidence in your long experience in the country that I feel it unnecessary to caution you against incurring needless perils in the prosecution of this service; at the same time I rely on you sparing no efforts to distinguish yourselves by success, and so as to earn for the honourable Company and their officers the approbation of Her Majesty's Government and the English public.

2. Private letter from Sir George Simpson, Lachine, to James Anderson, Fort Simpson, 18 November 1854.[1]

With this you will receive an official letter addressed to yourself and Mr. J. G. Stewart in reference to an Arctic Searching Expedition, whereof he and you are appointed to the joint command which will afford you a rare opportunity for distinguishing yourselves. The instructions are so full that it is unnecessary for me to add anything further on the organization or object of the expedition. As regards the charge of McKenzie River District you must make the best arrangements in your power on so brief a notice. Chief Trader R. Campbell[2] will naturally be your successor but if he be within reach it is desirable he should accompany you to Great Slave Lake in order to be at hand to act as a second officer for the Expedition in the event of any accident preventing Mr. Stewart proceeding upon it. If Mr. Campbell be not at Fort Simpson when you receive this, as you cannot spare time to send elsewhere for him, you must take with you one of the clerks who in a pinch could act as an assistant on the Expedition.

It is all important you should be at Fort Resolution some time before Great Slave Lake breaks up as the arrangement of the party after being assembled there will occupy several days and not one ought to be lost in the preliminaries after it is practicable to be under way on your journey towards the coast. If therefore you should not have made your appearance at Fort Resolution at the date the navigation opens, Mr. Stewart has orders to start without you in the chief command of the party, with post master McKay if he be then available, or some other person who may be at hand, as his second.

The accounts of this service should be carefully kept in the same manner as those of the Expeditions which have preceded. All outlay is to be charged as a transfer against the Montreal Department and detailed statements thereof transmitted from time to time to Lachine whence they will be presented to Government for adjustment.

In my official letter I have noticed the probability of your promotion this winter; it is always hazardous to speak with certainty of anything that is not a fait accompli, but I have good reason to believe your Chief Factor's Commission will be transmitted to you next spring.

Your various letters of this season have been duly received but I shall not address you on McKenzie River affairs to which after you receive [this] you will have no time to give attention. I will reply to your letters to your successor from the seat of Council next summer.

[1] HBC E.37/11, f. 6; BCA AB40 An 32.2, ff. 42–4.
[2] At this point at Fort Liard.

3. Private letter from Sir George Simpson, Lachine, to James Green Stewart, Fort Carlton, 18 November 1854.[1]

In the accompanying official letter to Mr. Anderson and yourself you are informed of your appointment to the joint command of an Expedition to be forwarded next summer to the Arctic Coast to make further search for the remains of Sir J. Franklin's party at the point where last seen by the Esquimaux. I am sure you will enter with alacrity on this important duty which will enable [you] to obtain reputation as an Arctic explorer & at same time entitle you to those rewards at the hands of the Government & the Company which distinguished service on this occasion will merit. At your father's advanced age it will no doubt be a disappointment to both of you that your visit to Quebec must be deferred, but I think I know you sufficiently well to be assured you would not on that account relinquish this opportunity of displaying yourself – an opportunity never likely to recur; and I believe your father would regret that you let it slip out of consideration for his feelings.

By my official letter (addressed to Mr. C. T. Anderson and yourself) you will see a party of men for the Expedition will be sent from Red River, and that Postmaster W. McKay is to accompany them. When that party reaches Carlton you must lose no time in your preparations, but making over your charge to any person Mr. Ballenden may send to relieve you,[2] or, if no one be sent, to your Assistant (Spencer), push on vigorously with them to Athabasca,[3] from whence you are to forward the dispatches for Mr. Anderson without loss of time.

You are left a full discretion as to the employment of the people who accompany you at Athabasca & Fort Resolution before the navigation opens, and as to all minor details and arrangements. You will be able to proceed from Athabasca to Fort Resolution some time before you can get through Great Slave Lake, but every thing should be perfectly ready to make a start the very first day the Lake breaks up, whether Mr. Anderson has arrived or not. The success of the Expedition depends very much on your forethought and judgment in the arrangements you make for providing against every possible contingency, and until Mr. Anderson joins you these arrangements must be made on your own sole responsibility, so that should any thing miscarry through defects therein your reputation will suffer. The construction of the canoes should claim your particular attention, and you must consider the chances of your wintering on the coast in 1855/56 and make provision accordingly, taking care your party are properly armed and supplied with ammunition, net thread, tobacco, and other matters which your experience will suggest. You should also decide where your reserve stores of provisions are to be deposited, so that you may know where to fall back upon them, and when the boat under McKay shall proceed to meet you on your return.

I hope Ooligbuck's party from York and Norway House, with the supplies they are taking on, may reach Athabasca soon after yourself.

You will understand you have carte blanche to draw on any post within reach for

[1] HBC D.4/75, f. 262; HBC A.8/17, ff. 238–9.
[2] From Fort Garry.
[3] Fort Chipewyan, at the west end of Lake Athabasca, headquarters of the Athabasca District.

supplies of goods or provisions, or for the services of the Company's servants; – nothing is to interfere with the efficient equipment and organization of the Expedition.

Your promotion next Outfit I think is pretty well secured. You will probably meet a Chief Trader's commission on your way back from the Arctic Coast and I trust your services on this occasion may earn for you a claim for promotion to the next grade at the earliest possible date.

4. Private letter from Sir George Simpson, Lachine, to John Ballenden, Fort Garry, 15 November 1854.[1]

On the 8th inst. I dispatched a packet of letters to your address via U/S Mail to Pembina, with instructions to forward an express to the Saskatchewan with letters for Mr. John Ross and (blank). I now write by the same channel in the hope of your receiving this in time to detain that express for a few days, until you are in possession of dispatches for McKenzie River of great urgency, to be forwarded from hence in the course of next week, & which are to be transmitted with all possible dispatch via the Saskatchewan to their destination. Dr. Rae's report has created great sensation in England and H. M. Government without a day's delay decided on the necessity of forwarding to the mouth of Back's Fish River an expedition to search for further traces of Franklin's party and have in a very complimentary manner placed the organization, management & entire conduct of the Expedition in the hands of the Company, whose officers & servants are to be employed upon it.

The whole business is placed in my hands without other instructions than the object to be attained, with an intimation that all expences connected with the undertaking will be cheerfully borne by the country – that H. M. Government will reward distinguished services [illegible] etc.

The Expedition is to be organized at Great Slave Lake in June next, descending from thence to the Coast in two canoes, with crews of 7 men each (and 2 or 3 officers).

Dr. Rae has recommended that the following men, who have served under him, should be engaged for this expedition vizt.

Thomas Mustegan	Steersman	Norway House
Murdoch McLellan	Mid'man	do.
John Fidler	Steersman	Red River
James Isbister	Bowsman & fish'man	do
John McDonald	Mid'man	do
Jacob Beads	Bowsman & carpenter	do
Charles Kennedy	do	do
Samuel Sinclair	do	do
Harry Fiddler	Mid'man	do
William Ouligbuck	Esqx. Interpreter	Churchill

I wish you to keep your eye on all of these men who reside at Red River, & have

[1] HBC D.4/75, ff. 233–5.

them available at a short notice, but neither to them nor anybody else would it be advisable to make known the contents of this letter until receipt of my next dispatches with the full details of the Expedition's arrangements.

P.S. As a party of men will have to be sent across from Red River to Athabasca this winter, to join the Expedition at Slave Lake in June, you can make provision for their journey at once, by having guides, dogs, sleds, provisions & all other requisites collected – & let all expences be charged to the Montreal Department, headed 'Arctic Searching Expedition,' transmitting statements to Lachine from time to time as has been usual with these expedition accounts ...

5. Letter from Sir George Simpson, Lachine, to John Ballenden, Fort Garry, 18 November 1854.[1]

The enclosed copy of an official dispatch addressed to Messrs. Anderson & Stewart will make you acquainted with the arrangements connected with the expedition to be employed next summer to visit the Arctic Coast at the outlet of Back's River, in order to follow up the search for Sir J. Franklin's party at the point where it was reported by the Esquimaux to Dr. Rae 40 persons perished in 1850. This important service, which excites the most lively interest in England & America has been left by H. M. Government in the hands of the Company and its officers; and as that circumstance has been publicly announced, I trust that no effort will be spared in the Country to carry out the arrangements with a zeal and completeness that shall redound to the credit of the Company's Service.

The following men who have served under and are recommended by Dr. Rae, you will have to engage on any terms that will be an inducement for them to join this Expedition, viz. John Fidler, James Johnston, John McDonald, Jacob Beads, Charles Kennedy, Samuel Sinclair & Henry Fidler. If any of them are not forthcoming you must replace them with the very best men for the work who can be found in the Settlement;[2] if accustomed to canoes it would be desirable.

With as little delay as possible after this reaches you, I hope not exceeding two days at the utmost, the people engaged by you, with four Iroquois canoemen who will be sent from hence,[3] should be dispatched for Carlton, where they will place themselves under the orders of Mr. Stewart, who will lead the party for the remainder of the journey to Athabasca. There is also to be attached to the Expedition Post master Wm. McKay (a son of James McKay, the old Saskatchewan guide, and now either at Egg Lake or Touchwood Hills) who will either accompany Mr. Stewart's party or that from York with Ouligbuck, the Esquimaux Interpreter, as may be most convenient, which will depend on where he may happen to be stationed; the York party will of course be later than Mr. Stewart, and follow the usual route from Norway House, via Moose Lake & Cumberland. You must make the best arrangements in your power to supply Postmaster McKay's place, and also if you can send some person to succeed Mr. Stewart at Carlton, who might take charge of the party

[1] HBC D.4/75, ff. 266–9; HBC A.8/17, ff. 243–5.
[2] The Red River settlement.
[3] Only three were, in fact, sent, in the care of James Bissett.

of servants from Red River for the Expedition. If you have no one to send, Spencer must take charge of Carlton until Mr. Sinclair[1] can make better arrangements.

You will forward by the Expedition servants to Athabasca an assortment of goods, suitable for trade with the Esquimaux, consisting of beads, files, knives, daggers, thimbles, fish-hooks, needles, axes, ice chisels, etc., also some tobacco, tea, chocolate and sugar, and you should see that the men are properly equipped for the service, in as small a compass as possible.[2]

Next summer it will be necessary to forward an additional boat with the Portage la Loche Brigade, laden with supplies for the Expedition, say pemmican, flour, tea, sugar, net thread, ammunition, 2 or 3 bales of blankets and warm clothing, and such other articles as your experience may point out as likely to be required in the event of the party remaining out two seasons. This cargo is to be deposited at Great Slave Lake, and held subject to the orders of Messrs. Anderson & Stewart.

With this general outline of the arrangements, aided by my instructions to Messrs. Anderson & Stewart, I leave you with all matters of detail for giving them effect as far as they depend on you, relying with confidence on the zeal & experience of yourself and all others of the Company's officers to do all that may be possible to ensure the success of the undertaking.

I transmit this package under the charge of Mr. James Bissett, Clerk of the Company's service, who will start from hence tomorrow, accompanied by four Iroquois Canoemen, via St. Paul[3] for Red River, where I trust they may make an early arrival. The letters for York & Norway should be sent forward immediately, so that the people and supplies to be forwarded from those places may reach Athabasca before the navigation opens, which can only be accomplished by great promptitude in making the arrangements, and diligence on the march. A duplicate packet will be dispatched in the course of a few days by mail via St. Paul, and so important is this service that you will be pleased to send expresses with the duplicate letters to Carlton & Norway House, in case of any accident to the originals.

Mr. Bissett is to return by the same route he goes up by.

6. Letter from Sir George Simpson, Lachine, to George Barnston, Norway House, 18 November 1854.[4]

The unexpected information respecting the fate of Sir John Franklin's party, which was conveyed to England by Dr. Rae, has created great sensation. Her Majesty's Government immediately decided, in accordance with the public sentiment on the subject, that an Expedition should be sent next summer to the mouth of Back's River, to make a search for traces of the party of whites, who, it is reported, perished there in 1850. The mode of carrying out this service, and all the arrangements connected therewith, have been left entirely to the Company, the Government undertaking to defray all expenses, and to reward distinguished merit on the part of the Company's people employed thereon.

[1] At Fort Edmonton, in charge of the Saskatchewan District.
[2] I.e. their outfits should be of minimum bulk.
[3] In Minnesota.
[4] HBC D.4/75, ff. 270–75; HBC A.8/17, ff. 241–2.

The scheme of the Expedition which has been decided on, is to organise a party of 2 officers, 12 canoemen and 2 Esquimaux Interpreters at Great Slave Lake in June next, who will proceed in 3 canoes down Back's River to the Coast, to explore the mainland and adjacent islands, and communicate with the Esquimaux, returning to Great Slave Lake the same season, if possible. The Officers appointed to this duty are Chief Factor James Anderson (a)[1] and Mr. J. G. Stewart. Among the men recommended by Dr. Rae are two who were with him, and now at Norway House, or the adjoining Mission Village, viz. Thomas Mistigan & Murdoch McLellan, and I have to beg you will immediately secure their services for this new Expedition by the offer of such wages as may be a sufficient inducement for them to join it. I fix no limits; you will understand the men are wanted and act accordingly. They should be in readiness to proceed to Athabasca in company with Wm. Ouligbuck and another Esquimaux Interpreter, whom Mr. McTavish, in the accompanying packet to his address, is instructed to forward to Norway House with the least possible delay. Ouligbuck's party will bring with them some supplies for the Expedition and from Norway House you will forward 3 or 4 pieces[2] consisting of goods suitable for traffic with the Esquimaux, say beads, files, knives, daggers, thimbles, fish hooks, needles, axes, ice chisels, and such other articles as your experience will suggest; also some tobacco, tea, and sugar.

If either of the two men pointed out by Dr. Rae are not forthcoming, you must endeavour to supply his place by some really good hand. Thomas Mistigan is reported to be a trustworthy, pushing fellow, and I presume would be qualified to have charge of the party from Norway House to Athabasca; but if you think otherwise, you must send with them Wm. Anderson[3] or any other good man about, in whom you have confidence, and who could expedite the march. At Athabasca they will find Mr. Stewart who will arrive there before them with a party from Red River, via Carlton & Rapid River, and under whose orders they will act.

In all the arrangements which devolve upon you, you have carte blanche to carry them out in the manner you think most likely to ensure success without reference to expense or the inconvenience that may be occasioned to the Company's service, as we are pledged to H. M. Government to use every exertion to carry out their views.

An extra boat is to accompany the Portage la Loche brigade[4] next summer, laden with provisions, clothing etc. for the Expedition, which are to be deposited at Great Slave Lake subject to the orders of Messrs. Anderson & Stewart.

On Mr. Anderson's withdrawal from McKenzie, Chief Trader R. Campbell[5] will succeed to the charge, unless his services be required in the Expedition, in which

[1] Where there were two (or more) men of the same name in the Company's service at the same time, the convention was to distinguish them (in Company correspondence, records, etc.) by letters of the alphabet, as here. James Anderson (b), cousin of James Anderson, the leader of the expedition, was then serving at Eastmain on the east shore of Hudson Bay.

[2] A pièce was the standard bale or package, weighing 90 lbs in which furs and supplies were packed.

[3] Then at Fort Churchill.

[4] The brigade of York boats which made the annual trip from Norway House via Cumberland House and Ile-à-la-Crosse to Portage la Loche to rendezvous and exchange cargoes with the brigade coming in the other direction from Fort Simpson with all the returns of the Mackenzie and Athabasca districts.

[5] Then at Fort Liard.

case Mr. Anderson is authorised to employ him therein, and make some other arrangement for the charge of the District.

7. Letter from Sir George Simpson, Lachine, to William MacTavish, York Factory, 18 November 1854.[1]

Her Majesty's Government have decided that an Expedition shall be forwarded to the mouth of Back's River next summer to prosecute the search for the remains of Sir John Franklin's party, at the point where, according to Dr. Rae's information, they were last seen by the Esquimaux. The organization and entire management of this Expedition is confided to the Company.

It is to start from Great Slave Lake in June next, under the command of Chief Trader J. Anderson (a) and Mr. J. G. Stewart, the party to consist of 12 men and 2 Esquimaux Interpreters, in three Canoes. We require for this expedition the services of Wm. Ouligbuck, Dr. Rae's Interpreter, who, I have to beg, you will summon from Churchill with all possible dispatch, and forward to Athabasca this winter, where he will find the Expedition party assembled. He should be accompanied by another Churchill Esquimaux, who could in any emergency also act as Interpreter. The main object of the Expedition being to communicate with the Esquimaux tribes on the Coast, it would be imprudent to trust entirely to one Interpreter.

Dr. Rae states he left a Halkett India Rubber boat at York, which you will forward to Athabasca by Ouligbuck's party; you will likewise forward by them for the use of the Expedition an assortment of portable goods suitable for presents to, and traffic with the Esquimaux, consisting of beads, files, Knives, daggers, thimbles, fish-hooks, needles, axes, ice chisels, etc., etc., say to the extent of 2 'pieces' or more if the party can convey them, bearing in mind the necessity of their making an expeditious march to Athabasca. To this assortment should be added a moderate supply of Tea, Chocolate, and Sugar, and any portable articles belonging to Dr. Rae's late Expedition, which you may think likely to be serviceable.

At Norway House Ouligbuck's party will be reinforced by two men for the Expedition, Thomas Mustigan and Murdoch McLellan,[2] their loads being increased by a supply of tobacco, and another assortment of articles for trade with the Esquimaux.

The Government desire that no expense be spared in carrying out this undertaking, and you will act accordingly, forwarding every thing your experience may suggest as necessary or desirable, and in order that Ouligbuck[3] may not be delayed on the march by being overloaded, you may send as large a party with him as may appear desirable, the employment of 2 or 3 extra men & dog sleds being of no consideration as compared with the gain of a few days on the journey to Athabasca.

Of the outlay connected with this Expedition, like those which have preceded it, accurate accounts must be kept, charging it against the Montreal Department for adjustment with Government & sending to Lachine from time to time detailed statements of the account.

[1] HBC D.4/75, ff. 276–9; HBC A.8/17, ff. 239–40.

[2] Both of whom had been with Dr Rae on his most recent expedition to Repulse and Pelly bays.

[3] At this point it was intended that Ouligbuck would first come south from Churchill via York Factory to Norway House before heading west to Fort Chipewyan.

8. Extract of letter from Sir George Simpson, Lachine, to John Bell, Fort Chipewyan, 18 November 1854.[1]

I beg to refer you to Mr. Stewart for information respecting an Expedition to be fitted out in the country next season, and placed under the joint charge of Mr. Anderson and himself for the purposes of making a search for traces of Sir John Franklin's party at the mouth of Back's River. The party are to rendezvous at Great Slave Lake in June, in time to start from thence for the coast as early as the navigation admits.

Mr. Stewart is to get two Canoes built at Athabasca, and two are also to be constructed at Fort Resolution, and I have to beg you will render him your best assistance in this as in all other arrangements connected with the expedition, bearing in mind that everything is to give way to this important undertaking. In the event of the services of any of the officers or servants in Athabasca District being required for the Expedition, Mr. Stewart is authorized to avail himself of them.

Mr. Stewart and a party of men will proceed to Athabasca this winter, and a second party with Ouligbuck the Esquimaux Interpreter will be sent from York and Norway House, and I hope will be with you some time before the navigation opens.

9. Extract of a letter from Sir George Simpson, Lachine, to George Deschambault, Ile-à-la-Crosse, 18th November 1854.[2]

Mr. Stewart will acquaint you with the arrangements that have been made respecting an Expedition to be forwarded by way of Back's River to the arctic Coast next summer, to search for traces of Sir J. Franklin's party at the place they were last seen, of which Mr. Anderson and Mr. Stewart have the command.

Mr. Stewart with a party of men engaged for this service in Red River, will proceed to Athabasca this winter, and a second party with Ouligbuck, the Esquimaux Interpreter, will also be forwarded thither from York. To both of these parties you will be pleased to render every necessary aid and facility for the expedition's prosecution of their march, it being very important they should reach Athabasca some time before the navigation opens. In whatever way you can facilitate the arrangements of the Expedition, you will be pleased to do so, as Her Majesty's Government have placed the entire management of this undertaking to the zeal and experience of the Company and their Officers.

10. Extract of a letter from Sir George Simpson, Lachine, to William Sinclair, Fort Edmonton, 18 November 1854.[3]

The services of Mr. Stewart being required on an Expedition to the Arctic Coast next summer, I have directed him on receipt of my dispatches to proceed forthwith from Carlton to Athabasca, making over his charge to any officer Mr. Ballenden may send from Red River to relieve him, or if none be sent, to his assistant Spencer, until you can make other arrangements.

[1] HBC D.4/75, ff. 280–81; HBC A.8/17, f. 245.
[2] HBC D.4/75, ff. 282–3; HBC A.8/17, f. 246.
[3] HBC D.4/75, ff. 284–5; HBC A.8/17, ff. 246–7. Sinclair was in charge of the Saskatchewan District, and hence was Stewart's superior.

Further details of the expedition will reach you from other sources; the Government has placed it entirely in the hands of the Company, and all their resources are to be made available for this important service. In whatever way you can render any assistance to the undertaking, I count on your cordial cooperation. The officers commanding the party, Messrs. Anderson & Stewart, have carte blanche to call upon every district within reach to contribute goods and people when required.

11. Contract signed by the Iroquois canoemen, 20 November 1854.[1]
We the undersigned have entered the service of the Honourable Hudson's Bay Company to be employed on an Expedition to the Arctic Coast to proceed immediately from hence to Red River Settlement & afterwards to obey all such instructions as may be given us by the Company's officers. For this service we are each to receive wages at the rate of Twenty five dollars ($25) per month.

J. Bte. Assanayunton, his mark.
Ignace Montour, his mark.
Joseph Anarize, his mark.

Signed at Lachine, Canada East,
the 20th November 1854, in
Presence of Edwd. M. Hopkins.

12. Letter of introduction for James Bissett from Sir George Simpson, Lachine, to: The Postmaster, the Agent of the American Fur Company and James W. Simpson, all of Saint Paul, Minnesota Territory, H. W. Libby, Mindota, Minnesota Territory, the Postmaster, Crow Wing River, Minnesota Territory, the Officer Commanding U/S Troops, Falls of St Anthony, Minnesota, H. Kittson, Pembina, Minnesota, and Colonel Dibble, Biddle House, Detroit, Michigan, 20 November 1854.[2]
The bearer hereof, Mr. James Bissett of the Hudson's Bay Co's service takes his departure hence for Red River Settlement via Minnesota Territory in charge of important dispatches from Her Majesty's Government to continue the search for Sir John Franklin's party at that point of the Arctic Coast where they were last seen – as reported by Dr. Rae. This service has enlisted the sympathy of all classes in America as well as England and I feel assured you will cheerfully render to Mr. Bissett & the voyageurs who accompany him every assistance in your power in the prosecution of their journey, it being of the utmost importance that the instructions of which he is the bearer should reach Red River at an early date.

Any drafts by Mr. Bissett on me for funds to defray the expences of his journey will be duly honoured.

13. Letter from Sir George Simpson, Lachine, to John Ballenden, Fort Garry, 20 November 1854.[3]
The three Iroquois sent from here are engaged at the rate of $25 per month. These extravagant wages cannot of course be taken as a standard for the men to be

[1] HBC A/C/10/H862.
[2] HBC D.4/75, f. 295.
[3] Ibid., f. 296.

engaged in the country, with whom the best bargains practicable must be made, bearing in mind that those named by Dr. Rae are to be had if possible. To induce the Iroquois to undertake the long winter journey from Montreal to Athabasca very high wages had to be offered; if any of them break down on the route, he should be at once sent back, his wages ceasing at the date of his return to Lachine. Their names are J. Bte. Assanayneton, Ignace Montour and Joseph Anariz.

14. Letter from Sir George Simpson, Lachine, to John Shepherd Esq., Deputy Governor of the Hudson's Bay Company, London, 19 November 1854.[1]

I have the honour to acknowledge receipt of your communication of 27th ulto conveying the instructions of Her Majesty's Government for fitting out two expeditions to be employed next summer on the Arctic Coast, one to procure information respecting HMS 'Enterprize' the other to prosecute the search for Sir John Franklin's party at the point where, as reported to Dr. Rae they were last seen by the Esquimaux. I trust I need hardly assure you that this important service has received my most anxious attention, and that I have lost no time in adopting the measures that appeared most advisable for carrying it out, which I hope may meet the approbation of Her Majesty's Government and the Governor and Committee...

...In order that you may be acquainted with the details of the arrangements, I transmit herewith copies of my instructions to Messrs Anderson and Stewart, who are to command the Expedition, and to Messrs Ballenden, Barnston, W. Mactavish, Sinclair, Bell, and Deschambeault, the Officers in charge of the various districts of the Northern Department from whence are to be drawn the Servants and supplies required for the service. The time we have disposable for collecting the people and supplies at Athabasca, so as to be ready to commence operations at the opening of the navigation, is inconveniently short, and they can only be brought together by great promptitude in making the arrangements, and extraordinary activity in carrying them out on the part of the Company's officers. Mr. Stewart, if not incapacitated by ill health or other contingency, being stationed at Carlton,[2] on the route between Red River and Athabasca, may be counted on as available, but I am not so sure that Mr. Anderson will be able to reach the rendezvous at Slave Lake in proper time; he is stationed at a considerable distance beyond that point, so that it will be late before he receives his instructions, and he will then have a long journey to perform to join the party.[3] I have engaged here three of our most skilful and trustworthy Iroquois canoemen,[4] who have usually formed part of my own crew for several years past, who, I think, will be a valuable acquisition; – they will be immediately forwarded from here to Red River, and from thence along with the people who are to be engaged there, and no doubt reach Athabasca in good time. I am not quite certain that Ouligbuck, the Interpreter, can be got forward sufficiently early, having to be summoned all the way from Churchill, but I am confident no exertions will be

[1] HBC D.4/75, ff. 286–91; HBC A.8/17, ff. 227–31.

[2] Fort Carlton, on the North Saskatchewan River, some 80 km north of the present city of Saskatoon.

[3] Fort Simpson, Anderson's base, lies over 500 km down the Mackenzie River from Fort Resolution, the proposed starting point for the expedition.

[4] Baptiste Assanayunton, Ignace Montour and Joseph Anarise.

spared in the country to bring together all the 'matériel' of the expedition at the proper date, and that if there be any failure it will arise from insurmountable difficulties that cannot be foreseen.

The engagement of canoemen in Canada[1] for this service is not suggested in the outline of arrangements prepared in England; my reason for employing these men, to whom it is necessary to pay very high wages ($25 per month) is the fear that, in consequence of the disuse of canoes in the interior,[2] it may be difficult to find men there competent to take the head of the craft in descending Back's River, the navigation of which is considered dangerous, and an accident to either of the canoes going down would defeat the whole expedition.

I have had much difficulty in making up my mind as to the proper persons to command the expedition. Mr. Anderson is so strongly recommended by Dr. Rae that he is named as the leader; but, as before said, I am doubtful he will be forthcoming in sufficient time. He has heretofore been employed in the charge of a district, and has no experience in Arctic travelling. Mr. Stewart possesses, in an eminent degree, many of the qualifications for such a command, being considered the most hardy, active, and persevering traveller now in the service;[3] it will be recollected that he has for many years distinguished himself in exploring the country on the banks of the Pelly River. These two gentlemen are probably the best qualified for this duty among the officers who are available, but they neither of them have any knowledge of the use of astronomical instruments, and their want of experience in the Arctic regions renders me less sanguine of the success of the Expedition than I should have been had Dr. Rae assumed the chief command. Under his able management we might have confidently anticipated that whatever it was possible to accomplish, would be performed, and even at this, the eleventh hour, I beg leave most strongly to recommend that he be requested again to undertake this duty, for which he would yet be in time, as by proceeding direct to Red River, and thence across to Athabasca, he would be at the rendezvous quite as early as Ouligbuck can reach it from Churchill.

It will be observed by my instructions that I have prepared Messrs. Anderson & Stewart for the contingency of having to winter on the Arctic Coast, as it is by no means improbable the first summer may, from the prevalence of fogs etc., be unfavourable for their operations, and, moreover, they may not immediately meet with Esquimaux to direct their search to the proper point, or to furnish information of the fate of the party seen in 1850, to procure which is an important object of the expedition. In this, as in every other particular which my long experience suggested, I have endeavoured to make my instructions as comprehensive as possible.

As the early receipt of my letters at Red River is of great moment, I have decided although at serious inconvenience to the business of this establishment, to forward our accountant[4] in charge of them, and, if the instruments come to hand, he will

[1] Simpson is referring here to his three Iroquois canoemen.

[2] In favour of York boats.

[3] This reputation is based on Stewart's performance in what is now the Yukon Territory, under the command of Robert Campbell.

[4] Mr James Bissett.

start tomorrow (Monday) morning, accompanied by the three Iroquois voyageurs, proceeding with all possible expedition via the Minnesota Territory. Duplicates of all the letters will be forwarded to Red River after the arrival of the next English mail, through the United States Post Office. As one Expedition only is to be fitted out,[1] I shall forward but one set of instruments retaining the other set here subject to future instructions.

The various changes which will have to be hurriedly made in the charge of McKenzie River District, Carlton, Egg Lake post,[2] and in this office,[3] will prove inconvenient to the Company's service, but, in accordance with the spirit of your instructions, I have regarded any considerations affecting the Company's interests of secondary importance to the efficient organization of the Expedition.

P.S. 20 November, 4 P.M. The instruments have not yet been received; they have already occasioned several days delay in dispatching the instructions to the country.

15. Letter from Lady Franklin to James Anderson[4], 15 December 1854.[5]

You will receive I am sure with kindness the earnest wishes of one who is most deeply interested in the important mission with which you are charged. May God strengthen and guide you in the execution of it. I am sure you will permit one a few observations as to our expectations from the task undertaken by the Hudsons Bay Company and intrusted to your charge. Their resources are now at your disposal and the range of your efforts will I trust have a wide scope and above all be unfettered by preconceived opinions. I ought in candour to tell you that Dr. Rae's report is not accepted by the public nor the Arctic officers in all its details with the exception perhaps of Sir George Back. The opinions of Sir E. Parry, Sir John Richardson, Sir James Ross etc. differ from Dr. Rae. The general opinion is that the ships are not crushed, but have been abandoned and pillaged on the coasts north of King William's Land, either on the E. side, or on Boothia or N. Somerset or to the West on Victoria Land – it is thought that the fugitive party seen by the Esquimaux were probably not the only survivors of expedition, but that a few stragglers may yet be found amongst tribes of Esquimaux and that different routes may have been taken by different retreating parties – also it is not considered clear that the position assigned by Dr. Rae to the catastrophe reported by him is correct. Everybody regrets, even Dr. Rae, I believe, himself, that he should have reported the shocking story of cannibalism – if true it should have been buried within his own bosom – there is but one feeling on this point. We trust that you will be satisfied with nothing less than the solution of all our doubts and that you will not trust too much to the report of the Esquimaux interpreter who having said a thing will say it again, because he has committed himself and may have an interest in saying it. I do not expect my dear husband to be amongst the survivors – if you should meet with his

[1] Since word had been received that HMS *Enterprise* was safe.

[2] I.e. to replace Anderson, Stewart and MacKay.

[3] To replace Bissett.

[4] Anderson received this letter at Fort Reliance on his return from the coast to Fort Resolution, on 7 June.

[5] HBC E.37/11, ff. 17–18; BCA AB40 An 32.2, ff. 18–20.

corpse which I think will be found wherever the ships are found, I beg you to bring me his locks of hair and I also entreat of you to bring me sealed up and directed to myself all the letters you can find addressed to him or me which may be supposed to have been in his possession. I feel that my dear husband's private letters and papers ought to be sacred from every eye but mine. In reading this, you must not attribute to me a want of confidence in your honor as a gentleman and a man of conscience and feeling. In your hands these cherished relics will be safe, but I wish you to give strict injunctions to all under you to observe the same precautions and that you should secure by peremptory orders that any papers or letters bearing his or my name outside or which may be supposed to belong to him shall be instantly sealed up and addressed to me. I think this rule should be observed in the case of all the other officers, but it is my especial request and injunction that it be done in the case of my husband. I shall give £700 reward to whoever brings me or forwards to me this packet. My husband took with him a bound quarto memorandum book in which he was to write his private journal – it had brass at the corners and a lock and key – this also I desire to possess and it will meet with the reward. The ordinary journals of the officers must of course be unearthed as they may enable you to guide your researches – but it is the private letters and papers I desire to be kept sacred from every eye but my own. My husband took with him a large quantity of letters and papers in his desk which he never intended anyone to see but me. I do what he would wish in giving you this injunction.

I think much more is required than an examination of the Fish River: the coasts E. & W. of it, the coast of Boothia as far N. as the Magnetic Pole or Bellot Strait and as much coast on the opposite side – also the passages across the isthmus into Regina Inlet and again the coast W. towards the Coppermine. I do not think with your utmost exertion you can do all,[1] but you can recommend what ought to be done. It is the prevailing opinion that a ship should be sent to Barrow St. and that sledge parties should be sent down Peel and Victoria Sound on each side to meet your parties. I do not know whether this will be done[2] – what you cannot do yourself, I beg of you to recommend should be done in some other way.

I write this letter under great suffering of mind and body, scarcely able to hold my pen. I believe Sir F. Beaufort of the Admiralty[3] will write to you to the same effect if he has time but we have been surprised as to the time and write now in extreme haste.

Trusting in your zeal and ability and commending you to the Almighty protection, I beg to assure you that you will have my heart-felt prayers and if you throw light on the fate of those dear to us and bring us back their precious relics you will obtain my lasting gratitude.

[1] An understatement! Bellot Strait lies some 500 km north of the mouth of the Back River – in a straight line, and the mouth of the Coppermine some 800 km west of it!

[2] It was not. In view of the disastrous outcome of Sir Edward Belcher's expedition (1852–4), during which four ships were abandoned in perfect condition, but beset in the ice, not counting HMS *Investigator*, abandoned by Captain Robert M'Clure in Mercy Bay, and *Breadalbane*, crushed by the ice near Beechey Island, and in view of the fact that this expedition had produced no new information as to the fate of the Frankin expedition, the Admiralty had abandoned the search.

[3] Sir Francis Beaufort, Hydrographer to the Navy.

16. Letter from Sophia Cracroft to James Anderson, 15 December 1854.[1]

I deeply regret that my aunt, Lady Franklin, should be unable as she fears, to address you with her own hand[2] owing to a sudden and most [illegible] and affliction in the loss of her sister Lady Simpkinson.[3] She requests me therefore to convey to you her warmest wishes for your safety in the execution of the important and solemn mission committed to you, and that you may be guided to its right and adequate fulfilment.

My aunt is anxious also to put before you the conclusions and anticipations of our chief arctic authorities, Sir John Richardson, Sir Edward Parry, Sir James Ross, Col. Sabine and others whose names must have been long known to you – and which opinions are those also of the leading officers who have been engaged in the search for my uncle's expedition. All appear to be of one mind in believing that the ships are probably not crushed or destroyed, but may yet be found within the area of Victoria Strait and Peel Sound, including with their shores those of King William Land, where Sir John Richardson believes they have been arrested.[4]

It is the discovery of the ships which would afford you that certainty as to the actual fate of the expedition of which as yet we have not even more than a glimpse; nothing more being proved by the communications brought home by Dr. Rae and by the relics than that some disaster of a fatal character had happened. Beyond this general assumption no one can confidently presume. There is certainly no evidence whatever that the ships are not in existence, nor any that the party reported to Dr. Rae were not a mere fraction of the main body. The manner of their death is disputed with about an equal division of the probabilities, and the precise locality of the catastrophe is doubtful. I mention these contested points to show you the better, all that is hoped from your efforts and it is a great relief to my aunt to find that Sir George Simpson takes an enlarged view of the work to be done and considers that it may certainly require two seasons.

No verbal communication with natives can afford you that positive evidence with which the discovery of the ships would reward your efforts. Alongside on the shore, would doubtless be found the burial place of those earliest taken, and on board those latest survivors who were unable to leave the spot and were left at last with none to bury them. In your hands, dear Sir, will be committed the solemn task of giving decent burial to those whom you may find here, as well as on the track of the fugitive parties! – You will do it as one hoping for the like office when you shall be taken from this world.

[1] HBC E.37/11, ff. 13–14; BCA AB40 An 32.2 ff. 22–7.

[2] Apparently Lady Franklin felt better later in the day, and was able to write a long and detailed letter to Anderson, just quoted, herself. The two letters appear to have travelled by different routes, this letter reaching Simpson before he left Fort Simpson, but Lady Franklin's not until he returned to Fort Reliance on his return from the coast.

[3] Lady Franklin's younger sister Mary (née Griffin). She is buried beside Lady Franklin in Kensal Green Chapel. See F. J. Woodward, *Portrait of Jane: A Life of Lady Franklin*, London, 1951, p. 364.

[4] In this, of course, Sir John Richardson was perfectly correct, as was demonstrated some five and a half years later when Lt William Hobson and Captain Francis Leopold McClintock found relics, skeletons and a single cryptic message during their search of King William Island in the spring of 1859, and as further confirmed by later expeditions. For further details see pp. 249–55.

Other duties too, will be yours on the discovery of any party as well as the ships – those namely, of collecting and securing for us (in whose behalf you will stand as administrator to the dying wishes of those hapless beings) all that will be most precious to us – your own heart will guide you in this work of charity. You will doubtless find records, journals (private, as well as those public ones, the examination of which may guide your future operations), letters addressed to them before they left their country, and from them, expressing their dying wishes to those dear friends whom they were never more to see in this world. I may particularly allude to a remarkable volume, square in form, and bound at the corners in brass, with a lock. This book was intended for my aunt alone and she would consider its restoration to herself as worthy of a special and ample reward to the finder. It is of course needless to suggest to you such immediate precautions as will secure the inviolability of these precious private documents – and yet Lady Franklin ventures to remind you that all may not and cannot possess the honorable feeling which would shrink from the examination of such private documents, whether found sealed or unsealed. Your party will not be composed of gentlemen of your equals who would instinctively recoil from searching into that which the writers intended only for the persons to whom they were addressed. She trusts therefore that you will issue such positive instructions to your subordinates as may compel the instant sealing up of all private documents whatever, whether addressed to the deceased members of the Expedition, or by them to their friends, and placing upon each packet the name of the person to whom it should be transmitted, unopened of course. The public documents will easily be distinguished from these and their examination may be important in deciding your future operations.

We fear you may have some difficulty in determining how far to rely upon the communications of the interpreter as Dr. Rae tells us that he cannot be trusted in the matter of food, nor where his personal interests are concerned.[1] It is believed by many, that when he ran away from Dr. Rae's party he did so in order to be guided by the Esquimaux party to the fountain head of greater treasures than any which had been delivered to Dr. Rae – that in fact he knew where the ships were, or at least knew where he could ascertain the spot.[2] Then again, we have his revelation as to

[1] What Rae in fact wrote was that he had been informed by a reliable source (before hiring Ouligbuck) that:

> young Ouligbuck could be perfectly relied on; that he would tell the Esquimaux exactly what was said, and give the Esquimaux reply with equal correctness; that when he had any personal object to gain, he would not scruple to tell a falsehood to attain it, but in such a case the untruth is easily discovered by a little cross-questioning. This description I find perfectly true (*Household Words*, 'The Lost Arctic Voyagers', 248 (23 December 1854), p. 434).

[2] This is a reference to an incident at Pelly Bay in April 1854, before Rae had heard the stories of the fate of the Franklin expedition, when he and his party encountered a group of Inuit who

> ...were extremely forward and troublesome, they would give us no information on which any reliance could be placed, and none of them would consent to accompany us for a day or two, although I promised to reward them liberally. Apparently there was a great objection to our travelling across the Country in a westerly direction. Finding that it was their object to puzzle the Interpreter [Ouligbuck] and mislead us, I declined purchasing more than a small piece of Seal from them, and sent them away...
>
> [Next morning, 21 April] our Interpreter made an attempt to join his Countrymen, fortunately his absence was observed before he had gone back very far, and he was overtaken after a sharp race of

the presumed cannibalism, an assertion which has met with universal reprobation and disgust, both at the conception and the publication of it. Mr. Dickens' powerful pen has been employed in this particular question in two nos. of his Household Words,[1] conveying a scorn and disgust which will be as effective then as his narrative of other starving beings, who yet never contemplated and never employed the 'last resource' alluded to by Dr. Rae. Mr. Dickens' just and deep feeling represents that which has been conveyed to us from all quarters at so gratuitous a revelation.

It has probably occurred to you as well as to many others that the unhappy party reported to Dr. Rae may have been the victims of violence and not immediately of starvation, and that thus was revenged by the native tribes, the murders committed by certain members of Sir George Back's party, since which no European had visited that region.[2] I am sure, dear Sir, that no (possible) prepossession will induce you to forego the exercise of that calm judgment and candid discrimination upon which so much depends in an undertaking like that entrusted to you. There is so little of the detail of the report communicated by Dr. Rae which can be relied on that a mere

---

four or five miles. He was in a great fright when we came up to him, and was crying like a child, but expressed his readiness to return, and pleaded sickness as an excuse for his conduct. I believe he was really unwell, probably from having eaten too much boiled Seals flesh, with which he had been regaled in the snowhuts of the Natives (Rich and Johnson, *John Rae's Correspondence*, pp. 273–4).

[1] 'The Lost Arctic Voyagers', *Household Words*, 245 (2 December 1854), pp. 361–5; 'The Lost Arctic Voyagers', *Household Words*, 246 (9 December 1854), pp. 385–93.

[2] Sophia Cracroft is alluding here to the rumours that some of Back's party, while exploring independently, had encountered and killed some Inuit, allegedly in self-defence. At the party's farthest north point (Point Ogle) Back sent three men (James McKay, George Sinclair and Patrick Taylor) equipped with a telescope, some distance farther, to reconnoitre ice conditions (King, *Narrative of a Journey*, II, p. 69). Once the expedition had returned to Britain, Back and King heard that there had been a violent confrontation between these three men and a party of Inuit. King reflected in retrospect that the men's demeanour on rejoining the party had been evasive, that they had been very nervous as the party approached the site at the mouth of the Back River where they had encountered the Inuit on their outward trip, and that it explained that the latter, previously so friendly, had fled in alarm when the party's boat reappeared. Back made a determined effort to confirm the story. He contacted William Morrison, Justice of the Peace in Stornoway (SPRI MS 395/77/1) and through him obtained a statement from Malcolm Smith, who was responsible for the rumour reaching King. Smith testified that according to Mackay the three men 'having observed some Swans swimming on a Lake, Commenced firing at them, which brought a Crowd of the Esquimaux upon them, who commenced shooting their arrows at them, when they retired; but finding themselves closely pursued, fired three shots which Killed three of the Esquimaux, when they immediately halted and they proceeded to join the rest of the party' (SPRI MS 395/77/2). Smith and John Ross, who was present when Mackay revealed this secret to them, decided not to inform King or Back until they were in Britain. John Ross (one of the artillerymen with Back's expedition) was interviewed by Major Jones at Woolwich Barracks on 26 December 1835; he agreed that Smith's version of what McKay had told them was essentially correct, 'but then McKay also said the whole affair was a jest, got up to impose on Smith' (SPRI MS 395/77/3). Back also wrote to George Simpson concerning the matter; the latter promised to enquire about the matter when he returned to Hudson's Bay (SPRI MS 395/77/3), but the results of that enquiry (if any) are unknown.

Thus the results of Back's investigation were frustratingly inconclusive. Given the detailed nature of McKay's story (as relayed by Smith), it has the ring of truth about it. If it is true and this violent confrontation between some of Back's men and the Inuit in 1834 did occur, it clearly is of direct relevance to how any Inuit who encountered the Franklin survivors some sixteen years later may have reacted. At the very least, they probably went to considerable lengths to avoid them, and certainly would have been reluctant to help them in any way. And if a large group of Inuit encountered a small group of stragglers, the Inuit may well have been tempted to seek revenge. On the other hand, there is not the slightest hint of such a violent encounter in the local oral traditions among the Inuit (Graham Rowley, personal communication, August 1997).

repetition of these, or the obtaining a few or even many more relics such as those he has brought home will be regarded as equally unsatisfactory, and my aunt ventures to impress this upon you in order to save you from disappointment as well as those who feel that now that the clue is given us we may hope for the final solution of the mystery.

The hearts and hopes of many are turned towards you, dear Sir, and you will think of this, I am sure, even more than of the honor and pecuniary reward which, if God blesses your efforts and spares your life, will crown your mission.

That you may be effectually guided and strengthened is the fervent prayer of Lady Franklin and myself.

17. Contract signed by Red River voyageurs, 25 December, 1854.[1]

We the undersigned have entered the service of the Honourable Hudson's Bay Company, to be employed on an expedition to the Arctic Coast, to proceed immediately from this place to any point that Messrs. Anderson & Stewart, who are in command of the Expedition, may think proper and to obey all such instructions as may be given us by the Company's Officers in command of us. For this service we are each to receive wages at the rate of Fifty Pounds Sterling per annum. To start from Fort Garry on Tuesday the 26 Instant on which day our wages commence.

[Witnessed & signed by: George Daniel, Ambroise Jobin, Alfred Laferté, Alexander Landrie, Francis Demren, Francis Desmarais, George Kippling, Jerry Johnstone, Donald McLeod, John Fidler, Henry Fidler, Edward Kippling].
Lower Fort Garry, Red River Settlement, 25 December 1854.

[1] HBC A/C/10/H862.

CHAPTER IV

GETTING TO THE RENDEZVOUS

INTRODUCTION

James Stewart

As soon as James Lockhart arrived at Fort Carlton with the three Iroquois canoemen and the voyageurs hired at Red River, on receipt of Sir George's letters of instructions James Stewart set preparations in motion. Using sleds and dogs, the party got under way on 6 February. Travelling by way of Big River and Green Lake, they reached the Company's post at Ile-à-la-Crosse where they were given a warm welcome by George Deschambeault. After a brief rest Stewart and his party got under way again on the 16th having, 'actually cleaned out Mr. D's store'.

From there the route lay north through Buffalo Narrows, across Peter Pond Lake, up La Loche River and Lake, then across Methy Portage (Portage la Loche) and down the Little Athabasca (now Clearwater River). The trip north down the Athabasca River was uneventful, and they reached Fort Chipewyan on 6 March, having made excellent time from Fort Carlton.

A week later Stewart started out again with ten men and seven sledges, bound down the Slave River to Fort Resolution. They reached their destination on the 19th. Here Stewart left seven men to push ahead with building the two canoes which Simpson had specified should be built here. Having made all the necessary arrangements with Bernard Ross[1] (in charge of Fort Resolution) Stewart started back south for Fort Chipewyan on 23 March, reaching his destination on the 28th. Stewart was understandably quite pleased and proud with his rate of travel (and rightly so) but it was at the expense of his dogs which were 'knocking up & no wonder, having come so far in so short a time, not to mention their coming from Carlton'.

[1] Bernard Rogan Ross was born on 25 September 1827 in Londonderry and died in June 1874 while visiting Toronto. On first joining the Hudson's Bay Company in 1843 he was posted to Norway House, and thereafter served at York Factory, Fort Simpson, Fort Norman and Fort Resolution. He was promoted to Chief Trader in 1856 and from 1858 until 1862 was in charge of the Mackenzie River District. He retired in 1871. In 1860 he had married Christina Ross, daughter of Chief Factor Donald Ross; they had three children. Ross was a keen amateur naturalist and made collections for the Smithsonian Institution, the Royal Scottish Museum and the British Museum. On the basis of birds he had collected, the Ross's goose (*Chen rossii*) was named after him in 1861. See H. Boosfield, 'Bernard Rogan Ross', in *Dictionary of Canadian Biography*, XI (1881–90), Toronto, 1982, pp. 52–3.

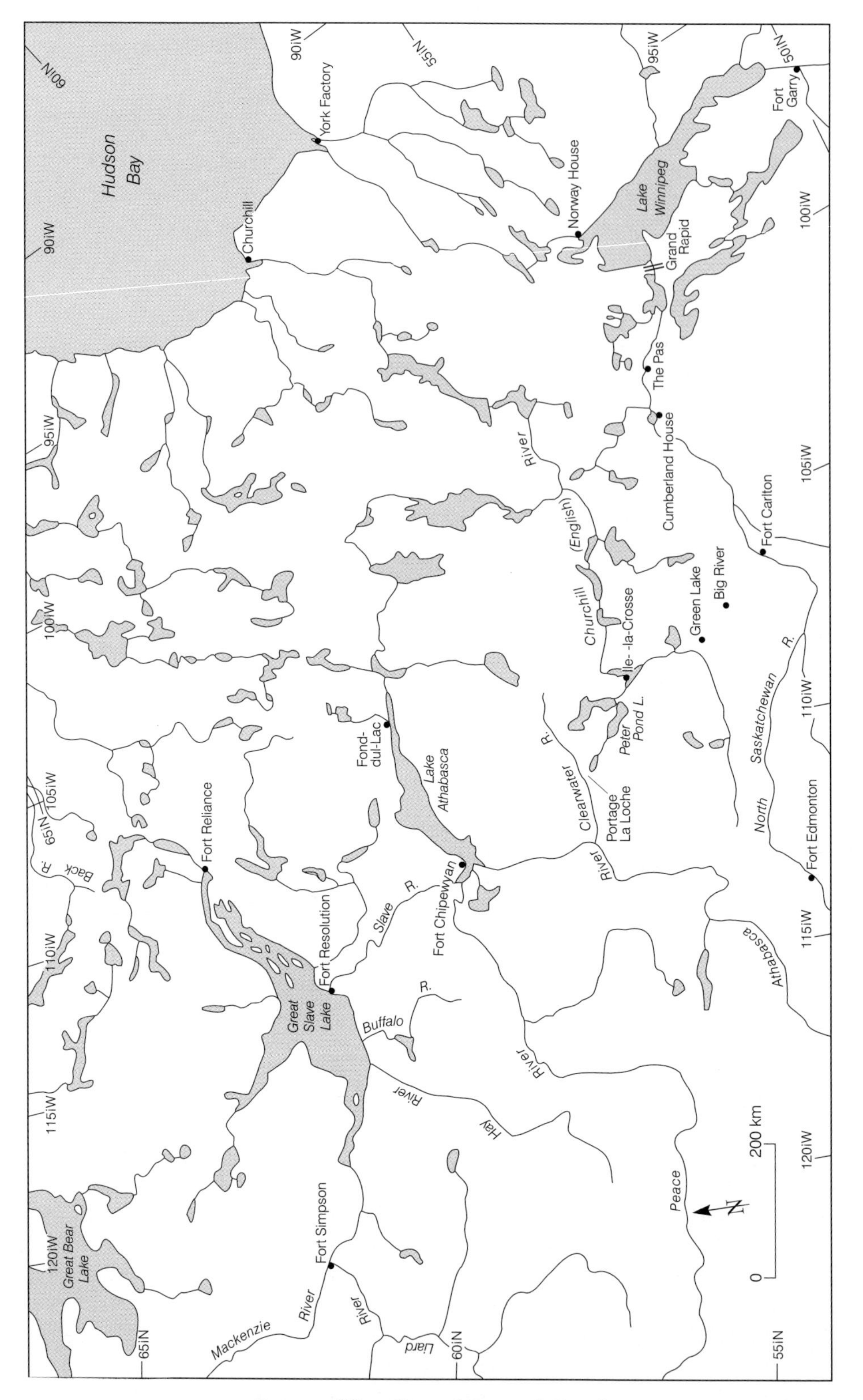

PLATE IV. Central Rupert's Land
showing how the various components of the expedition converged on Fort Resolution

On his return Stewart found that all the materials for canoe building had been assembled, i.e. wood, gum and birchbark. The job of building the two canoes at Chipewyan was entrusted to the senior of the Iroquois canoemen, Baptiste, who was delighted to be able to 'work at his favourite craft'. The cold weather delayed the start of the canoe building until well into April, however.

The men from Norway House (Thomas Mustegan, Murdoch McLellan and Paulet Papanakies) arrived in early April, along with the trade goods etc., forwarded from Norway House. To Stewart's growing concern, however, there was still no sign of the Inuktitut interpreters. And there was still no sign of them by the time Stewart, Lockhart and party started down the Slave River in their two new canoes on 26 May.

With the trees starting to bud and the mosquitoes already in swarms Stewart, Lockhart and party pushed off from Fort Chipewyan on 26 May. Having camped one night at Salt River, Stewart hired 'King' Beaulieu to act as interpreter for the group under James Lockhart, who would spend the summer at Fort Reliance, preparing winter quarters in case they were needed by the main party. Stewart and party reached Fort Resolution on 30 May, to find, not suprisingly, that Anderson had not yet arrived and, to Stewart's evident dismay that 'nor has Mr. Ross got our canoes even made though the wood is nearly ready & about all the birch which it is difficult to procure'.

The Inuktitut interpreter

It is convenient, at this point, to focus on the attempts by the Company's officers to provide the expedition with at least one competent Inuktitut interpreter. As we have seen Rae had recommended, and George Simpson had intended, that William Ouligbuck, who had been with Rae on his recent expedition to Pelly Bay and Boothia, and hence was already familiar with the Inuit stories concerning the Franklin expedition survivors, should be engaged as interpreter for the Back River expedition. Simpson had specifically instructed William MacTavish at York Factory to make arrangements for Ouligbuck to be sent south from Churchill to Norway House, along with another Inuk, to act as a back-up interpreter, to join the expedition (p. 15).

When the message reached Churchill Ouligbuck was away from the post, probably hunting and trapping farther north on the Hudson Bay coast, but he, and another Inuk known as Munro (who had also travelled with Rae, and whom William Anderson, the Chief Trader at Churchill, thought would be suitable as the back-up interpreter) were expected back at Churchill by mid-March. But by 7 April Ouligbuck had still not returned, despite the fact that Anderson had sent for him. Both Anderson and MacTavish were convinced that Ouligbuck was staying away because he had no desire to be involved in the Back River expedition.

By early April time was pressing and MacTavish instructed Anderson to send the Inuktitut interpreters across country to Fort Chipewyan, rather than south to York Factory and Norway House, from where they would have taken the winter sledge route via Cumberland House, Ile-à-la-Crosse and Methy Portage to Fort

Chipewyan. In the continued absence of Ouligbuck, Anderson had to find an alternate interpreter. His choice fell on William Oman.

Oman was the son of an Orkneyman (also William Oman) from either Stromness[1] or Sandwick[2] and of a Cree woman. He was born at Churchill in 1802 or 1803 and first appeared on the Company's books in 1820–21, when he was paid £15 as a labourer.[3] He continued to be employed by the Company for the subsequent forty years, being usually listed as a labourer, but also, on occasion, as a midman or a harpooner. He died at Churchill in 1868.

Oman regularly served as the post's Cree interpreter. For example an entry in the Churchill journal for 17 July 1842 indicates that Oman served the function of interpreter at church services for the Cree.[4] He possibly also had some command of Chipewyan since his wife, Margaret, was a Chipewyan. But of particular relevance to his new assignment was that he had close contacts with Ouligbuck and other Inuit, and was regarded as being the post's resident Inuktitut interpreter.

Oman left Churchill on 9 April, heading across country with two Chipewyan guides for Fort Chipewyan, a distance of some 950 km in a straight line, and substantially longer as Oman would have to travel, given the inevitable sinuosities of a winter trip across the rugged, lake-strewn terrain of the Canadian Shield. The three men were probably travelling on snowshoes, pulling toboggans.

Two weeks into the journey Oman was forced to give up, due to 'sore feet'. If he had had little previous experience of travelling on snowshoes and was trying to keep pace with his Chipewyan guides, a severe case of 'mal de raquettes', probably explained his sore feet. This was a common complaint among newcomers to the North, caused by the strain of walking on snowshoes on unaccustomed muscles and tendons in the feet and ankles.

Intriguingly one of the Chipewyan guides continued to Fond-du-Lac, the outpost attached to Fort Chipewyan, at the east end of Lake Athabasca, where he delivered Oman's letter of introduction. Since probably neither man could read, they would not have realized the pointlessness (and irony) of this achievement.

Oman returned to Churchill on 14 or 15 November, claiming that neither of his guides had wanted to escort him back any earlier and since, presumably, he was not confident of finding his way on his own.

What emerges clearly from this is that the failure to provide Anderson with an Inuktitut interpreter was not for want of trying on the part of the Hudson's Bay Company. Unfortunately, however, in this one critical area, the organizational skills of Sir George Simpson and his subordinates were foiled by a combination of circumstances beyond their control: Ouligbuck's apparent reluctance (for whatever reason) to participate in the expedition and hence his decision to deliberately make himself unavailable – and William Oman's unfortunate experience of being temporarily crippled on his undoubtedly strenuous winter trip across the rugged Canadian Shield from Fort Churchill to Fort Chipewyan.

[1] HBC A.30/11, f. 43^{v}.
[2] HBC B.42/f/2.
[3] HBC B.2398/g/4.
[4] HBC B.42/a/177.

In a letter to James Anderson the following summer, concerning plans to develop trade with the Inuit in the Anderson River area (east of the Mackenzie Delta) Sir George Simpson underscores the fact that obtaining reliable Inuktitut interpreters is a perennial problem; his very sensible advice to Simpson was 'to raise an Interpreter on the spot...'.

James Anderson

Having decided, sensibly, that there was little point in trying to reach Fort Resolution from Fort Simpson before the ice breakup had occurred on Great Slave Lake, Anderson waited at Fort Simpson until breakup had occurred on the Liard and the Mackenzie. He started with two canoes, loaded with supplies for the expedition from the supplies which had been stored at Fort Simpson in case they were needed for future arctic searching expeditions, and with ten men, on 28 May. He must soon have wondered if this start were not premature, since there were still large quantities of ice drifting down the river, and making either tracking or paddling extremely hazardous; on some days Anderson sensibly (although no doubt impatiently) decided to stay in camp to wait for the ice to clear. Hearing of the ice conditions from the Indians, Mr Miles, left in charge at Fort Simpson, had the sense to send a canoe with additional provisions for Anderson and his men. The result of the delays was that it was not until 11 June that Anderson reached the post at Big Island, which functioned mainly as a fishing station to supply Fort Simpson and other posts. Even on Great Slave Lake his progress was severely impeded by ice; he could make headway only by taking advantage of shore leads or by 'boring' through drifting ice, a procedure which led to serious damage to the canoes and, inevitably, to even more delay while this damage was being repaired. It was not until late on the evening of 20 June that Anderson and party reached Fort Resolution. Due to the very late date of breakup, it had thus taken Simpson twenty-three days for the trip from Fort Simpson; this should be compared with, for example, Lieutenant Pullen's trip over the same route at the same time of year in 1851; although also delayed by ice on Great Slave Lake Pullen's party took fifteen days for the same distance.[1] Here Stewart, with the remainder of the party, had been awaiting Anderson's arrival. Final preparations took only a short time, and once a northerly gale had slackened, the party pushed off in three canoes on the evening of 22 June.

[1] H. F. Pullen, ed., *The Pullen Expedition in Search of Sir John Franklin*, Toronto, 1979.

DOCUMENTS

James Stewart

1. The Journal of James Stewart, 6 February–30 May 1855.[1]

Tuesday, February 6th [1855]. Left Carlton House[2] with Mr. Lockhart on our way to the Ar. Sea to search for Sir Jn. Franklin's remains.[3] God grant us success & His providence & His protection in all our proceedings & preserve those I left behind me. Snowing hard all day. Wind N.E. & mild. Camped in this side of Lac qu'il doit permit.

Wednesday 7th. Mild during most of the [night]. Left camp @ 3 a.m. & camped at Grand Marais. Met the Northern Express. Took our letters & the men left at once [illegible] from Athabasca being nearly frozen & obliged to turn back was the cause of the delay.

Thursday 8th. A Beautiful day. Started late having lost a dog. Camped at La Grande Riviere.[4] All well. We are going very slowly.

Friday 9th. Started early & camped at the beginning of Lac Cruche. Sunny & mild.

Saturday 10th. Started early & camped within 2 miles of Green Lake. Fine, clear.

Sunday 11th. Beautiful weather. Arrived at Green Lake & started again; caught up with our men. All well.

Monday 12th. Started early & camped at Campement des Anglais. Fine weather.

Tuesday 13th. Left our encampment at 6 a.m. Met Mr. Jas. Finlayson at dinner time & camped at a small lake.

Wednesday 14th. Started early & arrived at Lac a la Crosse where we soon forgot our walking, by the hospitality of Mr. Deschambaults who received us in his usual kind manner.

Thursday 15th. Busy arranging sleighs & snowshoes etc.

Friday 16th. Got everything ready for starting tomorrow morning; we have actually cleaned out Mr. D's store.

Saturday 16th. Left Isle a la Crosse at 10 a.m. & bid goodbye to Mr. Deschambaults with regret. Beautiful weather. Camped at Pointe au Gravoir.

Sunday 17th. Started early & camped on the portage at Clear Lake. Cold – Wind North & clear.

Monday 19th. Started late & camped at the 2nd point in Buffalo Lake.[5] Francois Roy caught up with us at night with letters from my dear wife.

[1] PAA 74.1/137. The full title is 'Original diary of James Green Stewart, from 6th February to 4th November 1855 of the expedition sent to the mouth of the Back (or Great Fish) River by the Hon. Hudson's Bay Company 1854–1855'.

[2] Fort Carlton, on the North Saskatchewan River, about 80 km north of the present city of Saskatoon. Its primary function was to provide buffalo meat for making pemmican to fuel the Company's transport system of canoes and sledges. The fort was burned down during the Riel Rebellion of 1885 but has now been very faithfully restored.

[3] The party was travelling by dog sled.

[4] Now Big River.

[5] Peter Pond Lake.

Tuesday 20th. Started early & camped in Riviere la Loche. Fair & cold.
Wednesday 21st. Started at daylight & camped on Lac La Loche. Clear & cold.
Thursday 22nd. Arrived at Morrin's[1] at 10 a.m. & remained the rest of the day to rest the dogs; the same weather.
Friday 23rd. Started early & camped on a portage on the Little Athabasca river;[2] cold & clear; dogs going very slowly.
Saturday 24th. Started late & camped at La Bonne after a very fatiguing day's march. Beautiful weather.
Sunday 25th. Started early & came to a camp of Indians & camped late below Rapide a Pierre; the best day we have yet made.
Monday 26th. Started early; passed a camp of Indians at Pembina River & camped on the end of the first Portage.
Tuesday 27th. Started & camped at the end of the long reach between the Forts.
Wednesday 28th. Started early & camped below Pointe au Saline. Snowing all day; wind North; dogs low & so is Mr. Lockhart.
Thursday March 1st. Started in a snow storm & passed Red River [illegible] & camped below Pierre au Calumet. Clear & cold at night.
Friday 2nd. Started early & dined below Isle aux Freines; camped at Riviere au Barrier.
Saturday 3rd. Heavy snow in the morning & very mild. Camped opposite Piche's house. Hauling very heavy indeed.
Sunday 4th. Blowing & drifting all day & cold. Started late but notwithstanding made a good day. Camped at the upper end of the reach to the old Fort.
Monday 5th. Started early & camped at Lake Marmonance. Cloudy & cold.
Tuesday 6th.[3] —

2. Extract of a letter from John Bell, Fort Chipewyan, to Sir George Simpson, 30 March 1855.[4]

... The Expedition party arrived here on the 6th Inst. and every thing has been done towards the furtherance of this humane object. Our establishment being not large, we have had every difficulty in supporting them; we have been taken so unexpectedly, that we're quite unprepared for such a disbursement, reducing in every thing to the last item, thereby depriving us of the means of settling with our Indians in a satisfactory manner. There is not the slightest doubt that we shall be in debt to our Indians to no small amount of made beaver ensuing Spring, a very great error to fall into being injurious to the District...

3. Journal of James Stewart, continued,[5] Fort Chipewyan to Fort Resolution and back, 13–28 March 1855.

Tuesday 13th March. Left Athabasca with ten men & 7 sledges early this morning for

[1] This was at or on Portage la Loche (Methy Portage), the major portage across the divide between the Churchill and Mackenzie drainage basins.
[2] Clearwater River.
[3] One must assume that this was the date of Stewart's arrival at Athabasca (Fort Chipewyan).
[4] HBC D.5/40, p. 141.
[5] PAA 74.1/137.

Great Slave Lake.[1] It is quick work to be so far on our route seeing that it is only about 4½ months since Dr. Rae landed in England. Camped below Peace River[2] on a large island. Beautiful weather.
Wednesday 14th. Started early & camped a little above the Isles de Pierre. Fine clear weather; wind north.
Thursday 15th. Made a very good day. Camped below the Cassette portage;[3] the same fine weather; wind S.W.
Friday 16th. Started early; passed Salt River.[4] Smoked a pipe with Beaulieu[5] & camped 1 pipe below Hook's Island. Wind N.W. & cold.
Saturday 17th. Camped in Buffalo River. Cold with hard wind & a little snow.
Sunday 18th. Started early. Met Jno. Beaulieu in the Prairie portage & camped at this end of it.
Monday 19th. Arrived at Slave Lake at 3.15 p.m. Found nobody at the Fort;[6] we passed Mr. Ross[7] at the mouth of Buffalo river. A most beautiful day, our 7th from Athabasca.
Tuesday 20th.[8]
Friday 23rd March 1855. Left Slave Lake at 3 a.m. & camped at this end of the Rapid Portage; blowing from S.W. with snow.
Saturday 24th. Started late; dined at the Grand Détour & camped at Hook's Island. Fine cold weather.
Sunday 25th. Breakfasted with Beaulieu; took an Indian lad with us to take an Express to Isle à la Crosse & camped below the Cassette Rapid. Beautiful weather.
Monday 26th. Started at the usual time, and camped 3 reaches above the Isles de Pierre; rather warm today. Dogs knocking up & no wonder having come so far in so short a time, not to mention their coming from Carlton.

[1] Presumably the purpose of this trip was to make arrangements for canoes to be built at Fort Resolution and for other preparations for the expedition. It is not clear why this required such a large number of men and sledges. Anderson could not see the point either, and reprimanded Stewart for it in no uncertain terms in his letter of 6 May to George Simpson.

[2] Just below the confluence of the Slave and Peace rivers.

[3] The portage around Cassette Rapids, the most southerly of the four spectacular rapids on the Slave River just south of the present town of Fort Smith.

[4] A left-bank tributary of the Slave River which drains the Salt Plains, an area of largely barren salt flats with salt springs which provided an important source of salt for the whole of the Mackenzie basin.

[5] François Beaulieu, the patriarch of the Beaulieu 'clan' of Salt River, was born in 1771, son of Jacques Beaulieu and a Yellowknife woman. He was thus eighty-four years old when Stewart met him. He had accompanied Alexander Mackenzie on his journey to the Pacific Ocean in 1793 and had participated in Sir John Franklin's second expedition of 1825–7. For much of his life he supported three wives, but discontinued two of these relationships when he was baptized in 1848. For the later part of his life he held a monopoly from the Hudson's Bay Company for providing salt (naturally produced at the Salt Plains) for use at the various posts. He died in 1872 (L. H. Neatby, 'François Beaulieu', *Dictionary of Canadian Biography*, X (1871–80), Toronto, 1972, p. 38). In the summer of 1855 his son would play an important role in James Lockhart's support party which was entrusted with the tasks of building winter quarters at Fort Reliance and transporting two boats and supplies north to the headwaters of the Back River in case Anderson decided on a second season of searching.

[6] Fort Resolution.

[7] Bernard Ross, in charge of Fort Resolution.

[8] On this, and the subsequent few days, Stewart was no doubt engaged in preparations for the upcoming summer's expedition.

Tuesday 27th. Very warm & heavy walking. Camped at Peace River. Dogs very nearly done for.
Wednesday 28th. Arrived at Athabasca. All well. 6 days from Slave Lake.

4. Letter from James Green Stewart Esq. to Sir George Simpson, dated Fort Chipewyan, 29 March 1855.[1]

By the return of the duplicate Express I am enabled to inform you of our safe arrival at this place on the 6th Ultimo, and of my having been at Fort Resolution with 7 men, who were to commence at once to get the canoes in readiness, as well as 'traineau'[2] to carry them to where we find open water. Great Slave Lake breaking up very late we should lose a great deal of the short season by waiting for that event.

The wood, gum, bark, etc. are ready prepared here, and we only want warm weather to commence building the canoes at this place; no time has been lost in our preparations, and I think every thing will be ready here by the time Slave River breaks up.

It will be understood that the arrangements are made as if I was going alone, but I fancy Mr. Anderson will make but few alterations, except perhaps in the minor details; at present the plan of operations is as follows: On the breaking up of the ice in the River[3] we start for Slave Lake with two canoes, from thence getting them in two Traineau, with other sleighs for provisions etc.; we go with three canoes as far as possible or until we meet open water, by which means we hope to be on the Fish River about the beginning of June, follow the ice to the Sea, and commence the search according to the instructions, and return to winter at the East end of Slave Lake, where Back's old Fort Reliance was; for that purpose Mr. Lockhart will be left to collect provisions etc. and should anything prevent Mr. Anderson's coming Mr. B. R. Ross[4] will go with me to the sea.

On my way to Fort Resolution we fortunately fell in with the Yellow Knife Chief, to whom I gave instructions to collect his young men at the East end of Slave Lake to meet us there to receive orders, to hunt for the Expedition and render assistance in every possible measure.

Mr. Lockhart will also receive instructions to come with a party of Indians to the Fish River with supplies of clothing and provisions, as owing to the Expedition Goods having been taken down the McKenzie last year, we shall not have enough should the same weather take us en route. Fortunately 104 lbs provisions were left at Fort Resolution which will be quite sufficient for our wants during Summer; at this place provisions are so scarce that I have been obliged to send the men to fish for their maintenance.

Fourteen was the number sent from Red River, which makes two more men than the number mentioned in the instructions, but even that will not be sufficient for three canoes in such a river as the one we have to go through, consequently I have

[1] HBC A.12/7, ff. 516^{v}–517^{v}; HBC A.8/17, ff. 269–72; Great Britain, *Further Papers Relative*, 1856, pp. 22–3.
[2] Sledges – these, in the event were not required.
[3] The Slave River.
[4] Bernard Rogan Ross, clerk-in-charge of Fort Resolution. See p. 67 n. 1.

engaged two Orkneymen who volunteered at this place, which, leaving two with Mr. Lockhart for the summer work at Fort Reliance, will give us 14 for the sea, besides the Esquimaux Interpreters, which latter, I believe, not being used to work, need not be counted. They have not yet made their appearance, but the season will allow of their travelling for some time yet. Should any thing prevent their coming it will be a serious obstacle to our success, as it will impede our communication with the natives, who, I fancy, know more than they told Dr. Rae.

Canoes are rather frail craft for the sea and ice, but evidently best adapted for the work we have on hand; in order to make them safe they are to be lined inside with tarred bale cloth, so that should the bark be crushed or otherwise injured we shall have time to get ashore before the water gets through the cloth. I do not know whether this has been tried before, so that the Fish River will have the benefit of the first attempt.

Our wintering at all depends upon our success or whether another season would do any thing further than will have been done this season; should we be successful, or should it be found useless to prolong the search, you will, if I am alive, see me at Lachine next March, and should we winter will hear of us per winter Express.

No efforts shall be wanting to give us success and fulfil your wishes, but as so many abler men have been employed in this service without (with the exception of Dr. Rae) bringing any information or in fact doing anything in the way of clearing up the mystery of the lost ships, it will not be at all surprising should we fail also; let us hope otherwise.

5. Letter of introduction from George Barnston, Norway House, to 'the Gentlemen of posts in the Cumberland & English River Districts' carried by Paulet Panpoumakis [Papanakies], dated 26 February 1855.[1]

The Esquimaux[2] having to cross the country straight from Churchill to Athabasca to join the Arctic Expedition, I shall send off the remaining supplies to be furnished here for that Expedition immediately, and you are hereby called upon to lend every assistance in getting them speedily forward to Fort Chipewyan, as I by my former letters have previously advised You. To prevent difficulties in the District, Eighty Dogs are again charged by this Depot, to the Arctic Expedition, which will go forward with the Loads, but You will have to furnish extra Dogs & Sleds at each post to assist these. The time and labor of Your teams will be charged by You to the Expedition, for the period they may be absent on such duty. Any A. Ex. Dog breaking down or getting useless will have to be changed for another serviceable one, at whatever post that may be required. Please Examine the pieces, or Lading of the Sleds, as per way Bill herein enclosed, as they arrive at Your post, and forward this with the same way Bill, to the end of the Journey – Athabasca Lake.

[1] PAA 74.1/138. This letter was evidently presented to Stewart on the party's arrival at Fort Chipewyan, and retained by him. Papanakies had evidently been nominated leader of the group of voyageurs being forwarded from Norway House.

[2] This in fact was William Oman, who was travelling straight from Churchill to Fort Chipewyan via Fond-du-Lac, rather than via York Factory, Norway House, Cumberland House, Ile-à-la-Crosse and Portage La Loche, as previously planned.

A trusty Man, who can take care of property and a good guide, will have to accompany these sleds from post to post.

Depending upon Your strenuous exertions in rendering all the assistance as above and formerly required, I remain Gentlemen,

With much regard,
Your obedt. humble Servt.
Geo. Barnston

To be returned to Paulet Panpoumakiss
until he reach Athabasca Lake

6. Letter from James Stewart, Fort Chipewyan, to William MacTavish, York Factory, 15 April 1855.[1]

By the arrival of the men from Norway House[2] last evening I am in receipt of yours of the 11th Feby ult. as well as the supplies forwarded from York Factory and Norway House all of which reached this in as good order as could be expected from the state of the road. I was sorry that circumstances prevented your sending the interpreters but there is yet time for their arrival before the Expedition can leave this. Let us hope therefore that such an acquisition may not be lost. Every thing at Ft. Resolution and at this place is going on as well as we could wish. The canoes are begun and the first open water will see them afloat. The Expedition is much indebted to the gentlemen in charge of Posts for furthering the Company's aims in every way in their power, notwithstanding the limited means at their disposal.

7. Letter from James Stewart, Fort Chipewyan, to George Barnston, Norway House, 15 April 1855.[3]

By the safe arrival of Thomas Mustegon, McLennan & Paulet I am in receipt of your letters dated 11th & 26th Feby last. The supplies they brought were in as good order as could be expected from the state of the roads, and we are much indebted to the Gentlemen along the road for forwarding them in such an Expeditious manner. The Esquimaux interpreters have not yet arrived but there is yet time for them to come from Churchill via Fond du Lac as the ice is as fast still as in the middle of winter. I hope that we shall not be obliged to start without such an important requisition. The Canoes at this place or Fort Resolution are already begun and will I trust be afloat the very first open water.

With the exception of provisions the supplies sent in last summer have all been taken to Ft. Simpson but with the little we can get at Ft. Resolution, this place and those from Y.F. & Norway House we shall have a sufficiency for one season.

Paulet if not required will go out by the boats as you desire with many thanks for your letters to the gentlemen along the route (which has been well attended to).

[1] PAA 74.1/139.
[2] The voyageurs from Norway House under the leadership of Paulet Papanakies.
[3] PAA. 74.1/141.

8. Journal of James Stewart, continued, Fort Chipewyan to Fort Resolution, by canoe, 26–30 May, 1855.[1]

Saturday 26th May 1855. Left Athabasca in two Canoes accompanied by Mr. Lockhart en route to G.S. Lake. We are afloat at last & started on our journey to the Arctic Sea. Blowing hard from N.W. & cloudy. The trees are now beginning to bud.

Sunday 27th. Started at ½ past 5 a.m. & camped at the lower end of the Cassette. Cloudy & warm with a little Southerly wind now & then. Mosquitoes thick.

Monday 28th. Passed all the portages in Slave River; rained a little in the afternoon. Camped at Salt River.

Tuesday 29th. Wind North all day & cool. Started early after embarking Beaulieu's son & family who is engaged as interpreter for Fond du Lac & Jambe de Bois, a Yellowknife guide for Fish River who I had told to meet me in the spring.

Wednesday 30th. Arrived at Ft. Resolution and found the ice in the Lake as hard as in winter. Mr. Anderson has not yet arrived nor has Mr. Ross got our canoes even made though the wood is nearly ready & about all the birch which it is difficult to procure.

William Oman, Inuktitut interpreter

9. Extract from Fort Churchill Journal, 6 April 1855.[2]

... As Oman is preparing to start for the Athabasca there being no appearance of the Esquimaux as yet & the season advancing it will be getting late for crossing the Rivers on ice.

10. Letter from William Anderson, Fort Churchill, to John Bell, Fort Chipewyan, 7 April 1855 (concerning the Inuktitut interpreter).[3]

Having received orders from William McTavish Esquire, C.F. to send two Esquimaux Interpreters across Country to Fort Chipewyan for the Expedition but being unable to procure any after waiting untill nearly the latest date that the guide can make sure of getting to Fort Chipewyan, before the opening of the navigation, I have been drove to the necessity of sending William Oman, our Fort's Esquimaux Interpreter, & I hope that he will reach Fort Chipewyan in time to join the Expedition & will be able to Explain to the Esquimaux [whatever] the gentleman in Charge may want.

The guide who accompanies Oman only knows the way to your out-post on the borders of Athabasca Lake[4] where they will be able to get a guide or whatever they may want to enable them to get to Fort Chipewyan as quick as possible.

I may also mention that I have promised the Chipewyan guide, that he will get

[1] PAA 74.1/137.

[2] HBC B.42a/187, f. 78^{v}.

[3] HBC B.42/b/61.

[4] Fond-du-Lac at the east end of Lake Athabasca.

to Norway House by your boats & from there to York Factory by the first opportunity.

11. Letter from William Anderson, Fort Churchill, to 'the Gentleman in Charge of the Trading Post, north end Athabasca Lake [Fond-du-Lac]', 7 April 1855.[1]

Please furnish the bearer William Oman with a guide, Provisions & whatever else he may require to enable him to get to Fort Chipewyan before the opening of the navigation as he is required to join the Expedition that is to leave Slave Lake in June.

I Beg you will not detain them at your place as William Oman's services is required for Esquimaux Interpreter for the Expedition.

12. Letter from William Anderson, Fort Churchill, to William MacTavish, York Factory, 9 April 1855.[2]

I am sorry to inform you that Oulibuck & Munro have not yet come in or no other Esquimaux has visited this since Mr. Beddome left here.

After waiting untill near the latest date that the guide could make sure of getting to Fort Chipewyan before the opening of the Navigation, & as you said that an Esquimaux interpreter was essential to the Expedition, I was at the necessity of asking Oman & after an evenings consideration he agreed to go at 40£ a year & started this morning with the full determination of getting to the Athabasca in time for the Expedition.

The guide now says it will take them 28 days to the out-post[3] & 12 days more to Fort Chipewyan. So that still leaves time enough for their getting there before the 1st June even should they have to get Canoes at the lake.

There is very little chance of the Esquimaux being here now before the 6th May to the seal hunt, so please let me know if I shall try to keep Oulibuck & Munro & send them to York Factory when they come here or to keep Oulibuck only, as I am now certain that they are stopping away on purpose, as they are 3 weeks past the time that they promised to be here allowing no message had been sent.

There was 12 distant Chipewyans here in March but only 2 young lads that had been at the little house,[4] & they went there from the northward & did not know the way from this side, but their Father knows the place well, & it is him that Oman & Nardarl-yousa is going to take as his tent is in their way & the sons promised that they would stop at the same Place Fourteen days to wait for the Esquimaux......

There is yet some of the most distant Chipewyans that is expected here. If Oulibuck arrives before them and think that he will be able to get to the Athabasca, he shall be sent if possible.

13. Extract from Fort Churchill Journal, 9 April 1855.[5]

Wind North to NW. Blowing strong with a little snow & drift. Oman & Nar darl

[1] HBC B.42/b/61, f. 34^{v}.
[2] Ibid.
[3] Fond-du-Lac.
[4] Ibid.
[5] HBC B.42a/187, f. 79.

yousa started this morning for Fort Chipewyan being provided with 20 days provisions & 5 lbs Gunpowder with an Equal proportion of Ball & Shot etc., and all other necessarys to enable them to get into the Athabasca with out delay...

14. Extract from letter from William MacTavish, York Factory, to Sir George Simpson, 20 May 1855.[1]

...Mr. Anderson ... succeeded in getting messages delivered to them by Esquimaux, but as neither Ouligbuck or Munro came into the post there can be little doubt that they disliked the service...

15. Extract from letter from William Anderson, Fort Churchill, to William MacTavish, York Factory, 20 June 1855.[2]

...On the 27th May five Esquimaux visited this... I also made enquiry at them about Oulibuck; they said that he was gone far off and they did not know when he would be here. Munro they said was also far away & had been starving all the winter, which is very likely the cause of his not coming in here.

16. Extract from letter from William Anderson, Fort Churchill, to William MacTavish, York Factory, 28 July 1855.[3]

Our Esquimaux Chief ... arrived here last evening... The Chief can give us no information about Oulibuck & says that he has not heard any thing of him since he left his hut last winter, but I trust that Buck[4] will yet come in here this summer & I will try to keep him at the house...

17. Extract from letter from William Anderson, Fort Churchill, to William MacTavish, York Factory, 26 August 1855.[5]

... Munro, the Esqx. came in here along with the whale hunters, says that he was very near dead the last winter from starvation. Oulibuck did not come; he remained inland at a place where the deer passes. I have spoken to Munro & the Chief, and both of them has promised faithfully to come in here along with Oulibuck, on the first ice. I also told Munro that it was likely he & Oulibuck would be wanted for another Expedition & Munro said that he was willing to go again...

18. Extract from Fort Churchill Journal, 14 November 1855.[6]

... 40 of the distant Chipewyans arrived with William Oman just as we settled with the last one of the others...

19. Extract from letter from James Hackland to William MacTavish, York Factory, 7 January 1856.[7]

Oman also arrived with the dist Chipewyan on the 15th Nov. having stopped all the time with them, he gave up with sore feet his 15th day from here, 23rd April 1855,

[1] HBC B.239/b/104b.
[2] HBC B.42/b/61, f. 35v.
[3] Ibid., f. 37.
[4] Ouligbuck.
[5] HBC B.42/b/61, f. 37v.
[6] HBC B.42a/188.
[7] HBC B.42/b/61, f. 39.

when the two Indians proceeded for the Athabasca out post,[1] which Nar,darl,yousa did reach. Ee,so,ey gave up 2 days from the house and got another Indian to go in his place with Nar,darl,yousa, which after he had given up the letter returned to Oman who wanted to return here but none of the Indians would accompany him.

20. Extract of letter from Sir George Simpson, Norway House, to James Anderson, Fort Simpson, 14 June 1856.[2]

I am glad to observe by your letter of 16 February that you are directing your careful attention to the extension of the trade with the Esquimaux. We are very sensible of the importance of this effort and rely on your zeal & perseverence in its prosecution. We would willingly aid you with the services of an Interpreter could one be found, but they are very scarce, & you must have observed how difficult it is to obtain the services of such Interpreters, by the abortive results of the efforts to find one for your Arctic Expedition.

Your cousin at Eastmain is in as great trouble as yourself from a similar cause & we have tried to procure him an Interpreter from Labrador, but so far without success. You must not rely on us in this affair, but endeavour to raise an Interpreter on the spot, by getting an Esquimaux boy to live at one of the Company's posts to learn English & to teach others his own language. This subject of the Esquimaux trade is of much interest and I shall be glad to hear further from you upon it hereafter...

James Anderson

21. Letter from James Anderson to Sir George Simpson, 25 March 1855.[3]

March 31st. I had written this far when the express arrived on the 29th about mid day. I am not easily put out of the way but certainly this appointment was a bit of a 'stunner'. I can only say that I am deeply gratified for the honor you have conferred on me, and that all the energy I possess shall be put forth for the accomplishment of the objects of the expedition and for the honor of the service.

All the supplies are here excepting the pemmican and flour and I could do nothing if I went to Resolution except getting the canoes built and engaging some of the Fond du Lac hunters,[4] while there is much to arrange here. I have sent minute instructions to Stewart as well as to Ross to get these matters arranged. I shall leave here as soon as the Upper McKenzie is open with the remainder of the supplies and proceed to Big Island which I shall leave for Resolution as soon as the south end of Slave Lake is navigable; the north end[5] breaks up later; we may think ourselves fortunate if the ice will permit us to leave by the middle of June.

I trust that a copy of Back's journal has been sent – I read it long ago. The river is,

[1] Fond-du-Lac.

[2] HBC D.4/76A, f. 831.

[3] HBC E.37/10; BCA AB40 An 32.2, ff. 117–22.

[4] Anderson is referring to the 'Fond-du-Lac' of the East Arm of Great Slave Lake, not the outpost of Fort Chipewyan at the head of Lake Athabasca.

[5] Here Anderson is actually referring to the East Arm.

I believe, a bad one and the ice-encumbered sea will not be very favorable to our frail craft. Putting aside detention from ice I see no difficulty in accomplishing our mission in one season; by Collinson's voyage[1] and Rae's discoveries the field of search is materially narrowed. Unless we receive certain intelligence from the Esquimaux I think I shall divide the party at the mouth of the River and search both sides of the Gulf at once as far as Cape Britannia and Point Ogle. I see no prospect at present, without great detention and loss of time, of being able to convey sufficient provisions for a party to pass the winter on the coast with any prospect of safety but I trust that there will be no occasion for this. I shall engage Fond du Lac hunters to accompany us and make caches of provisions along the route; this party I will put under the command of a son of Beaulieu's who I have told Stewart to hire; he is an excellent hunter and will serve also as interpreter.

Regarding the astronomical instruments, I shall leave them, with the exception of compasses, both Stewart and myself being perfectly unacquainted with their use.

Stewart and everyone else seem to have a perfect fever of useless excitement. His letter gives no information, not even the number of men he has brought down; he can't, however go far wrong with Ross[2] at his elbow.

I have sent for Campbell to arrange everything with him; he will then go up and close the trade at Liard as far as possible and leave Harrison to send off the returns. I must keep Harrison now. Campbell will reach here early in June; Mr. Mills can conduct affairs until his arrival. I will leave everything here in apple-pie order for Campbell, and Ross can conduct the summer business.

I don't know who you intend to appoint to the charge of this most valuable district, but Campbell is certainly not capable of undertaking it. If I get back all safe – and you wish it – I have no objection to remain a year or two longer in this quarter if my health remains good. If so let me know by the very earliest opportunity.

This express must leave tonight. I have not shut an eye since the arrival of the packet and am nearly done up.

P.S. I hardly know how my right leg will stand a long walk; the veins are swollen as big as your thumb (varicose) but I suppose that I shall rub through it.

Herewith you will receive sundry documents brought by Hardisty's Rat Indians from the Enterprise;[3] how easy it would have been for Collinson[4] to have accomplished our task.

The winter has been exceedingly mild.

[1] Captain Richard Collinson (in HMS *Enterprise*) had wintered in Walker Bay (south-western Victoria Island) in 1851–2 and at Cambridge Bay in 1852–3. His sledge patrols had searched the western, southern and part of the eastern coast (as far north as Gateshead Island) of Victoria Island without finding any definitive traces of the Franklin expedition. See: Richard Collinson, *Journal of H.M.S. Enterprise, on the Expedition in Search of Sir John Franklin's Ships by Behring Strait, 1850–55*, London, 1889.

[2] Bernard Ross, in charge of Fort Resolution.

[3] Collinson had sent dispatches south from HMS *Enterprise*, wintering at Camden Bay on the arctic coast of Alaska, with a group of Gwich'in Indians who visited the ship. They had been forwarded by William Hardisty at Peel River post.

[4] This is a recurring theme of Anderson's (with some justification). He felt that Collinson had missed a golden opportunity for searching King William Island and Adelaide Peninsula by sledge parties from his wintering base at Cambridge Bay in the spring of 1853.

22. Letter from James Anderson, Fort Simpson, to Sir George Simpson, 30 March 1855.[1]

I beg to state that your despatch of 18 Novr. last appointing me to the first and Mr. J. G. Stewart to the second command of an Expedition to ascertain the fate of Sir John Franklin and his unfortunate companions, reached me yesterday at mid-day.

(2d) I need not say that I feel highly honored at being selected to perform this duty and that all the energy and ability I possess shall be put forth to perform your instructions and not discredit the service.

(3d) Mr. Stewart is at Slave Lake getting canoes built; he will also engage Fond du Lac hunters and make other arrangements. All the A.S.Ex. supplies[2] are here excepting the pemmican and flour which are at Resolution and there are many arrangements to make. I shall leave this immediately the Upper McKenzie breaks up with these supplies, and await the opening of the navigation along the south shore of Slave Lake at Big Island, when I shall immediately proceed to Resolution, which I shall reach long before the north end of Slave Lake is navigable.

(4) The chief difficulties to be overcome are the navigation of the River and the ice-encumbered sea with our frail craft, but with care this may be safely accomplished. As great a quantity of provisions as possible – with due attention to our rapid progress – shall be taken and a party of Fond du Lac hunters be engaged to accompany us, and await our return about the head of the River,[3] by which time I anticipate that they will have gathered a large stock of provisions. If not detained by ice I see no difficulty in accomplishing our task in one season.

(5) I shall have the pleasure of addressing you previous to our departure from Resolution, when I shall be able to give you an account of the arrangements that have been made. This Express leaves immediately otherwise it will not reach Athabasca before the Slave River becomes dangerous.

23. Letter from James Anderson, Fort Simpson, to James Stewart, Fort Resolution, 30 March 1855.[4]

The Southern Express arrived here yesterday at Mid-day, by it I received Sir Geo. Simpson's instructions regarding the Expedtn. we have the honor of being appointed to conduct.

2d. Sir G. Simpson appears to be apprehensive that I may arrive too late at Fort Resolution to take the command, being probably unaware that the southern portion of the Lake breaks up eight to ten days earlier than the northern portion. Were there the least risk of such an event, I should of course now take my departure, but as there is not, I shall remain here till the McKenzie breaks up, and ascend the River with 2 Canoes or a boat, in which I shall embark the supplies for the men, articles of Esquimaux Trade, Ammunition, etc. By adopting this plan I shall save much time, trouble and expense, and moreover have an opportunity of seeing C. T. Campbell with whom I have much to arrange.

[1] HBC E.37/10; BCA AB40 An 32.2, f. 137.

[2] The supplies for possible use by arctic searching expeditions. See p. 29.

[3] The Back River.

[4] HBC E.37/10, f. 111.

3d. You have of course already commenced the construction of the canoes. I trust that the Chipewyan model has not been followed, as their canoes can stand no swell. It will be requisite to procure a large supply of Gum[1] and Wattap,[2] with Bark for repairs. The Gum should be well cleaned, cooked and blazed. The Iroquois will know how to do this. Some extra paddles should also be made, as we shall be in a country where wood is a rarity, the requisite Agrets[3] will be brought up from here.
4. On the arrival of the Fond du Lac hunters, six at least should be engaged for the purpose of hunting for the Expedition; the most skilful should of course be chosen, and if their conduct be good, they shall be most liberally remunerated. You will also hire King Beaulieu at the rate of £40 per annum while employed in the service, as Interpreter and Hunter, and to look after the party of Indian hunters, who I intend shall meet us on our way back. I consider this man as indispensable and he must not refuse.
5. Two men must be sent immediately on the receipt of this to Chipewyan, who will return in Canoe as soon as Slave River breaks up, with the following supplies belonging to Government, Viz. 1 Keg Gunpowder, 1 Bag Ball, a bag Shot and also the Tobacco if any be left. If Mr. Bell[4] can exchange the Bag of Shot for a Bag of Ball it would be more serviceable for us. I need hardly add that this amn. must not be touched on any account, except it may be to furnish a small supply to any Indians who promise to hunt for us.
6. I send up a little tea for the men, and you may tell them that they shall be equipped properly with the requisite clothing etc.
7. In addition to the above supplies belonging to Govt. you will also request 60 pairs Moccassins from Chipn., half of them topped and large enough to hold a sock.
8th. On the last ice[5] Ten men including two of the Iroquois are to be sent to Big Island to meet me and this will also be the means of saving some provisions.
9. In all your arrangements you will consult with Mr. Ross who can of course afford you much valuable information, and who in the event of your absence has permission to open this letter and act on its contents.
10. I can only spare 1 Keg of G'powder from here with lead, so it is absolutely necessary for the success of the Expedition that the ammunition at Chipewyan be brought down. I believe there is nothing else requiring immediate notice, and with best wishes.

24. Extract from letter from James Anderson, Fort Simpson, to Eden Colvile,[6] Fort Garry, 1 May 1855.[7]

You may imagine that I was rather surprised at receiving a dispatch from Sir G.

[1] Spruce gum, made from spruce resin with an admixture of moose fat to make it less brittle, for waterproofing the seams of the canoes.
[2] Spruce roots, for repairing the canoes.
[3] Accessories for the canoes, camping and cooking equipment etc.
[4] John Bell, at Chipewyan.
[5] I.e. just before breakup.
[6] Governor of Rupert's Land, based at Fort Garry.
[7] HBC E.37/10, ff. 105–6.

Simpson on the 29th March appointing me to the command of the A.S. Expedn. I of course feel highly honoured at this mark of confidence, and assure you that all the energy and ability that I may possess shall be put forth to accomplish the objects of the Expedn.

Mr. Stewart and the entire party with the important exception of the Esquimx. Interpreters reached Chipewyan with the Express and are now I suppose at Resolution.

Sir George under the impression that the dispatch would reach me much later than it did, or perhaps being unaware that the N.E. end of Slave Lake broke up later than the Southern End, directed me to throw up this charge and proceed to Resolution immediately, but as I could have done nothing there but superintend the building of the canoes and engage some Couteaux Jaunes[1] hunters, which Mr. Stewart can do, and as there are many arrangements to make here, I decided on remaining here until the Upper McKenzie breaks up when I shall leave for Big Isld. and coast along the south shore of Slave Lake as soon as the ice will permit me, taking up at the same time the requisite supplies which are all here except the provisions which I fortunately left at Resolution last fall. I shall reach that post long before the Upper end of the Lake is navigable. It would have been all but impossible to have transported the supplies here to Resolution by trains[2] and would have occasioned immense trouble & expense, without forwarding our progress an hour. By remaining here I shall be able to arrange the Expedn. supplies completely, have had an opportunity of communicating with C. T. Campbell (who will come here on the 10 June to assume this charge), and without any detriment to the affairs of the Expedition, be able to leave the affairs of this most valuable District in such a condition that Mr. Campbell will have no trouble.

Stewart has plenty of pluck and endurance; as to his foresight & prudence I have my doubts. The sending up 3 Iroquois Boutes[3] was an excellent idea of Sir George's, as such a thing as a good Canoeman is almost unknown in the North. Among the other men are some of Rae's old hands. I wish Old Sinclair who was a steersman to Back had been one of them. Sir George mentions that charts have been sent up, as well as astronomical instruments. The latter will be left as neither I nor Stewart understand the use of them. However it is of little consequence as the ground is already laid down and we can't well lose our way. I trust that Back's Journal has been forwarded.

Canoes are certainly not the craft for navigating an ice-incumbered sea, but with our short notice we cannot do better.

I have directed Stewart to engage Couteaux Jaunes hunters. I shall take them as far as they will consent to go and they will hunt back making caches en route in case we may require them. I shall leave Resolution with 3 Mos. provisions which with the supplies will load 3 Canoes. We shall be able to dispense with one of them when we get among the portages. About 2 days from Slave Lake we shall bid adieu to all wood.

The winter was very mild, but the spring so far has been cold & lingering & I am praying for a change. Lakes Clinton Colden and Aylmer are said to break up as

[1] Yellowknife Indians.

[2] Dog-sleds.

[3] Bow-men.

soon as the Upper end of Slave Lake. I trust that such is the case. If however we can reach the sea by the end of July it will be early enough. There is little prospect of the inlet at the mouth of Fish River being free of ice before the month of August. Collinson (from whom we heard via the Youcon), says that the ice at Cambridge Bay began to break up on the 25 July and that he was able to leave on the 5th August. Since I have been in this quarter we have seen no ice in Slave Lake on our way to the portage.[1] Last year it was free about the 15 June, but in '48 the boats were detained 12 days by ice and on the 11 July their bows were nearly cut through by young ice.

If we find Esquimaux at the mouth of the River of course my plans will be determined by the information I expect to receive from them. If they should have the Despatches[2] and the sea be free from ice our task will be speedily accomplished. Should the inlet be still fast I shall proceed with a party & the Halkett Boat on foot to Point Ogle and Montl. Island leaving 10 Men with the Canoes to follow if the ice breaks up. If we see no Esquix. nor find any traces of the missing party at Pt. Ogle & Mtl. Isld., if the ice will permit it I shall send Stewart with half the party to the Wesd. as far as Ellice's R. and proceed myself first to Simpson's Cairn at Pt. Herschell, King Wm. Isld,[3] which I shall examine as a likely spot to find a notice of the party, and then proceed round the Isld., if possible, in search of Esquimaux, the two parties to rendezvous at Pt. Ogle, cross to Cape Britannia & return along the East side of the inlet to the mouth of the River. This is all that could be done in that quarter, and to perform it we should require 3 weeks of open sea and moderate weather. I only mention these as plans floating in my mind; no one can fix on any definite plans for an expedition in the Arctic Regions; there are so many obstacles that may intervene, which no human prudence or foresight can overcome. The River is a bad one and we may possibly meet with accidents tho' I shall risk as little as I can.

From neither Rae in 1851[4] nor Collinson having heard anything of the missing party or their vessels from the Esquix. of Victoria Land it is evident that the vessels were not lost on the Eastern side of Victoria Strait (though the piece of wood found by Rae most likely belonged to one of them), they were therefore in all probability lost on the W. coast of Boothia between Bellots Straits & the Magnetic Pole. Is it not strange that the party did not strike for the Fury's stores where relief was certain?[5]

[1] On his annual trip to Portage la Loche to make rendezvous with the La Loche brigade coming from Norway House.

[2] Journals or documents from the Franklin expedition

[3] A large and conspicuous cairn on the south-west coast of King William Island, built by Thomas Simpson and Peter Dease during their impressive trip along the arctic coast by boat in the summer of 1839. See: Thomas Simpson, *Narrative of the Discoveries on the North Coast of America; effected by the Officers of the Hudson's Bay Company during the Years 1836–39*, 2 vols, London, 1843.

[4] In the summer of 1851, operating from a base at Fort Confidence on Great Bear Lake and travelling by boat, Dr John Rae searched most of the south and east coasts of Victoria Island. See: Rich and Johnson, *John Rae's Correspondence....* Much of this coast was searched again by Collinson in the summer of 1852 and the spring of 1853. See: Collinson, *Journal of H.M.S. Enterprise*.

[5] Anderson's reference here is to the major depot of supplies left by Captain William Parry at Fury Beach on the east coast of Somerset Island, when he was forced to abandon HMS *Fury* (which had been driven ashore by the ice) in the summer of 1825 (see: Parry, *Journal of a Third Voyage*). Although Sir John Ross's party had used some of the stores during their wintering at Fury Beach (after abandoning *Victory*) in the winter of 1832–3 (see Ross, *Narrative of a Second Voyage*), substantial amounts of these stores were still left, although rumours were current that the whalers might have used some of them. Of significance

In the event of the failure of the Expedition this season Sir George has suggested that all, or a portion of the party, should winter on the coast to resume the search in the spring. This is impracticable: our craft will with difficulty convey the provisions etc. for the summer trip, and even half rations for a party of 6 would require 36 pieces Pemmican and at least 10 or 12 of other supplies. By the time we return to the mouth of the River, the most likely place for a winter station, the deer[1] will have migrated to the Soud. and the fisheries have finished. Moreover we know nothing of the resources of that portion of the Country to justify such an attempt. I correspond intimately with Rae and know his opinions about wintering on an unknown coast. When Sir J. Richardson proposed to send him to winter on the coast,[2] he told me that he should refuse unless fully provisioned. His wintering at Repulse Bay is not a parallel case. There he brought his Boats laden with supplies and provisions alongside a well known deer pass[3] and Fish Lakes, and even then did not decide on wintering till the 1st Septr. when he found that the deer hunt had not failed. Had it failed he had ample time to retreat to Churchill by open water.

I believe I must now have tired you with this arctic gossip. If we, as I trust we shall, prove succcessful in our entreprise, I shall have to go straight to England, I suppose via R.R.[4] I was considered once a good walker, and certainly for ten Years had rather more than enough practice. I don't know how my right leg will stand a long tramp now, the veins being swollen and knotted (varicose). However I suppose that I shall rub through it.

I shall leave the Distt. I think in fine order; the returns will be excellent. The R.R. traders who are overrunning the country even threaten to invade this quarter.

As I may not have another opportunity, I beg to recommend to your notice two most able officers in this Distt. as being worthy of promotion, namely Messrs. Bernard R. Ross[5] and Wm.L. Hardisty.[6] I perceive that Mr. Stewart, who is junior to either of them, and whose activities and services are far inferior, is to be promoted next spring; this will create much dissatisfaction. Mr. Stewart was 4 Years a supernumerary, then second at an inland post, and only placed in charge of a post, Carlton, this Year.

Mrs. Anderson was confined of another Boy on the 14th Ult. Both she and my

---

in this context is the fact that Captain Francis Crozier, captain of HMS *Terror* on Franklin's expedition, had been a midshipman in HMS *Hecla* under Captain Parry when *Fury* had been abandoned and when this large depot of stores was cached.

[1] Barren-grounds caribou (*Rangifer tarandus groenlandicus*), which with the exception of a few scattered bands, migrate south of the treeline in winter.

[2] Presumably during the expedition of 1847–9 during which they collaborated, searching the coast from the Mackenzie Delta to the mouth of the Coppermine for traces of the Franklin expedition, but from a wintering base at Fort Confidence on Great Bear Lake (south of the treeline) and not from a base on the arctic coast. See Richardson, *Arctic Searching Expedition*; Rich and Johnson, *John Rae's Correspondence*.

[3] A location, such as a narrows in a lake, where migrating caribou could be easily hunted.

[4] Red River.

[5] Of Fort Resolution.

[6] Of Peel River (Fort McPherson).

children are well. I hear excellent accounts of those at Red R. Pray excuse this hurried letter; I am indeed much occupied. I shall only close this at Resolution. Mrs. A. unites with me in sincere wishes for the health & happiness of yourself and Mrs. Colvile...

25. Extract from letter from James Anderson, Fort Simpson, to Sir George Simpson, 6 May 1855.[1]

I have now everything ready for a start, men's equipments, equipments for Indian hunters, and assortt. of iron works etc. for the Esquimaux and agrets[2] and supplies for the Expedition. A bastard canoe I have here is thoroughly repaired and I expect another larger one from Liard of superior quality immediately after the disruption[3] of the Liard River.

As I have some leisure I have determined on devoting it to giving you my present opinions etc. regarding the expedition. I informed you in my last that I had determined on not proceeding to Resolution till the Upper McKenzie broke up, when I would leave here with the requisite supplies and creep along the S. shore of Slave Lake as soon as the ice would permit me. All of the supplies are here with the exception of most of the flour and pemmican which I fortunately left at Resolution. With the means at my command and at that late season it would have been all but impossible to have rendered supplies undamaged to Resolution, but admitting that it could have been done, it would have been incurring an enormous expenditure of provisions without forwarding our progress one hour as the lower end of Slave Lake breaks up at least a fortnight before the upper end. By remaining here I have been able to make my arrangements more completely, have had an opportunity of communicating personally with Mr. Campbell, and as there is a good canoe maker at Liard, of getting a canoe made which I think will be superior to any that can be made at Resolutn. or Chipn.

The idea of starting on ice from Resolution[4] of course occurred to me, but when I came to consider that the canoes from Chipewn. would only reach there about the 1st June, when the ice would be rotten and as far as the islands in motion, large crevasses forming early in the spring, I soon perceived the impracticability of transporting the canoes and some 60 pieces over the ice at such a late season. Moreover we should have had to await at the upper end of the Lake for at least the partial disruption of the ice on the lakes in the interior, and in the meantime our provisions would have been disappearing.

When I last addressed you there was every appearance of an early spring; the rivers were then all covered with water, but since the beginning of April the weather has been very cold and everything prognosticates a late season; the Liard which broke up on the 1st inst. for the last 3 years is still fast. Last year I believe that the whole of Slave Lake was free by the 20th June; a boat from Resolution got along the South shore and reached here on the 12th June last spring. I have never seen ice on

[1] BCA AB40 An 32.2, ff. 123–35.
[2] Accessories for the canoes, i.e. spare poles and paddles, sails, birchbark, gum, etc.
[3] The breakup of the ice.
[4] I.e. transporting the canoes on sledges.

Slave Lake on going out to the portage.[1] This is however not an unusual circumstance; in 1849 the boats were nearly cut through by young ice on the 11th July.

It is very probable that I may take a different route from that of Cap. Back to Artillery Lake. I will however say more on this head when I hear from Mr. Ross who is making sundry inquiries for me.

I should have wished to reach the sea in the beginning of August not later than the 5th – but I can hardly hope to accomplish this owing to the backwardness of the season, but even if we could get there by the 10th much may be accomplished if we have a clear sea and tolerable weather. We must however reach the mouth of Fish River on our return not later than the 1st Septr.

Should we see Esquimaux at the mouth of Fish River and recover the journals or despatches of the party[2] from them, and if as it is supposed – the party perished at Pt. Ogle and Montreal Isld. our task – if the inlet be navigable – will be speedily accomplished. As it is possible that the despatches may have been deposited in the graves I shall open them. If we neither see Esquimaux nor discover traces of the party at Pt. Ogle or Montl. Island – should the ice and weather permit it – I shall send Stewart with half the party to trace the coast to the westwd. towards Ellice's River, while I shall strike across to King Wm's Island and proceed straight to Simpson's cairn at Pt. Hershell which I shall search for a record and, if possible, go round the island in search of Esquimaux – the 2 parties to rendezvous at Pt. Ogle and if nothing should have been discovered cross over to Cape Britannia and return to Fish Rr. along the E. coast of the inlet. This is all that can be done in that quarter, and to do it we shall require 3 weeks of clear sea and moderate weather. Such are the ideas now floating through my brain. I don't mention them to you as fixed plans, for no one can fix a plan for an expedition in these regions where obstacles may arise and accidents occur, not to be overcome or avoided by human determination, prudence or foresight.

In the event of our failure, I consider at present that it will be impracticable for even a portion of our party to winter on the coast. Our craft will barely convey sufficient provisions etc. for the summer trip. Now, a party say of 6 persons would require 36 pieces of provisions for half rations for 12 months, and at least 14 pieces of other supplies. To live they would require to be stationed in the immediate vicinity of a deer pass or well ascertained fisheries. Again by the time we return to the mouth of the river – the most likely place for a winter station – the deer will have migrated to the Sd. and the salmon fishery be over. A good supply of ammunition cannot be supplied from the Distt. We are also unprovided with alcohol or lamps for spring tripping though this last difficulty might be overcome by the cumbrous substitute of tallow wicks. I correspond intimately with Rae and know his ideas on 'wintering on an unknown coast'. When Sir J. Richardson contemplated sending him to winter on the coast he told me 'he should refuse unless fully provisioned.' Rae's wintering at Repulse Bay is not a parallel case. On his first trip his provisions and supplies for a party of 13 persons amounted to 140 pieces; these he brought

[1] On his annual trip south to the rendezvous at Portage la Loche with the returns from the Mackenzie District.

[2] The Franklin party.

alongside a good deer pass and fish lakes; on his second trip he had a boat load for 6 or 7 persons and did not determine on wintering till the 1st Septr. when he had ascertained that the deer hunt was successful; had it failed he had ample time to retreat to Churchill by open water.

I shall take a party of Couteaux Jaunes hunters as far as they will consent to accompany us; they will act as guides as far as the river[1] and hunt back, making caches en route in case we may require them. I shall endeavor to leave L. Aylmer with 2 months provisions. About every 50 or 60 miles a bag of pemmican will be deposited en cache for our up trip.

Except for proceeding as far as Point Ogle I consider that this route abounds with more difficulties and is the most unpromising and dangerous of any to the Arctic Sea. The inlet at the mouth of Fish River is, I fear, choked with ice late in the season. Stewart has not forwarded Back's journal – surely it has been sent – and I have only a general recollection of the route, as I read the work long ago. Rely on it, however, that whatever can be accomplished without sacrificing the lives committed to my care shall be done.

A boat with the requisite supplies will meet us either at Fort Reliance or where I shall decide on wintering. I shall, however, distribute a portion of the party – for the sake of saving provisions – at Rae and Resolution. A pedestrian trip from Slave Lake from its immense distance from the point of search would be a hopeless task and if it be decided on that we are to continue the search from this quarter we shall require to be furnished with alcohol lamps for cooking and some rifles (say 6) with their agrets.[2] Half of the party will then be left at some fit place to procure provisions for passing the winter, while the others carry on the search.

On reading over Rae's despatch attentively – one paragraph of which I predict will bring a storm about his ears[3] I think from the selection of the articles procured by the Dr. from the Esquimaux and the quantity of amn. stated to be in possession of the party, that the vessels were not destroyed suddenly but may have been crushed at winter quarters like the Fury.[4] From neither Rae in 1851 nor Collinson in '53 having heard anything of the vessels from the Victoria Land Esquimaux, I suspect they were lost on the W. coast of Boothia between Bellot's Straits and the Magnetic pole – if such were the case is it not strange that Franklin did not strike for the Fury's stores where relief was certain. What an opportunity Collinson had of sweeping over the whole field of search!!

In my opinion a small expedition should be sent out to winter at the Fury's stores and survey both sides of Victoria Straits, returning by the Isthmus of Boothia.[5] Some

[1] The Back River.

[2] Accessories such as ramrods, flints, etc.

[3] One assumes that Anderson is referring to Rae's relaying of the Inuit reports that the Franklin expedition survivors had been engaged in cannibalism – in which case his remark is very perspicacious.

[4] HMS *Fury*, driven ashore by the ice and wrecked at Fury Beach, eastern Somerset Island, during Parry's third expedition of 1824–5.

[5] This, effectively, was what Captain Leopold McClintock's expedition in *Fox* did in 1858–9. From winter quarters at the east end of Bellot Strait sledge parties led by Captain McClintock and Lt William Hobson searched the entire coastline of King William Island and found numerous relics and skeletons as well as the only message ever found which threw light on the fate of the Frankin expedition. See F. L.

of the party might still be found living among the Esquimx. and records would undoubtedly be found at their winter quarters.

I shall be very anxious till I hear of the arrival of the Esq. interpreters; would it not have been better to order them to come via Cariboo Lake[1] and Fond du Lac and Athabasca.

I will now say a few words about the Distt. Mr. Campbell arrived here on the 14th ulto. and left again for Liard on the 16th; he will come down here after closing the business at Liard about the 10th prox. Mr. Mills will take charge ad interim – he has behaved extremely well and been very active and attentive to his duties. Everything will be left here in such a state that Mr. Campbell will have little else to do but put the furs in the boats. At his request I have made out several sheets of memo and have given him the best advice I can. He feels doubtful about being able to transact the writing part of the business. Mr. Mills will be able to give him much assistance in this respect – as will Mr. Ross for the summer and fall business. I do sincerely trust that this most valuable and improvable Distt. will not be neglected.

In my last I mentioned that if you wished it and my health was good that I had no objection to remain a year or two longer in this quarter. I however now plainly perceive that this service will in all probability require more than one season and were I so fortunate as to procure the despatches[2] this year, on no earthly consideration would I trust them to another, so that my service had better not be relied on.

25th. This letter will be something like a journal. Your duplicate despatch and advices from Y.[3] and No. Ho.[4] reached me on the 12th and a few hours after the Liard broke up and cut across the McKenzie.[5] The Upr. McKenzie broke up on the 22nd and is still drifting thickly. The canoe from Liard arrived yesterday; it is well made and the bark superb, but the Dr's suggestion[6] regarding shortening the canoe has spoilt it (Note – this canoe was 5 feet too short and as crank as a small camel). The canoes ought to have been full sized North canoes to take the requisite supplies. I shall leave directly the McKenzie is clear of ice probably in 2 days.

The news from Y.[7] is most disturbing; the Esq. interpreters are not forthcoming; a quantity of iron works, tea, sugar and tobacco have been forwarded by 3 trains from Y. and No. Ho.; the latter will be acceptable but all the other supplies forwarded I had already arranged here with the sole exception of real harpoons. I regret that so much unnecessary trouble and expense have been incurred.

McClintock, *The Voyage of the 'Fox' in the Arctic Seas: A Narrative of the Discovery of the Fate of Sir John Franklin and his Companions*, London, 1859.

[1] Reindeer Lake. This was probably the route taken by William Oman and his Chipewyan guides, if indeed Oman got that far.

[2] Records of the Franklin Expedition. Anderson is suggesting that he would then have felt obliged to convey the records to Montreal (or even farther) himself.

[3] York Factory.

[4] Norway House.

[5] The ice breakup at the confluence of the Liard and the Mackenzie is usually very spectacular since the former river usually breaks up before the latter, and some spectacular ice-jams commonly result.

[6] Dr Rae's suggestion that for manoeuvrability the canoes be made 5 feet shorter than the standard *canot du nord* which was 25 feet in length.

[7] York Factory.

Mr. Barnston[1] has forwarded 3 first rate men – they will be required. Stewart has now forwarded me a list of the men from R.R.[2] Several of them are not first rate – one of them a £60 pounder – old George Kippling is quite unfit for such a service. Had Sinclair (or McKay) Back's steersman been sent he would have been of great service. At the same time I received Back's journal and find that our difficulties will be greater than I expected. I heard also from Stewart; he had been down to Resolution with 11 men and 34 dogs. He remained there 8 days and returned to Athabasca with 10 dogs and 3 men. You can calculate what all these men and dogs require to feed them, and all the nets and lines at Resolution produced on an average 6 fish a day. I don't at all approve of this parading about with dogs and carioles, consuming provisions unnecessarily. Fortunately I write by the Express to Mr. Ross to send all the expedition men and dogs (& three of his own) to Big Island, there to await my arrival; had I not they would have cleared out the stores at Resolution. Stewart proposed to me to start on the ice and told Ross to get sledges made 18 ft long, 6 inches high and 2 inches broad for the canoes!! I have already given you my views on this subject. Stewart tells me that the bark of his canoes is so bad that he has been obliged to line them with sheeting; the plan is very creditable to his inventive powers, but I have little faith in its efficacy. I fear that he has been in an unnecessary hurry, for you know as well as I that there is abundance of splendid bark at the Embarras Rr.[3] and also at the Bark Mountains[4] only two or three days from Chipn. Canoes are frail enough for our work, but bad bark canoes are useless. Ross had the woodwork of his canoes ready, but only 2½ fms.[5] of bark. I know that Resolution is a bad place for that article – my interpreter, who was born there, tells me that no good bark is to be found within 8 days of Resoln. It is most fortunate that I have provided 2 canoes.

Mr. Ross has collected some most valuable information for me. He says that by following an Indian route to Artillery Lake we may reach there long before the ice on the upr. end of Slave Lake breaks up. It leaves the lake about the Mountain and follows a chain of small lakes which open early. We may then probably find the large lakes open along shore and by making occasional portages reach the river[6] long before they break up (Note – I found afterwards that this information was altogether incorrect). It is, however, very evident that we cannot reach the sea at the earliest before the 10th August. He has already sent off a party of Couteaux Jaunes hunters to meet us en route with provisions.

I had the honor of receiving a letter from Lady Franklin written by her niece, Mrs Cracroft. As a composition it is beautifuly written and portions of it most pathetic; my prediction[7] regarding the cannibal part of Rae's report is verified with a vengeance. I can assure you she works him up a trifle and informs me that 'Dickens has employed his powerful pen in 2 numbers of his Household Words on the

[1] At Norway House.
[2] Red River.
[3] One of the channels by which the Athabasca River reaches Lake Athabasca.
[4] Birch Mountains, south of the Peace-Athabasca Delta.
[5] Fathoms, i.e. 15 feet of bark.
[6] The Back River.
[7] That it would cause a storm of controversy in England.

subject'. This is a nice recompense for all Rae's sufferings and privations. She seems to expect that we shall reach the spot where Franklin's vessels were lost – very reasonable!

26. Journal of James Anderson, Fort Simpson to Big Island, 28 May – June.[1]

Monday, May 28th 1855. I took my departure with 2 canoes and 10 men, with supplies for the expedition, a little after midday. Ice still drifting in the upper Mackenzie. We broke one of the canoes near the Green Island.[2] It drifted so thickly that we were compelled to encamp at 7½ p.m. at the head of the Island. The water appears to have risen very high in the river; appearance of several dykes.

Tuesday, 29th. Detained by ice till 8.50 a.m., when we left and reached the point below Rabbitskin River,[3] where we were compelled to encamp, the ice drifting very thickly. In the midst of this B. Le Noir came drifting in a small canoe. He says that the river is free as far as Couteaux Jaunes R., but impracticable for even a boat to ascend. The people shot a few ducks and rabbits.[4] A few drops of rain fell and the sky was overcast all day. Got a fresh stock of duck eggs.

Wednesday, 30th. The ice detained us till 10½ a.m. We got many knocks and rubs, but reached Spencer's River at 8½ p.m. Saw Babillard's son and old Le Noir and son. Got a few fish, 2 geese, a beaver and a piece of bear from them. They had hunts varying from 40 to upwards of 100 M.B.[5] The birches and poplars began to put out their leaves. The weather was warm today. Previous to leaving Ft. Simpson the highest the therm. reached this spring was 62°.

Thursday, 31st. After gumming the canoes,[6] embarked at 4½ a.m. Obliged to take the paddle[7] owing to the quantities of ice on the beach. Experienced some heavy showers, accompanied by thunder. We had much trouble with drift ice, but managed to reach a little above the stream when we saw the ice coming down full channel, evidently from the little lake. By using our best exertions we managed to get our canoes out of the water 5 p.m., just as the ice came down with tremendous force, sending huge boulders up the bank like skittle balls. The canoes suffered much today. On one occasion a mass of ice tumbled from off the bank, sent a wave into the canoes and broke the paddle of one of the men. A few inches more and we should have been all smashed into a thousand pieces. As it was, we escaped, except an Indian, who was hurt by the handle of the broken paddle being driven into his side.

Friday, June 1. Detained all day by ice. Immense quantities have passed; about 3 a.m.

[1] HBC E.37/3. The full title is *Journal of Chief Factor Anderson of the H.B. Co., of a Journey from Fort Simpson, McKenzie River, to the Mouth of the Great Fish River, via Great Slave Lake etc.*, 1855.

[2] 21 km from Fort Simpson.

[3] 35 km from Fort Simpson.

[4] Varying hare (*Lepus americanus*) (A. W. F. Banfield, *The Mammals of Canada*, Toronto, 1974).

[5] Made Beaver, the standard of exchange used in the fur trade. In other words the men had taken an amount of furs (of various species) equivalent in value to 40–100 Made Beaver, at the current rate of exchange.

[6] The seams of a birch-bark canoe were sewn with split spruce roots and rendered watertight with spruce gum, made from spruce resin with some moose fat added to make it less brittle when it hardened. The gum was usually applied hot. See K. G. Roberts and P. Shackleton, *The Canoe: A History of the Craft from Panama to the Arctic*, Toronto, 1983.

[7] Instead of tracking the canoes.

the waters rose with a sudden rush, bringing down immense fields, portions of which were shoved with tremendous force up the bank. Fortunately I caused the baggage and canoes to be carried high up before the men went to sleep. Still one of our canoes had a narrow escape. The ice though still (8.30 a.m.) drifting thickly, is getting a little clearer. This is a bad place for hunting. Nothing has been killed today by the hunters. Weather warm.

Saturday, 2nd. Still detained by ice. Cloudy with some slight showers. The Big Island boat arrived at 11 a.m. Took out its crew and sent the Simpson people back in it, except two Indians. Mr. Clarke was a passenger. The ice is drifting thinly this evening, and I am in hopes that we shall be able to leave in the morning.

Sunday, 3rd. Cloudy all day. Just as we were preparing to leave, a canoe arrived from Simpson which Mr. Miles was kind enough to send with some provisions upon hearing the state of the ice from the Indians. Of these I took 1 bag of pemmican, 22 bags grease, 25 tongues, 1 bag potatoes and sent back the remainder. We left rather too soon, as we broke both canoes with ice, and were compelled to put on shore to repair them. It was tough work getting up to the head of the line.[1] The water is high, which precludes tracking, and the current very strong. Both canoes were nearly upset in rounding fallen trees. And the old canoes had a most narrow escape of being crushed by a floe of ice. Saw 5 Indians with excellent hunts, and a boy of 12 years old who had killed 70 MB in martens. The lowest their men had was 80; the others 100 and upwards. Encamped late in the little lake opposite Point au Foin. Men much fatigued after this hard day's work. It was one continued stretch.

Monday, 4th. A beautiful, calm, warm day. Vegetation has made considerable advances the last 2 or 3 days. We left the encampment at 4 a.m., and encamped at 7 and a half p.m. – the canoes requiring considerable repairs – at a pipe[2] from a small lake close to the 'Ecours'. Saw only a few pieces of ice until we encamped, when we saw a considerable quantity. I suppose from the small lake. Saw old Bedeau and the Grand Noir. The men who are unaccustomed to the paddle complain of sore arms and breasts.

Tuesday, 5th. Left early, but were stopped by a large body of ice (or rather a stream of drift ice, apparently much broken) supposed to have come from the channels, about Big Island. Did not unload till sunset in hopes of a passage cleaving. In the evening a heavy gale arose. We are encamped in the little lake opposite Lop Stick Point. The weather warm. I need not say the pain and vexation I feel at these repeated detentions. However, I could do nothing were I further advanced. Slave Lake is still firm,[3] but the appearance even of advance is consolatory.

Wednesday, 6th. Left at 1 a.m. Stopped by ice at the Island at 7.30 a.m. until 5 p.m. We then managed to cross among the drift ice and reached Charleson's fishery, where we were again compelled to encamp by our enemy at 8 p.m. Stopped at an

[1] From this point, about 10 km below the Trout River confluence to the foot of the Green Island rapids, 100 km downstream, the Mackenzie is very fast-flowing. Canoes were normally lined (tracked) up this section of the river – hence the name.

[2] Fur trade canoes normally stopped for a few minutes every hour or so to allow the men to rest and smoke a pipe. Hence a 'pipe' became a unit of time – and hence also of distance.

[3] I.e. the ice on the lake was still solid.

Island where we saw many of the small forked-tailed, black-headed tern.[1] They had just begun to make their nests, but had laid no eggs. Saw one of these birds drive off a crow.[2] Gooseberries[3] in flower. Very warm and clear until the evening, when it was overcast. Mosquitoes troublesome.

Thursday, 7th. Detained here the entire day by ice, drifting so thick that we can't see the water. It is all smashed into separate 'candles'.[4] Very sultry. Thunder at some distance, a few drops of rain fell here, but heavy showers falling to the northward. Saw a grasshopper, strawberries in flower.[5]

27. Letter from James Anderson, en route from Fort Simpson to Fort Resolution, detained by ice, to Lady Franklin, 7 June 1855.[6]

The touching and beautiful letter written in your name by your niece, Mrs. Cracroft[7] reached me on the 12th ultimo. with the duplicate of Sir George Simpson's despatches and tho' much hurried and harassed I cannot refrain from replying briefly to it.

Should I be so fortunate as to recover the records of the hapless party, you may rely on it that any of a private nature shall be held inviolably sacred. My second in command, Mr. Stewart, is a gentleman by birth and education, and would, of course, scorn to read what was not intended for his perusal, and with one or two exceptions the remainder of the party can neither read nor write. Regarding the volume in which you are so much interested, I fear that the brass ornaments on it may have led to its destruction.

I regret sincerely that the paragraph you allude to in Dr. Rae's despatch,[8] should have wounded your feelings so deeply. He is an intimate friend of mine, and I know that no one has been more anxious to afford relief to your gallant and unfortunate husband and his associates. I am sure that no one will regret the circumstance more than he.

I have seen none of the speculations alluded to by you regarding the destruction of the vessels and the fate of their crews. The conclusions I have arrived at are that the vessels were crushed – tho' not immediately destroyed – by the ice when at winter quarters. Had they been wrecked suddenly, it is not probable that such a selection of articles as Dr. Rae brought home could have been saved. This must have occurred on the Wt. coast of Boothia between Bellot's Straits and King Williams

[1] Arctic tern (*Sterna paradisea*). See W. Earl Godfrey, *The Birds of Canada*, Ottawa, 1966, pp. 190–91.

[2] Almost certainly a raven (*Corvus corax*), which Anderson invariably identified as a crow. However the range of the Common crow (*Corvus brachyrhynchos*) does extend north to Great Slave Lake and the upper Mackenzie River (see Godfrey, *The Birds of Canada*, pp. 275–6), so in this case at least, this may well be the species in question.

[3] *Ribes oxyacanthoides*.

[4] 'Candling' is the process whereby lake and river ice decays just prior to breakup. Intact masses of ice break into candle-like separate crystals from 2 to 3 cm in diameter and 10–30 cm in length. Dr Terry Prowse, Canadian National Hydrological Institute, personal communication.

[5] *Fragaria glauca*.

[6] BCA AB40 An 32.2, ff. 98–101; extract at SPRI MS 1100/4.

[7] This was Sophia Cracroft's letter of 15 December; Lady Franklin's letter of the same date reached Anderson only on his return to Fort Reliance in September.

[8] The allegations of cannibalism among the Franklin expedition survivors.

Land; had they been wrecked on Victoria Land, Rae in '51 or Collinson in '53 would have undoubtedly heard of it from the Esquimaux in that quarter. What a noble opportunity the latter had of clearing up the mystery. There is one thing, however, that I cannot understand. How was it that a party wrecked in Victoria Straits did not direct its steps to the Fury's stores where relief was certain?

I will now say a few words regarding the present expedition. I regret to say that we set out with very unfavorable prospects. The season is a very late one. From the brief notice given us birch bark canoes must be used and those not of the best quality. They cannot bear hard usage and there is every probability that they will be much injured before we reach even the Thlewycho,[1] one of the most dangerous rivers known; if we reach the sea in safety I shall be much surprised & don't expect it. You may then picture to yourself the risk we shall incur in navigating an ice-encumbered sea in such craft & mention these things lest you expect too much from us and be disappointed. All that can be accomplished, without absolutely throwing away the lives of the party committed to my care, shall be done.

Speaking for myself personally it is from no greed of lucre, nor hope of fame, that I undertake this task but solely from a sense of duty; had it been for any other purpose I should have declined the appointment as my health is not very good.

As we can barely take down enough provisions for the summer trip, we shall be unable to winter on the coast. Should they be required, I am getting boats built and making other arrangements for carrying on the search another year. At the same time I conceive that the route by the Fish River, from the great difficulty in getting down supplies, the imminent risk of accidents occurring, and the probability of finding the inlet at its mouth choked up by ice, is one of the most unpromising that can be followed.

In my opinion a small vessel with about 20 men – including officers and 3 Esquimaux interpreters – should be sent to Prince Regents Inlet and make the Fury's stores their starting point for an ice journey; they should be in two parties, one to explore both sides of Victoria Straits to King Williams Land and the other to explore the east coast of Boothia to Lord Mayor's Harbour. This might be accomplished in one season. There are many officers well adapted for such an enterprise.

Pray excuse this hurried production – believe dear Madam that I sympathize earnestly in your long protracted sufferings and that I shall put forth all the energy or ability I may possess to clear up this mystery and to fulfil the instructions I have received.

28. Letter from James Anderson, en route from Fort Simpson to Fort Resolution, detained by ice, to Robert Campbell, Fort Simpson, 7 June 1855.[2]

It is highly probable that the objects of the Expedition cannot be accomplished this season. I shall therefore make my arrangements for another.

I beg therefore that you will get Grey to build 2 Boats this Summer, they are to be Clinker built and must carry <u>at least</u> 45 pieces and a Crew of 8 Men (7 Oars & a

[1] The Great Fish, or Back River.

[2] HBC B.200/b/32, ff. 104–5.

Steersn) without being too much sunk[1] to run Rapids, the planks to be ½ Inch thick. The main points are that they draw little water and pull & sail well. I enclose a plan of Dr. Rae's Boats and you can also refer Grey to Dr. Richardson's Boats[2] in the work you were so Kind as to lend me and which I now return. The Keels must be plated with sheet Iron & strong stem and stern plates. The Bows should be sheathed with Tin or copper a little above and about 8 Inches below the Water line. As for the Rigs, with our crews I think the Common square sail will be the best. 12 Oars will be required for each boat also Oaken. 2 Gals Keg's prepared **A** Tar, a new Main line, Tracking lines & Pins, Oil Cloths, Grafting Saw. Plane, Chisel, Caulking Irons, Gimbles, 4 Sheets Tin & Tacks, Nails for repairs, also 2 planks per Boat. Should you not have the Proper Rope for Rigging & haulyards you can get some from Chipewyan.

These Boats must be ready by the time you return from P. La Loche.[3] 20 pces. Pemmican and 2 Bags Flour are to be embarked in each. They should reach the place where they will leave Slave Lake by the 1st September and endeavour to get to the headwaters of the Thleeychodese Rr.[4] by the 15th December. They will be met in Slave Lake by Mr. Lockhart and a party to show the road & assist them. They will return thence in Canoes; when they reach there a cache must be made secure against the attacks of Wolverines. The Pemmican to be raised from the ground and well covered with Oil Clothes, the Sails and Agrets[5] will be placed in the same Cache. You will require to furnish 2 Steer, 2 Bow, 4 White & 6 Indian Middlemen. I would suggest Bouvier & N. Sauvé for the Steersmen.

I shall still require the Canoes I requested you to get made. The Fort Rae Voyagers can take them to the Expedition Fort.[6] Oars should be fitted on them when 2 Men and a Steersman for each will be a sufficient crew.

I shall leave a Memorandum of their Ladings at Resn.[7] One article will be Birch for 12 Sledges, which should be raised this summer. Another will be Iron Runners for Sleds to be cut from Old Saws 1½ Inch broad with countersunk holes pierced at proper Nails; some of these might be made at Chipewyan.

I think that the Planks sawed last Winter for weather boarding would answer (for a boat at least) for the Boats; they are ¾ Inch thick but might be reduced with the Plane. If you get the Planks sawed it will be advisable to season them; it might be done in a short time by putting up 2 stoves in one of the Men's houses.

You will require to have 4 Men at least expressly for the Boats. I would suggest Gray, Brough, Swaney & Gilb. Williamson. You may tell them that they will get a gratuity if they perform their task to my satisfaction.

[1] I.e. not to have too deep a draft.

[2] The boats used by Dr John Richardson during his search of the coast between the mouths of the Mackenzie and the Coppermine in 1848. The book in question is probably Richardson's narrative: Richardson, *Arctic Searching Expedition*.

[3] From accompanying the brigades with the Mackenzie District furs to the annual rendezvous at Portage la Loche.

[4] The Thle-wee-choh, Great Fish or Back River.

[5] Accessories, i.e. paddles, poles, etc.

[6] Fort Reliance.

[7] Fort Resolution.

If the Boat Shed be already down, by putting all hands to the work it might be reerected before you leave.

29. Journal of James Anderson, continued, Mackenzie River.

Friday, 8th. Another warm day, thunder at intervals with a few drops of rain. Still detained by ice, running full channel. Saw a dragon-fly and some yellow butterflies.
Sunday, 10th. At 3 p.m. we managed to cross the river amongst the drift ice, and put ashore for supper at 9 p.m., after which we continued our route. Very warm, sultry. About 6.30 p.m. the sky to the N.W. became of an inky colour with long streamers like waving hair hanging like a fringe. The sun shone through this as if a hole had been cut in the cloud. This shifted gradually round the compass, accompanied by violent squalls and heavy showers of hail and rain. We had some narrow shaves in the ice, and the tracking of the rapids was execrable.
Monday, 11th. We marched all last night, got up the Batture Rapids[1] about daylight, when we were within sight of the Ile aux Bouleaux. We were again stopped by drift ice, but managed to get on by dint of wading and hauling the canoe through the small channels and afterwards by keeping along shore, which was shoal and full of stones. We reached the point before reaching the Big Island Fort at 5 p.m. There we were obliged to unload as the channel is choked with ice.[2] Sent all hands to the Fort except my servant. Saw some Big Island Indians and one belonging to Resolution, who is waiting for the disruption of the ice to go there. The men marched[3] 26 hours, except during the time they put on shore to sup and breakfast.
Tuesday, 12. The ice cleared sufficiently about mid-day to cross over to the island, along which we found a channel and reached a point on it about 15 miles from B. I. at 6 p.m. Here we found our road barred by ice. Encamped on a nasty swampy point. Set a short net which yielded by sunset 3 fine trout,[4] 8 W. fish[5] and 12 red carp.[6] Gave 1 oz. tea out this evening.

30. Letter from Sir George Simpson, Norway House, to James Anderson, 14 June 1855.[7]

I have very great pleasure in offering my congratulations on your advancement in the service. Of the probability of your promotion I advised you in the winter and I now hand enclosed your commission as a Chief Factor accompanied by a covenant which you will be pleased to execute in the presence of two witnesses and return to my address at Lachine by the first conveyance that may offer.

[1] Just above the present settlement of Fort Providence.
[2] In the fair copy of his journal Anderson reports: 'I there found 8 of the Expedition men who were awaiting my arrival – to these I added 4 men and left the Fort Simpson people at the post with orders to return immediately to that Post (HBC B 200/a/31, f. 1). The eight expedition men had been sent to Big Island by Stewart at Anderson's suggestion, so that they would not consume Fort Resolution's reserves of food. Since Big Island was primarily a fishing station, it had abundance of food.
[3] The verb 'to march' was used for any type of progress, including paddling.
[4] Lake trout (*Salvelinus namaycush*).
[5] Lake whitefish (*Coregonus clupeaformis*) or broad whitefish (*C. nasus*).
[6] Longnose sucker (*Catostomus catostomus*).
[7] HBC E.37/11, ff. 19–20; BCA AB40 An 32.2, ff. 51–3.

As I presume this letter will reach you only on your return from the Arctic coast it is useless for me to offer any suggestion on the subject of the Expedition; I can only say the report of your proceedings will be most eagerly anticipated both in this country and in England. I hope you have written me before taking your departure from Fort Resolution, detailing your arrangements and the proposed plan of operations, especially as to the contingency of having to remain out next winter. A boat laden with provisions has been forwarded along with the Portage la Loche Brigade, and will proceed to the further end of Great Slave Lake, or to such rendez vous as you may have appointed. I regret to learn that the services of Ooligbuck, the Esquimaux interpreter, have not been secured, but I trust William Oman and an Esquimaux who was to have accompanied him from Churchill, reached Fort Chipewyan in time to join the party before taking its departure. Oman is reported as being qualified for the duty of interpreter, in which capacity he was employed at Churchill.

The temporary charge of McKenzie River has been given to Chief Trader R. Campbell but by the Minutes of Council you will observe that in the event of your return from the Expedition next autumn you are to resume your position as head of the District.

In the uncertainty that exists as to your movements and when my letters may reach you I shall not reply to your communications received during the past winter and spring under dates of Septr 26 and 28 November last. It is unnecessary to say much upon the chief [illegible] to which you advert as your views respecting the re-establishment of Frances Lake etc. have been adopted and that measure [illegible] at all events?

With reference to the transfer of the Government supplies to the Company, I presume a large quantity of what was in Depot for the use of Arctic Expeditions will be applied to the outfit of the party employed under your orders this season; the remainder, especially the articles that will deteriorate by keeping, should be assumed by the Company at such rates as you may think fair and reasonable; it was considered that no person was so well able as yourself to decide what those rates should be, and accordingly both the Government and the Company are willing to abide by your adjustment of the matter. I would merely suggest that you should keep in mind that some articles which may be valuable for the purposes of an expedition and in certain localities, may be of little use to the Company for their trade in McKenzie River; so that in valuing such articles you should not look so much to their cost or intrinsic value, as to what they may be worth at the posts where they may be in depot. With this remark I leave this affair entirely to your discretion.

Wishing you every success on your present enterprise and looking forward to your next communication.

31. Letter from James Anderson to Sir George Simpson, 6 May 1855, continued.[1]
15th June. 15 miles from B. Island[2] detained by ice. I left Simpson on the 28th ulto.,

[1] BCA AB40 An 32.2, ff. 135–6.
[2] Big Island.

and have been detained every day by ice and sometimes 4 days on one spot. I can assure you that travelling against a strong current with ice drifting thickly is no joke. We had many narrow escapes. I reached this spot on the 12th and am blockaded by floating ice; further out the ice appears unbroken. This late season will increase our difficulties immensely. We shall have so much working among ice that our canoes will be much injured before we even reach the river; by the time we have got through its 84 rapids etc. they will not be in good order for a sea voyage and for our return trip. I sincerely trust that the Indn. route spoken of by Mr. Ross is practical, as from the description Back gives of his ascent of the Hoar Frost River at a low stage of water and almost light, I should imagine that it would be all but impracticable during high water and with ice drifting down it.

In case they may be required for another season I have directed 2 boats to be built and rendered to the head of Fish Rr. next fall; they will take up 45 pieces of provisions, which will be put en cache securely, and made several other arrangements. It is of no use repining but you can conceive how distressing this detention is to me.

32. Letter from James Anderson to Eden Colvile, 1 May 1855, continued.[1]

15 June. 15 Miles from Big Isld. Detained by ice.

I left Simpson on the 28th Ulto. with 2 Canoes laden with supplies & have been detained by ice almost every day since leaving, and I can assure you that travelling against a strong current in canoes with ice drifting is no joke. Our canoes were broken and we had many narrow escapes. I reached this spot on the 12th and am completely blockaded by floating ice. The season has been a late one and this will increase our difficulties immensely. I have received Back's Journal and find that I underrate the obstacles we have to surmount. We shall have so much working among ice that our Canoes – not of the best quality – will be much damaged before reaching the head of the River. We have then to descend it with its 84 Rapids & falls. By the time we have reached its mouth our craft will not be in very good order for a sea voyage and for our Return trip. It is probable that I may follow an Indian instead of Back's Route to the Lakes. It strikes off about the 'Mountain'. The advantages of this route are that it is through small lakes which are open before the Upper end of Slave Lake, and we shall avoid the ascent of Hoarfrost River, which by the description Back gives of it in the fall when the water is low, must be almost impracticable when the water is high and the ice drifting down it.

Though I do not expect to find the large Lakes open we will probably find a passage alongshore. I have no hopes of reaching the sea before the 10th August. I shall then have 3 weeks before me and if there be a clear sea and moderate weather, much may be done in that time. I should be at the mouth of the River by the 1st Septr. at latest. I have directed 2 boats to be built and brought up to the head of the River this autumn, with about 45 pieces of provisions, which will be put en cache there, and made other arrangements for another season should they be required. Tho' in my opinion this route offers the greatest obstacles and least chance of success of any that could be adopted, except for exploring the inlet at the mouth of the

[1] HBC E.37/10, ff. 106–7.

River. In the first place there is the immence difficulty and risk of getting down supplies, and in the second the great probability of finding the inlet choked with ice. Rae on his first trip took 140 pieces of supplies; these he rendered to his wintering ground without a portage. If Victoria Straits & the Coast of Boothia are to be explored, the starting point for ice journies should be from the Fury's stores.[1]

From the selection of articles Rae bought from the Esquimaux and the quantity of amn. etc. said to have been saved, it is evident that the vessels could not have been suddenly wrecked. I suspect that they were crushed, tho' not instantly destroyed at Winter quarters. Pray excuse these disjointed remarks. I am so annoyed by this detention and so anxious to get on that I cannot fix my mind to letter-writing. Rely on it, however, that everything shall be done & risked (without absolutely throwing away the lives of the party entrusted to my care) to fulfil my instructions. J.A.

33. Journal of James Anderson, continued, Big Island to Fort Resolution.
Wednesday, 15th. Detained all day by ice. The water rose and drove us to another encampment. Obliged to take up the net as the ice was covering it. It yielded 32 fish, chiefly whitefish. Some marsh flowers are in bloom, such as the large buttercup.[2] It is blowing fresh from the N.E. The land here is evidently encroaching on the lake. The process is first drift wood, then a sediment of mud, moss then springs up, and grass and marsh plants. Willows take root and when the ground is a little raised, birch. Beyond that we see spruce. The leaves of the birch here are just appearing and the grass is 18″ high, though the ground is frozen 6″ from the surface.[3]
Thursday, 14th. Heavy rain with wind all night. The ice is packed against the beach so that we cannot even set a net.
Friday 15th. Calm and cloudy. A most gloomy day. Ice as yesterday. This perpetual detention is most distressing, but it is useless repining.
Saturday, 16th. Very warm with a slight shower. Foggy. Mosquitoes dreadfully thick. About 5 p.m. we managed to get off from our beastly swampy encampment. We found some lanes of water, and bored through much drift ice till we reached near De Marais Islands,[4] where we could get no further, the ice being hard and in close pack. At the same time the fog was impenetrable. It was an awkward situation. We bored away into the bay and suddenly came on one of the islands and afterwards managed to reach the last one by sunset. Canoes rather damaged. Saw several fields of ice still white and hard. Very cold in the evening. Set the net.
Sunday, 17th. Left very early in hopes of finding a clear road. We were soon, however, undeceived, as after pushing through much drift ice, and injuring the canoes much, we were brought to a standstill by thickly packed ice in the bay at a short distance from Pt. des Roches.[5] Foggy with some showers of rain. Were the wind to blow off shore I think we could get on, as the ice is in pieces and moving. Mosquitoes awful. The net only produced 2 fish.

[1] At Fury Beach, Somerset Island.
[2] Probably the marsh marigold (*Caltha palustris*).
[3] This area lies within the discontinuous permafrost zone, hence beneath a layer of variable depth which thaws every summer, the ground may well remain frozen perennially.
[4] Iles des Marais, along the south shore of Great Slave Lake opposite the east end of Big Island.
[5] On the south shore of Great Slave Lake, about 21 km from Hay River.

Monday 18th. Got off at mid-day and after 5 hours' hard labour in getting through the ice reached Pt. des Roches. Got a few gulls' eggs. Weather warm with thunder. Mosquitoes awful. Set the net, the bay beyond the point quite blocked up.

June 19th. At 7 p.m. yesterday a slight land breeze drove the ice from round the point and left a channel. The net was instantly raised and we started. The channel, however, only extended a mile. We then began to bore through the ice, and at last found a fine open channel, which, with many bars of ice, took us to Hay River. Afterwards we bored through a great deal of ice, with occasional lanes of water, and reached here after being 23 hours on the water. (6 p.m.) We are encamped on a stony isle about 2 miles from the Sulphur Springs.[1] Much to my surprise the lake [ice] here seems much stronger than towards Big Island. The floes seem unbroken, white and hard. We shall require a breeze of S.E. wind ere we can start, as it is impossible to get through such ice. The canoes suffered much damage. We have been troubled with perpetual fogs for the last three days. Much of the ice yesterday and today covered with sulphur.

Wednesday, 20th. About 5 a.m. a breeze sprung up which cleared the channel outside, it ran far out, but I could not see whether it approached the Presque Islands. I, however, determined on venturing, and after breaking some ice we fortunately reached the Presque Isle, after which we got pretty clear water to Les Isles aux Mort.[2] A head wind put us ashore on one of the Les Isles Brulés for 3 hours. We then started and reached the house[3] about 10.30 p.m.

34. Letter from James Anderson to Sir George Simpson, 6 May 1855, continued, at Fort Resolution.[4]

22nd [June]. I arrived here[5] at last on the night of the 20th.[6] I found this bay which is under the influence of the Slave and Buffalo Rrs. quite free,[7] but outside the ice still fast. Last night was a gale as well as most part of today, and I trust that it has cleared our way; I start in half an hour. Stewart's canoes are better than he gave me reason to expect; they are however awfully heavy – the maitres like boat gunnels and the verrangues(?) too weak. The Esquimaux interpreters are not forthcoming; this is most distressing. I must conclude. Accept my sincere thanks for many acts of kindness towards myself. Rely on me for doing what can be done.

35. Letter from James Anderson, Fort Resolution, to Sir George Simpson, 22 June 1855.[8]

I beg to state for your information that the Expedition is now ready to take its

[1] Near Sulphur Point, about 69 km west of Fort Resolution.

[2] In the fair copy of his journal Anderson ascribes this open water to the influence of the waters of the Buffalo and Slave rivers entering the lake (HBC B 200/a/31 f.1).

[3] Fort Resolution.

[4] BCA AB40 An 32.2, f. 136.

[5] Fort Resolution.

[6] The 21st was spent in giving the men their advances and in arranging the loads of the canoes, according to the fair copy of Anderson's journal.

[7] The discharge of these two rivers having locally broken up the lake ice earlier than elsewhere.

[8] HBC E.37/10, ff. 112–13.

departure either this evening or tomorrow morning. This bay which is under the influence of Slave and Buffalo Rivers is quite free from ice, but the body of the Lake is still fast. The weather latterly has been very warm, and a strong Gale last night & today will I hope break open a passage for us.

I intend to adopt a different route from that followed by Sir G. Back. We shall leave Slave Lake about the 'Mountain' and follow a chain of Lakes with 7 portages to Artillery Lake. The advantages to be gained by following this route are, that it opens earlier than the Fond du Lac route, and we avoid the ascent of Hoar Frost River, which at this season must be impracticable (See note at end).

I regret to say that the Esquimaux Interpreters have not made their appearance; this is most distressing as I shall not be able to accomplish my instructions satisfactorily.

It will be impossible for a party to winter on the Coast, as the means at my command will not permit me to take down any supplies for such a purpose, the Canoes being hardly able to convey the provisions and supplies required for the summer trip.

I have directed two boats to be built, and to be brought to the head of Fish River next 'Fall', besides making other arrangements for another season, in case they may be required.

I have to acknowledge myself under great obligations to Messrs. Stewart, Ross and Lockhart for the zeal and activity they have displayed in forwarding the objects of the Expedition; the men seem to be in excellent spirits, and were the Esquimaux Interpreters here I should have strong hopes of accomplishing our task satisfactorily.

Note.

The above was the information I received from Mr. Stewart. I had not then had time to examine the Guide hired by him, according to my orders, but being wind bound at Pt. des Roches I examined the man, and found that he had never been further than Clinton Colden Lake, and that the only route he was acquainted with was one which struck off on the E. side of the Bay on which Fort Reliance is placed and by means of 7 or 8 portages & small lakes he would fall on Artillery Lake at the Rat Lodge. At Fort Simpson I had questioned a rather stupid Dog Rib Indian (Timbré) regarding the practicability of the Mountain Route which was evidently shorter and by which Artillery & Clinton Colden Lake as well as the Fond du Lac of Great Slave Lake would be avoided. He told me it was dreadfully bad and gave (what I afterward found) to be a pretty correct chart of the Route. As I found the Guide hired by Mr. Stewart to be worthless, I enquired if either of the 3 Yellow Knife Indians, were acquainted with this route, and found that two of them were. On examining them I found their statement to agree with those of Timbré's, and I then and there determined to follow this route, which alone caused the success of the Expedition that season. Had we followed the other Route we should never have reached the sea. The Canoes would have been worn out before reaching the head of the River, and most of our provisions would have been expended. J.A.

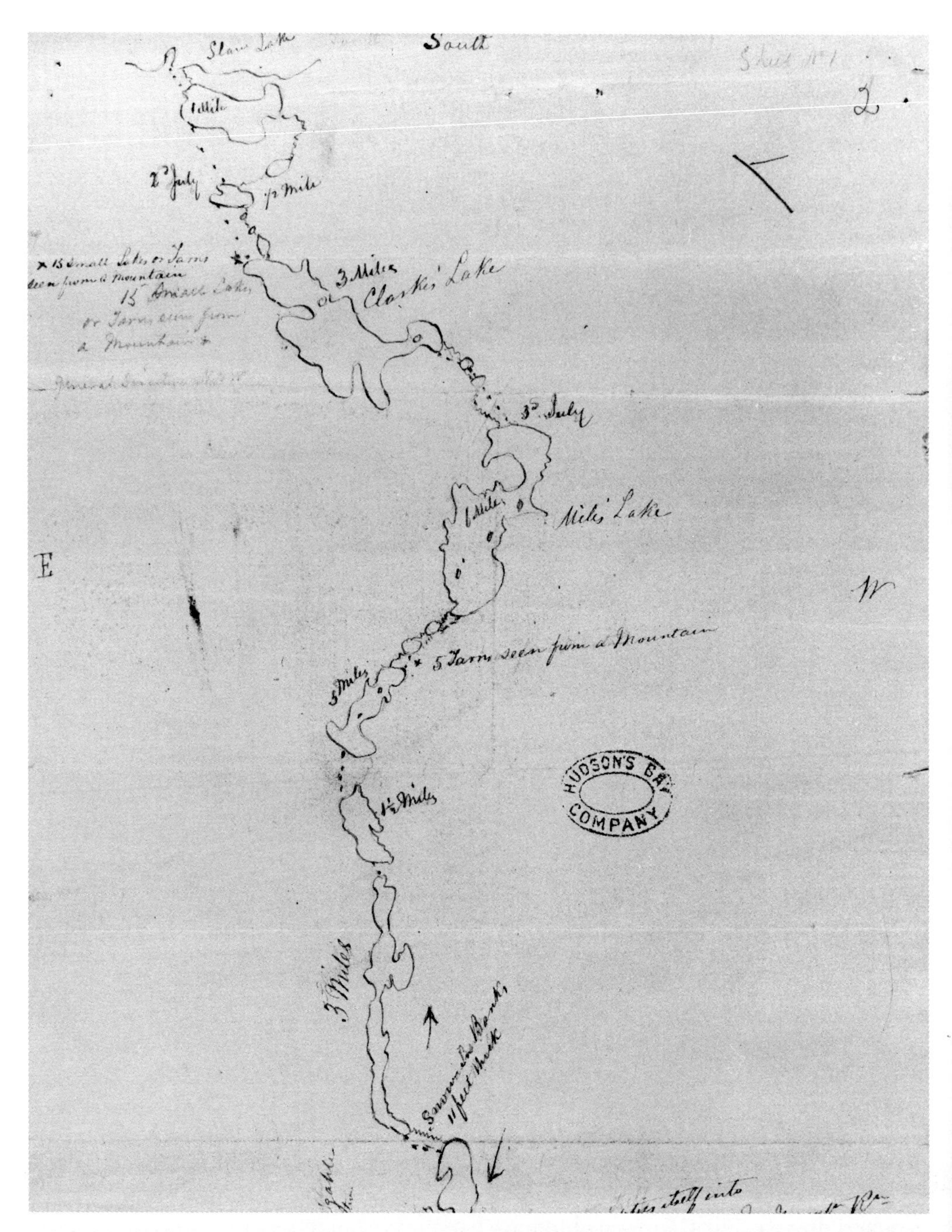

PLATE V. Anderson's draft of his map of the southern portion of the 'Mountain Portage'
South is at the top of the map
(Hudson's Bay Company Archives)

CHAPTER V

THE JOURNEY TO THE ARCTIC OCEAN

INTRODUCTION

In general it is appropriate to allow James Anderson's journal to speak for itself, but at the same time certain aspects of his canoe trip down the Back River (and back) merit comment.

Firstly the party did achieve a certain amount of primary exploration and survey. Faced with the knowledge that the Hoarfrost River, by which Captain George Back had travelled north from the East Arm of Great Slave Lake, had proved to be extraordinarily difficult, involving an endless sequence of rapids, waterfalls and difficult portages, and that the eastern end of the East Arm and Artillery and Clinton-Colden lakes would still be ice-bound, Anderson opted for a third route (previously unknown to Europeans), known as the 'Mountain Portage', which would lead to Aylmer Lake, by-passing these various obstacles. Starting north from the East Arm some distance west of the mouth of the Barnston River, it led north-eastwards via a series of lakes and relatively easy portages and converged with the Barnston River above its lowest, rapid-strewn and almost impassable section. Thereafter the route led north and east via Campbell, Ross and Mackenzie lakes, then over an easy divide and on via Margaret, Back and Outram lakes to the western end of Aylmer Lake. Although possessing no earlier training in surveying, Anderson produced a detailed map (at a scale of approximately 1:316,800) of his route from Great Slave Lake to the head of the Back River, which is preserved in the Hudson's Bay Company's Archives[1] (Plate V).

One point which must be stressed is that the summer of 1855 appears to have been unusually cold, with very late breakup of the lake ice and early freeze-up. Anderson and his party found Aylmer Lake still largely ice-bound as late as 8–11 July. Even as late as 24 July they found ice 2–3 feet thick on Lake Garry and Anderson remarks on the strange sight of the men cutting their way through the ice at such a late date. On the return trip it was snowing again by 3 September (on Clinton-Colden Lake) and by the 9th, on Artillery Lake, the ground was white with snow.

On reaching the sea at the head of Chantrey Inlet the party found itself in the

[1] HBC E.37/29.

unenviable position of attempting to navigate the waters of the inlet in birchbark canoes amongst drifting sea ice. They had a very narrow escape while making the three-hour crossing from Victoria Headland to Montreal Island on 1 August, when the canoes were almost crushed by the drifting pack ice. Finally, near Point Pechell, finding the sea totally ice-covered Anderson and his party were forced to leave the canoes and search Point Ogle on foot (although there was sufficient open water in Barrow Inlet to permit a party to cross to Maconochie Island in the Halkett boat); this island was the farthest point reached by the expedition.

Despite these handicaps Anderson and Stewart completed the round-trip to the ocean in an amazingly short time, a tribute to the skill of their crews and especially to those of the three Iroquois canoemen. They left Fort Resolution on 22 June, had reached their farthest point (Point Ogle) by 7 August, and were back at Fort Resolution by the evening of 16 September. With its endless array of rapids and its large lakes, on which one may be windbound for days, and on which finding inlets and outlets may be extremely difficult due to the intricacy of their shorelines, even today the Back River is considered to be a very challenging wilderness river by canoeists and kayakers. And they, almost without exception, make the trip in only one direction – downstream –, having prearranged to be picked up by plane or boat from the river's mouth.

In his journal James Stewart kept a careful note of his party's progress as compared to that of Back twenty-one years earlier. Admittedly Anderson and Stewart had the advantage of having a copy of Back's map to guide them (although Anderson's comments about it are invariably derogatory), but nonetheless their rate of travel was impressively faster than his. Comparing only their progress down the river and back, Back started down the river on 28 June and reached his farthest point (west side of Point Ogle) on 15 August, for a total travel time of forty-nine days; Anderson and Stewart travelled the same route downstream in twenty-eight days (12 July to 8 August). It took Back thirty-three days to reascend the river, as against the twenty-three days spent by Anderson and Stewart. By any yardstick, theirs was a superlative performance.

Undoubtedly the results of the expedition were constrained by the lack of a fluent Inuktitut interpreter. But by using mime and taking advantage of the (probably limited) knowledge of Inuktitut possessed by Mustegan and McLellan (which they had picked up while with Rae, from the Inuit of the Repulse and Pelly bays area), Stewart and Anderson were able to confirm from the Inuit they met at the mouth of the Back River, the stories which Rae had derived from the Pelly Bay people of the final days of the Franklin expedition. These reports were further confirmed by the various articles, incontestably originating with that expedition, which they found in the possession of the Inuit or in Inuit caches. On Montreal Island they found a location which provided clear evidence of a boat having been cut up. Much to the regret of all concerned they found no skeletal remains and no documents. One can only regret that shortage of time and heavy sea ice prevented them from pushing a little farther west from their turning point at Maconochie Island. It lies only some 7 km east of Starvation Cove where in early June 1879 the local Inuit told Schwatka, Gilder and Klutschak of having found a boat, skeletons, a profusion of articles of clothing and equipment as well as books and papers;[1] the Inuit reported that they had broken the boat up for its wood, but that the bones were still scattered around. When the group visited the spot in June there was nothing to be seen, since everything was still covered by snow, but when Klutschak returned to the site in November 1879 he was able to gather together the skeletal remains and bury them beneath a large cairn in the shape of a cross.[2]

It seems very likely that had Anderson and Stewart reached this same spot twenty-four years earlier, they would have been able to deduce a great deal of information from the skeletons, clothing and equipment, and it is even possible that some trace of the expedition's records might have been recovered. That they did not reach that spot was clearly not for want of trying. Both Anderson's and Stewart's accounts make abundantly clear that they had pushed to (and probably beyond) the limits of safety, in terms of time, ice conditions and the deteriorating state of their canoes.

DOCUMENTS

1. The Journal of James Anderson, 21 June–23 September 1855.[3]

Thursday, 21st. Gave the men their advances and prepared for starting, calm and warm.

Friday, 22nd. Last night and most part of the day blowing a gale from sea[4] which has undoubtedly cleared our road, as we can see large bodies of ice with the naked eye driven in. In the evening[5] we made a start and encamped a little beyond the small channel.[6] Mosquitoes awful.[7]

[1] Lieutenant Frederick Schwatka, Colonel W. H. Gilder (correspondent for the *New York Herald*) and Heinrich Klutschak (artist and surveyor) were three of the four white members of a small, but remarkably successful American expedition which visited the area in 1878–80 in search of remains or records of the Franklin expedition. Accounts of the expedition written by all three individuals have been published, namely: William H. Gilder, *Schwatka's Search: Sledging in the Arctic in Quest of the Franklin Records*, New York, 1881; Heinrich W. Klutschak, *Overland to Starvation Cove: With the Inuit in Search of Franklin 1878–1880*, W. Barr, trans. and ed., Toronto, 1987; E. A. Stackpole, ed., *The Long Arctic Search: the Narrative of Lieutenant Frederick Schwatka, U.S.A., 1878–1880, Seeking the Records of the Lost Franklin Expedition*, Mystic, CN, 1965.

[2] Klutschak, *Overland*, p. 133.

[3] HBC E.37/3.

[4] Off the lake.

[5] At 11 p.m., according to Stewart. He further noted that the flies were thick and the weather warm. His observations gave their latitude as 61°11′30″ (PAA 74.1/137).

[6] About a mile from the fort (HBC B.200/a/31, f. 1).

[7] Anderson set off with three canoes manned by fifteen expedition members and four Copper Indians. 'Their ladings amount to 24 Pieces each and consist chiefly of Provisions with a supply of Ammunition, Nets, a selection of Goods for the Esquimaux, Halkett Boat, etc. Double sets of Paddles, Poles, tracking lines etc. have also been provided. The Canoes are of a good size and model but the Bark is very indifferent, and they are too heavy – mine requires 6 men to carry her. We have however everything essential and I should be well pleased if it were not for the absence of the Interprr.' (HBC B.200/a/31, f. 1).

CREWS

1. Baptiste } Bow
2. Ignace } Bow
3. Joseph } Bow
4. Thos. Mustegan } Steer
5. Alfred Laferté } Steer
6. John Fidler } Steer
7. Mur. McLellan
8. Hen. Fisher
9. Edward Kipling
10. Don McLeod
11. George Daniel
12. Joseph Bouché
13. Will Reid
14. Paulet Papanakies
15. Jerry Johnson
4 Copper Indians

Saturday, June 23rd. Left at 3 a.m.,[1] but could not get beyond Rocky Island, owing to strong head winds. The Resolution canoes exceedingly leaky. Set two nets in the evening.
Sund. 24th. About 4 a.m. the wind lulled and we made a start but it soon rose again and we were driven ashore at Pt. des Roches[2] (whence we take this traverse)[3] where we remained all day. The nets set yesterday produced only 6 fish. They cannot be set now owing to the enormous quantities of drift-wood which line the shore
Mon. 25th. Unable to move from this encampment; blowing to a heavy gale all day with no appearance of its abating. An Indian here says that his band follow a road from near the Mountain to Lake Aylmer. It is through a chain of small lakes with many portages – six of them long ones. I wish to follow this road, but unless I can get additional information, shall adopt another which they all represent as longer but perfectly safe and with few portages. This falls on the east[4] of Lake Artillery near 'Rat Lodge'.[5]

[1] At 5.30 a.m. according to Stewart (PAA 74.1/137).
[2] Grant or Rocky Point.
[3] I.e. make an open-water crossing.
[4] In fact the portage route starts at the southern tip of the lake.
[5] This latter route is the relatively easy portage route from the head of Great Slave Lake to Artillery Lake known as Pike's Portage; it lies east of and parallel to the Lockhart River with its impassable falls and rapids.

By the time he was writing the fair copy of his journal Anderson was much more familiar with the various options and his entry for this date is more informative:

> I examined the Indian guide engaged by Mr. Stewart; the Road he proposes to adopt leaves the Lake near Fort Reliance and by means of a series of small lakes and 10 portages falls on Artillery Lake close to the Rat Lodge. I fear from the lateness of the season that the N.E. end of this Lake [i.e. Great Slave Lake] is still fast [i.e. ice-bound] and Artillery Lake, Clinton Colden & Aylmer lakes break up still later. If we adopt this route the Expedition will undoubtedly fail this season. This Indian is unacquainted with the 'Mountain Route.' I know this was condemned by Sir G. Back, still

Tues. 26th. Detained still by wind.[1] This delay is most distressing.[2] The men shot a goose, some ducks and gulls. I was in hopes that the Esquimaux Interpreter might have overtaken us here. Had this occurred I would not have regretted the detention.

Wed. 27th. The wind fell a little after 4 a.m. and we started immediately. Just after making the traverse,[3] it began to blow from the N.E. harder than ever, but we felt little of it among the numerous islands of the Simpson Group. But in making some of the traverses, the canoes shipped water. The evening is delightfully calm and serene. We are encamped about 8 miles from Pt. Keith[4] at 8.30 p.m. The view from a high rock near our encampment is of extraordinary beauty. On this rock is a nest (last year's) of a fishing eagle[5] composed of sticks, hay and moss. Set two nets. Saw some Canada geese[6] with their young ones. I may here add that Back's description is generally correct and that I do not intend to repeat his descriptions. I however think that he has estimated the height of the rocks too highly.

Thurs. 28th. A fair day with one or two showers; wind rather strong ahead. Back mentions that the rocks are from 200 to 2,000 feet in height; the *highest* estimate that both Mr. Stewart and myself have formed is 500 feet and this only in one or two instances; his descriptions otherwise are correct. The cut rocks (trap) strikingly resemble those of Nipigon Bay, Lake Superior.[7] Three peaks indistinctly seen by Back between Pethenent and the East coast, I perceive as portions of a considerable island. There are many islands along the coast not noticed in the map. Many plants are now in flower, but they are all to be found in the valley of the McKenzie. I have therefore collected only a few of the rarest. We left our encampment at 3 a.m. and encamped at 9 p.m. at the end of Tal-thel-la (a strait which does not freeze during the winter),[8] on an island called the 'Bag'. Our nets provided 8 white fish and a very fine trout. They were set again to-night. We met with a little ice in this strait and I fear we shall be stopped tomorrow, as it appears unbroken in the distance. I saw an eagle's nest.[9] The young eagles were peering out over the edge.

Frid. 29th. Young ice formed last night and we could not leave till the sun had some effect on it and the old ice, which when cemented together is as strong as ever. We embarked at 6 a.m. and after breaking through some ice, put in shore at a high

---

it is our only chance. One of the other Indians is, however, acquainted with this Route. He gives an awful account of it; he says that there are 24 Lakes & 25 Portages, mostly long ones, high Ranges of mountains to cross etc. This Road falls on the large River at the head of Lake Aylmer and he has never been farther than the mouth of the River (HBC B.200/a/31, f. 1–1^v).

[1] Stewart determined the latitude as 61°39′N and the magnetic variation 42°W (PAA 74–1./137).

[2] But Anderson ruefully welcomed it, since it should have broken up some of the ice (HBC B.200/a/31 f. 1^v).

[3] To Caribou Island.

[4] South-west tip of Keith Island.

[5] Osprey (*Pandion haliaetus*).

[6] *Branta canadensis*.

[7] Anderson's observations on the geology are very perceptive. Both the area of the East Arm, especially Kahochella and Pethei peninsulas, and the area around Nipigon – Thunder Bay, are dominated by diabase lava flows which give a striking terraced appearance to the landscape.

[8] Taltheilei Narrows, between Pethei Peninsula and the northern mainland; it does not freeze due to currents..

[9] These were ospreys, golden eagles (*Aquila chrysaetos*) or bald eagles (*Haliaetus encocephalus*).

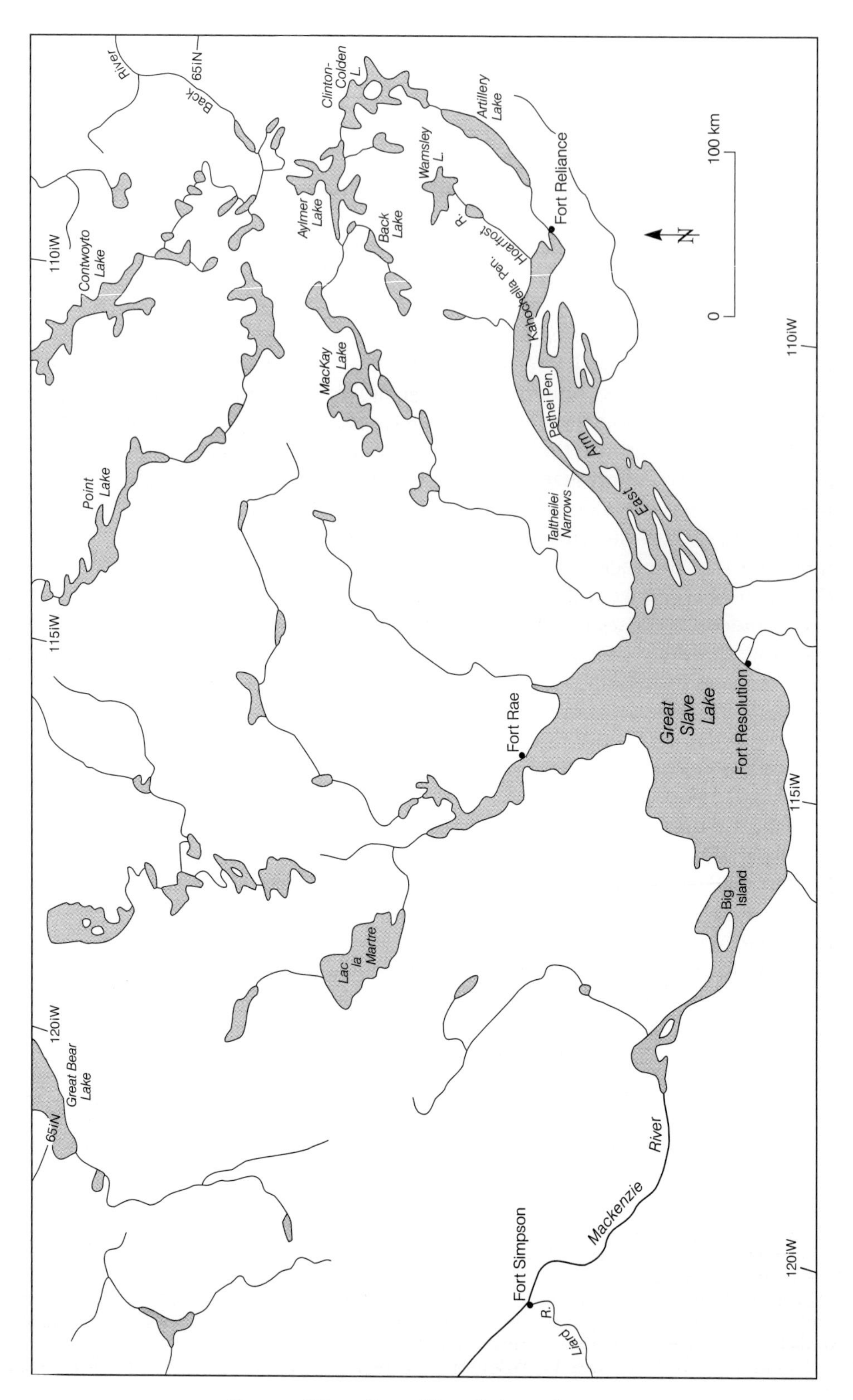

PLATE VI. Great Slave Lake and area

rocky island where we remained until 12 o'clock then made a move but after proceeding 2 or 3 miles, put in shore again as the ice was still too strong. Started again at 3 p.m. The ice was now breakable and we found occasional pools of water. We managed to reach a small stream about 15 or 16 miles from the 'Mountain Portage', a road leading to the Barren Lands and L. Aylmer. Back rejected this route as impassable, but as it is the only chance we have of reaching the Thlewycho[1] in time to descend to the sea, I have determined on adopting it. The head of the lake is still firm and the other lakes (Artillery, Clinton-Colden, etc.) will probably be still unbroken. This mountain route is a chain of small lakes with many portages. Our nets produced nothing. It was curious to see the men at this date on the ice chopping a road. Mr. Stewart took a meridian altitude when we breakfasted which gave 62°47′11″.[2]
Sat the 30th. Calm and clear. The ice froze in a mass last night and we could not attempt to leave before 2 p.m. It is thicker than what we saw yesterday and bore the men easily; it was from 1 to 2½ feet thick. By dint of chopping and pushing pieces apart, we made about 3 miles when it became so thickly packed that I could not venture to proceed further without risking the destruction of the canoes. We encamped at 5 p.m. within sight of our last encampment. The men went hunting but nothing was killed except a goose and a white partridge;[3] the latter had only half its plumage changed. At 6 p.m. Ther. in the air shaded 59+, in the water near the shore 39+. Our nets yielded nothing. On account of the ice none were set tonight.[4]
Sunday, July 1. The wind arose (N.E.) rather fresh, and by driving away the ice permitted us to leave. We made about 2 miles and were again driven ashore till half past 3 when by breaking through some ice we got paddling till 9 p.m. (breaking occasionally through ice) when we were brought to a stop by an impenetrable pack opposite Kahoochellah or Rabbit Point.[5] The wind blew very fresh from 2 to 7 o'clock, and has broken up the ice which had not previously moved. The rocks on the mainland (Wy) are higher than any we have seen, the ascent is sloping. I think the highest does not exceed 700 feet. Most of the rocks are in a state of disintegration. They appear to be of granite and trap.[6] The process is easily seen; the rocks are in layers of 5 feet thick. The upper layer is split into quadrangular pieces. Water enters into these cracks, freezes and splits off the outside one, so that at last the whole of the under layer, which is perfectly rounded and smooth, is covered with these blocks. In process of time the angles are worn off and they have much the appearance of boulders. This may explain why boulders – apparently – are found on high mountains – without having recourse to either water or ice. The islands are apparently of trap and resemble very much those in Nipigon Bay. They have many

[1] The Great Fish River, known now as the Back River.

[2] Stewart noted that the weather was very sultry and the temperature 82°F (PAA 74.1/137).

[3] Most likely a Willow ptarmigan (*Lagopus lagopus*), but possibly a Rock ptarmigan (*L. mutus*) since the latter species penetrates well south of the treeline in winter.

[4] Stewart's journal entry for this date includes the information: 'The scenery is nothing but barren hills for the most part rocks, covered here & there with moss. A little wood is also to be seen in the valleys, sufficient for cooking at any rate' (PAA 74.1/137).

[5] Possibly Gibraltar Point but it is not easy to determine where this was located on Kahochella Peninsula.

[6] I.e. diabase.

peaks with a cut face to the north.[1] The water is of immense depth even close to shore. Only a few ducks and geese are seen and a chance gull and a few small birds. I have not seen the Cypres (Banksian Pine)[2] since leaving Resolution. We passed two insignificant streams today.

Monday, July 2, 1855. Obliged this morning to make a portage ½ mile previous to embarking, after which we met only two bands of ice. We embarked at 3 a.m. and reached the 'Mountain Portage' at 8½ a.m. We passed one insignificant stream about 2 miles from the portage. Another falls into the bay where the portage commences.[3] This portage is an ugly business – it is almost a continual ascent for over 1500 feet. In the first place a portage of about half a mile is made to a pond of about a mile on length, which I have named .....[4] Another portage[5] is then made over the mountain of about 3 miles to a small lake now named ...... The whole of the loadings with the canoes were rendered by 10 p.m. and the men are now laughing over their day's work!!! The general direction of our route today about NNWs. Lat. of the head of the portage 63°46′19″, by meridian observation of Mr. Stewart. Moostigues or sand flies and mosquitoes dreadfully annoying.

Tues. 3rd. The men only got to bed about 1.30 o'clock last night. I therefore allowed them to sleep till 6½ a.m. We crossed a small lake (about ½ mile across) and made a portage to another lake about three miles in length.[6] From the top of one of the highest mountains, perhaps 1,000 feet above the level of Slave Lake, I had a fine view of that body of water (there seems still to be a good deal of ice in it) and counted no less than 15 small lakes or tarns. The interior is inconceivably rugged and desolate. The mountains are riven in every shape. Only a few dwarf spruce and birch are to be seen and scarcely even a bird to enliven the scene. Labrador Tea[7] is in full flower and some berries are nearly full size.[8] The first portage was about a mile in length, and of course, from the steep ascents and the ruggedness of the country, very fatiguing. We then made two short portages and crossed 2 small tarns. We then made a portage of about ¾ mile which, tho' it had some steep ascents, was less rugged than the others. It is thickly carpeted with Reindeer moss,[9] and from their vestiges appears to be a favourite haunt of those animals.[10] This brought us to a lake[11] where we encamped at 7¾ p.m., as the men, though in good spirits, seemed pretty well done up with their last 2 days exertions. Set 2 nets, as the lake is said to abound in Trout.[12]

[1] The peninsulas and islands of the East Arm consist of a striking series of cuestas with steep, often sheer, north-facing slopes and gentle south-facing dip slopes.

[2] Jackpine (*Pinus banksiana*).

[3] This route appears to have begun at the small stream entering Great Slave Lake about 8 km east of the mouth of the Waldron River. It then headed generally north-east to join the Barnston River above Barnston Lake, thus avoiding the ferocious series of falls and rapids on the lower course of that river.

[4] On his sketch map Anderson named it Sandy Portage Lake; otherwise Thai-Koh Antetti.

[5] Each of these portages involved four trips (HBC B.200/a/31, f. 2).

[6] Which Anderson named Clarke's Lake.

[7] *Ledum groenlandicum*.

[8] Probably crow berry (*Empetrum nigrum*) the earliest berry to ripen in this area.

[9] Caribou lichen (*Cladonia rangiferina*).

[10] I.e. Barren-grounds caribou (*Rangifer tarandus groenlandicus*).

[11] Named Miles Lake by Anderson, otherwise They-gee-yeh-too-ey.

[12] In his entry for this date Stewart noted: 'The wind has favoured us much though blowing hard, by keeping away the flies which would devour us no doubts without the least worry seeing that it is not

Wed. 4th. Began to load at 3 a.m. Our nets produced nothing. We made 8 portages today, most of them short, and about 35 miles of lake water.[1] The lakes are getting longer and the height of the mountains is diminished. Wood is fast disappearing.[2] The whole country is clothed in Reindeer Moss, and is evidently much frequented by those animals.[3] It is now utterly lifeless, with the exception of a very, very few birds, such as robins,[4] loons[5] and eagles. The water in the lakes is of crystal purity. They are said to abound in fine trout and W. Fish;[6] we, however, have caught none. We passed through a lake about 7 miles in length,[7] which empties itself into Slave Lake by a very rapid river (unnavigable).[8] A little to the N.E. of the mountain at the head of this lake we found banks of snow still 10 feet thick. A little before encamping we passed through a large body of water, broad and 10 miles in length. Another lake empties itself into it by a fine fall of about 50 feet in height.[9] (Spent up to this date 3 bags Pem'n.,[10] 2 bags flour. Opened one of each at midday today, 4th) It pours through a door-like cut in the rocks. We encamped a little beyond this at 7¾ p.m. Set the nets. Weather is very warm and mosquitoes and sand flies dreadful; a slight breeze today gave us some relief. I shall for the sake of reference name all the lakes we run through, but not those I see from high mountains; they are innumerable, of all sizes, and at every elevation. Saw some old Indian encampments, last year's, of 11 lodges. Lat. of the portage where snow was seen by a M. ob.[11] of Mr. Stewart's 64°4′52″. The general direction of our route is (compass) a little to the W. of North.

Thurs. 5th. Began to load at 3 a.m. We are very unlucky: the nets set last night produced nothing. We made 6 portages – two of them about ½ mile each in length, the others short – and about 47 miles through lakes;[12] two of these are 12 and 13 miles in length, two of 5 and 7, 2 others very small. We are now encamped about half way in a large lake full of islands;[13] we saw divers and gulls in it as well as white partridges[14] in their brown garb, and traces of marmots[15] are also seen at our present

---

every day they get the chance of such a meal'. He also noted that they saw a whisky jack (or Canada jay (*Perisoreus canadensis*) (PAA 74.1/137).

[1] Pruden's, Harrison and McFarlane's lakes.

[2] I.e. they were approaching the treeline.

[3] I.e. caribou (*Rangifer tarandus groenlandicus*).

[4] American robin (*Turdus migratorius*).

[5] Conceivably any of Canada's loon species; the range of all three northern loons, the yellow-billed loon (*Gavia adamsii*), the arctic loon (*Gavia arctica*) and the red-throated loon (*Gavia stellata*), all extend this far south, while the fourth loon, the common loon (*Gavia immer*) reaches its northern limits in this area. See W. Earl Godfrey, *The Birds of Canada*, Ottawa, 1966, pp. 9–14.

[6] Whitefish.

[7] Barnston Lake.

[8] The Barnston River.

[9] Named Rae Falls by Anderson (HBC B.200/a/31, f. 2v).

[10] Pemmican.

[11] Noon observation.

[12] In the sequence in which they occurred, Anderson named them Campbell's Lake, Ross' Lake, Hardisty's Lake and Mackenzie's Lake. They have since been renamed respectively MacLellan Lake, Misteagun Lake, Beirnes Lake and Anarin Lake, i.e. after members of his party.

[13] Margaret's Lake.

[14] Here again, these could be either Rock or Willow ptarmigan.

[15] Probably the Arctic ground squirrel (*Spermophilus parryii*) rather than a true marmot; but Anderson can be forgiven for this mistake since it appears that he did not see the animal itself.

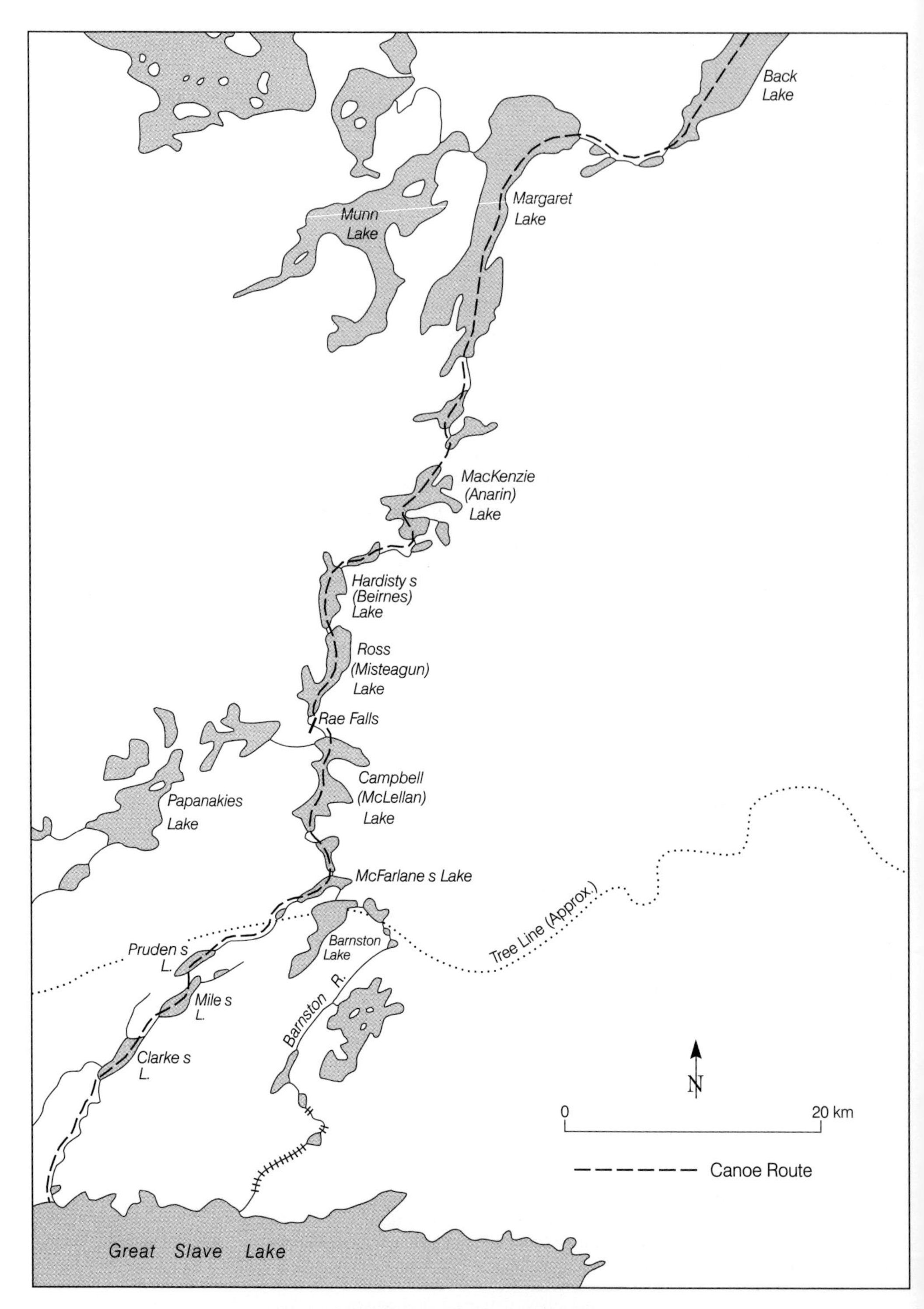

PLATE VII. The 'Mountain Portage'
the expedition's route north from Great Slave Lake to Aylmer Lake

encampment. A fine salmon trout[1] and a pike[2] were taken; the one with a line, the other shot. The appearance of the country is less savage. The mountains (granite) now rise gradually and rarely exceed 100 to 200 feet in height. Their rounded summits are covered with moss and debris of rock; the same process of disintegration is going on with the next layer. Some gravelly islands and sand hills were seen. Wood is getting rare, indeed; we cooked breakfast with a kind of heath today;[3] it burns well. The weather is excessively warm, but an aft wind tempered the heat and helped us on our way. It also kept down the mosquitoes and sand flies a little. In the evening, however, they were in clouds. Set the net again. Encamped at 9½ p.m.. Men rather tired. The canoes are very heavy, particularly mine. It takes 6 men to carry her. Our route today was crooked but the general direction is N.N.W. Compass.

Fri. 6th. Began to load at 5½ a.m., having given the men a little extra sleep. The Indian took us into a Bay yesterday evening and we lost ¼ hour in getting to the proper road. The remainder of the lake was free from islands; in some parts we had a clear horizon; it is a splendid body of water. Some rocks were still covered with ice, and patches of snow were seen throughout the day. It is evident that the ice has only lately broken up. This lake is 23 miles in length and perhaps 8 or 10 in breadth in most parts. The water from the lake runs towards Lake Aylmer. We ran the canoes down two short pieces of river, but the pieces were carried as they were both shallow. This brought us to the largest lake we have yet met with.[4] We encamped on it after making about 30 miles. The mountains are now gently sloping hills; some sand hills were seen in both lakes. Wood is very scarce. A patch of moderate sized spruce was, however, seen in this lake, but with this exception, it is about 2 ft. or 3 ft. in height; the trunks are shaped like carrots. At this encampment the trees are like walking sticks (the largest) and about 1½ feet in height.[5] We shall leave even this tomorrow. A marmot was seen and 6 white grouse,[6] with 2 Canada geese (moulting) killed. We were alarmed a little before encamping by seeing our road apparently barred by ice; fortunately we found a passage round it; it was a broad belt traversing the lake. One of our best men is sick; he has injured his testicle in some of the portages. Weather extremely warm. Flies as usual. Encamped at 8½ p.m.

Sat. 7th. Left at the usual hour. Made 3 portages. They together measured 5½ miles of bad road; and 17 miles of lake way. This brought us to a small lake communicating with the river falling into L. Aylmer.[7] Encamped at 8 p.m.; men tired. At the last portage but one we saw a clump of small spruce about 16 inches in height.[8] A few grouse were shot. Nets set. These lakes abound in fine salmon trout.

[1] Lake trout (*Salvelinus namaycush*).

[2] Northern pike (*Esox lucius*).

[3] Probably arctic heather (*Cassiope tetragona*).

[4] Still named Back Lake, the name given by Anderson.

[5] Anderson's party took advantage of these last spruce trees: 'Each canoe was provided with a large bundle of "attisoirs" for gumming' (HBC B.200/a/31, f. 2^{v}).

[6] I.e. ptarmigan.

[7] Named the Outram River by Anderson; the name is now also applied to the lake which this river drains, but which Anderson did not name.

[8] But were obliged to use heather for cooking (HBC B.200/a/31, f. 2^{v}).

Sun. 8th. Left our encampment at 5½ a.m. The canoes are well arranged. Took up the nets, which yielded only 2 trout. Got into the river at 6 a.m. and reached the mouth at 7½ a.m. Ran 6 good rapids. Except at the mouth of the river we found L. Aylmer fast;[1] along shore however, and the bays, afforded a passage. After paddling about 30 miles[2] we found our passage barred. Broke a piece along shore, but at last the ice began to drive on shore and we were compelled to encamp. The whole of the lake to the Nd. and Eastward is full of unbroken ice.[3] All hands were on it, chopping away, though the weather is very warm. In a shallow bay in this lake we surprised a whole shoal of splendid salmon trout; three or four were captured by the men with their hands.[4] The Cariboo tracks appeared to be fresher than those hitherto seen. The rocks in this part of the lake are chiefly sandstone fit for the finest grindstones and some granite.
Mon 9th. This day has been employed battling against ice. By making portages (3) of about 2½ miles in total length,[5] chopping and pushing ice aside, we rounded a deep bay and reached a point about 3 miles in a direct line from our encampment of last night. We were again stopped by ice and a similar day's work is before us. Wind as usual N. and cold. It froze hard last night and began to freeze at 9½ p.m. when we encamped. One of our canoes narrowly escaped destruction by being nipped between two fields of ice. They actually met, but by shoving poles under her the ice went under her bottom. All the canoes slightly damaged, notwithstanding all our care. A Canada goose shot today. One of the Indians injured his foot by letting a bag of Pemn. tumble on it! Our sick man still unable to work. Ther. 39 air; 34 water.
Tues. 10th. Wind N.N.E. and piercing cold. The ice all frozen in a solid mass, and to give it time to soften we left only at 10 a.m. The whole day was spent in breaking through ice and making portages; of the latter 4 were made, say 1½ miles. We are obliged to round all the bays; some of them are very deep. I really think that we have not made ten miles of direct distance. We are now in a bay, the N. and N.E. portion of which is formed of sand hills, and is, I trust, the Sandhill Bay of Back.[6] We have still much ice to break through before reaching the bottom. The men, notwithstanding their working among ice and water, are in famous spirits, and many a joke and laugh is raised at the expense of those who run a risk of breaking through weak portions of the ice.[7] In general it is about 2½ to 3 feet thick and sound, except close along shore. Encamped at 10½ p.m. Unable to set the nets.[8]

[1] Ice-bound. Lake Aylmer had been named by Back after Lord Aylmer, Governor-General of Canada. See Back, *Narrative of the Arctic Land Expedition*, p. 139.

[2] Along the north shore (HBC B.200/a/31, f. 3).

[3] From 2½ to 3 feet thick (HBC B.200/a/31, f. 3).

[4] In the fair copy of his journal Anderson expands on this comment: 'All hands were in the water in a moment, and after a ludicrous scramble 4 were captured' (HBC B.200/a/31, f. 3).

[5] These portages were over rocks and boulders and the canoes received some damage despite every care (HBC B.200/a/31, f. 3).

[6] So named by Back during his reconnaissance in August 1833 for the sandy eskers or sand hills. See Back, *Narrative of the Arctic Land Expedition,* p. 140.

[7] Anderson and Stewart were also probably the source of some entertainment due to their habit of taking a daily swim, despite the water temperatures. Stewart reports: 'Mr Anderson & myself have not omitted a day without plunging into the water which is dreadfuly cold though very refreshing, it helps to deaden the bites of the Mosquitoes' (PAA 74.1/137).

[8] In his journal entry for this date Stewart noted: 'For the last two days the weather has been cloudy & very cold & damp for the season with showers of rain at intervals, though in this respect we cannot complain having had most beautiful weather since leaving Athabasca Lake in May' (PAA 74.1/137).

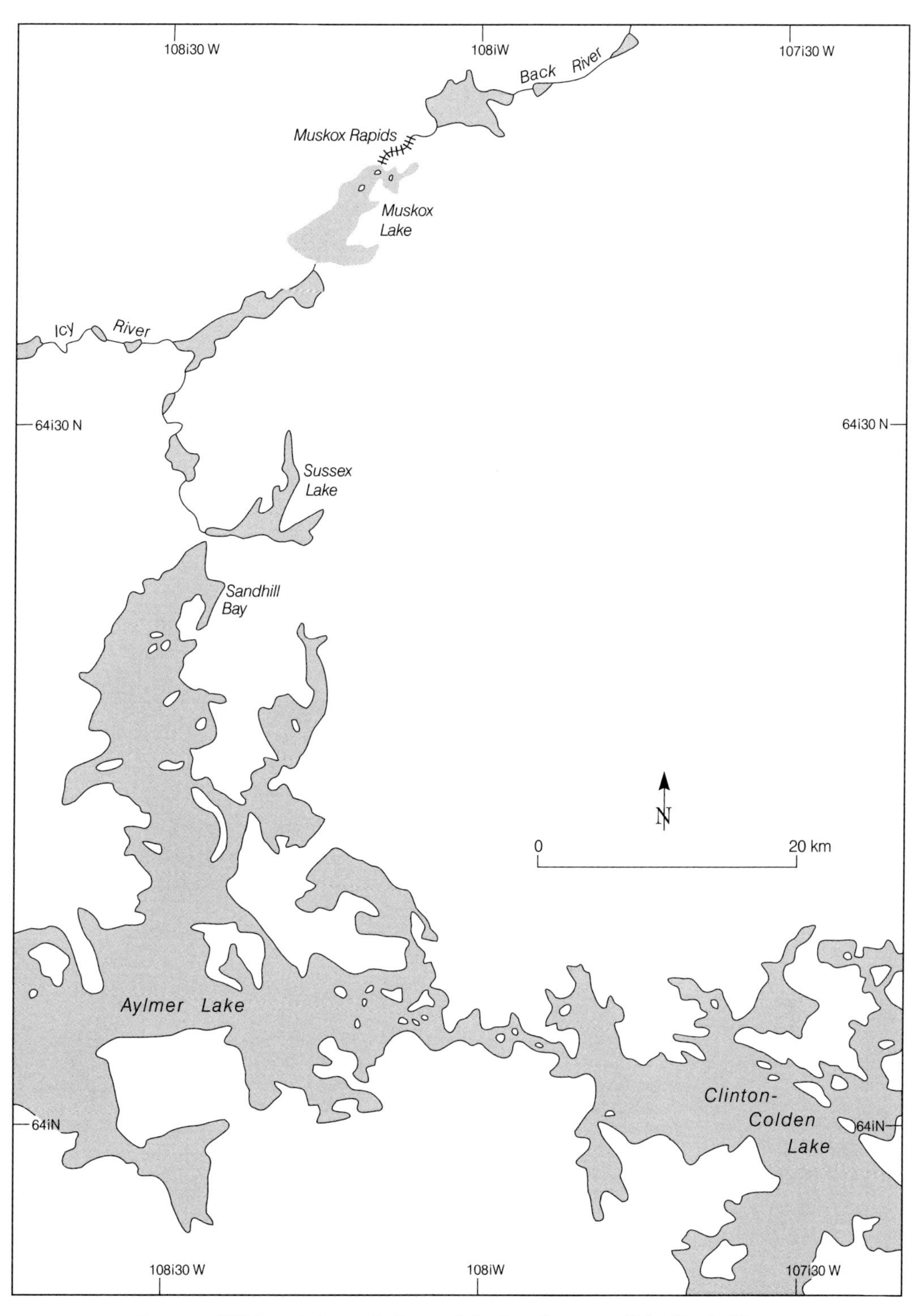

PLATE VIII. Aylmer Lake and the headwaters of the Back River

Wed. 11th. Wind moderate and variable. Cloudy with occasional showers. Left our encampment at 11 a.m., having waited to allow the ice to soften a little. Just before starting a crack appeared at the next point across to the other shore; along the side we were on was choked by ice, and though the risk was great I was determined on attempting it. Fortunately the wind was very light and after a sharp paddle we got safe through. We then had 4 hours of uninterrupted paddling, when ice again barred the road. Another crack appeared in the ice which we immediately entered and re-crossed to the opposite side. We were as nearly crushed as possible; 2 canoes only succeeded in crossing; the third had to retreat and take a passage across higher up. We then, with the exception of a decharge, reached the bottom of what we considered Sand Hill Bay of Back.[1] All our Indian guides were ignorant of this particular portion, having come either from the river falling into this lake, or from Clinton Colden Lake overland.[2] On mounting a high sand hill we immediately recognized Sussex Lake[3] from Sir G. Back's admirable drawing. The river running from it is nearly dry, and we are now cutting across to an elbow of the river by a chain of 3 ponds and 4 portages. The first one is made. Rocks granite, with occasional sand hills. Some of the rocks nearly white, with plates of talc. In some of the bays yesterday sandstone appeared. I never saw regions so destitute of animal life. Since leaving Slave Lake we have seen a white wolf and a marmot, some divers, perhaps 20 Canada geese, as many gulls, a few plover, some bands of grouse and a few small birds. One Indian has lamed himself and our sick man is still hors de combat. Fortunately, notwithstanding the dreadfully severe labour they have undergone the others are well and full of spirits.[4]

Thurs. July 12th The day commenced by making 3 portages and traversing 3 small tarns, which brought us to the river,[5] which is at present nearly dry. The distance from the Lake (Aylmer) is about 2 miles of portage, and 1 of lake. We then crossed it and made another portage of 1 mile to a small lake,[6] after crossing which we made 2 more portages – the river being still almost dry – of ¼ and 1 mile. We then

[1] Hauling their boat on runners across the ice of Artillery, Clinton-Colden and Aylmer lakes Back and his party had reached this point on 27 June 1834. See Back, *Narrative of the Arctic Land Expedition,* p. 141.

[2] Anderson explains this more clearly in the fair copy of his journal: 'None of the Indians had ever been here, having gone to Musk Ox Lake overland from Outram River or Clinton Colden Lake' (HBC B.200/a/31, f. 3^{v}).

[3] Named by Back after H.R.H. the Duke of Sussex, the expedition's patron. See Back, *Narrative of the Arctic Land Expedition,* p. 143.

[4] Stewart's description of the landscape, and his assessment of the trip thus far at this critical point reads as follows:

> It being cloudy we are not able to take observations today but owing to the low state of the water we may have that opportunity tomorrow as the portage is much longer than it must have been when Captn. Back passed. The scenery here is a little diversified by the sandhills & green colour of the moss. At a distance one would fancy it was grass; the dwarf Birch too adds a little to the deception. No deer have as yet been seen. It seems nothing can remain here except mosquitoes. Even deer hasten past as soon as possible. Our canoes have not yet suffered much. Though we have been a great deal in the ice, great care has been taken of them. 38 portages have already been made without damage. It is to be hoped they will stand the trip. Ice is our great enemy and we ought now to have passed it till we reach the sea (PAA 74.1/137).

[5] The Great Fish or Back River.

[6] Back and his companions found these small lakes still ice-bound and had to sledge their boat across the ice. See Back, *Narrative of the Arctic Land Expedition,* p. 298.

encamped at 9 p.m.; men very tired and several lame. Mr. Stewart and I went on ahead to view our road and determine on the best places for portages – two are before us, 1 short and the other long. Saw 2 white wolves and had a long shot at one of them. A grey wavy[1] was killed today. Our Indians are still ignorant of the route. We are guiding ourselves by Back's Journal; his description of the route is so minute and correct that it is needless for me to say anything. The wind was strong from the N.W. and very cold. No mosquitoes tonight; they were in clouds this morning.

Fri. 13th. The men were so fatigued that I gave them an extra hour's sleep. We made 2 portages, one of quarter the other 1¼ miles over the angular debris of rocks; 4 men were so lame as to be unable to carry. We then proceeded across the little lake and Muskox Lake.[2] Back's descriptions are excellent. I think he under-estimated the distance between the portage and Muskox Lake. Icy River[3] was past. The Island particularized by Back in the small lake is no longer conical; the middle is sunk, and the N.W. and S.E. ends raised like a saddle. The white rocks (are of gneiss) very little decomposed; the middle is in a complete state of disintegration. The rock first splits into squares by ice, then the angles are decomposed by the atmosphere, and they assume the appearance of boulders, and eventually are entirely decomposed, forming round spots of gravelly earth a little higher than the moss which surrounds. The rocks may be seen everywhere in these regions in all stages of decomposition. At the head of Muskox Rapid[4] we found a few Copper Indians. We purchased some meat from them[5] and encamped a considerable distance down the Rapids. The entire ladings were run, except at one place where a decharge was made.[6] From this encampment a sick man[7] and four Indians will return. The former and one of the Indians proceed to join Mr. Lockhart;[8] the others will join their relatives at Clinton Colden Straits. The Expedition will now consist of 14 men, Mr. Stewart and myself. This will leave only 4 men for one canoe, and 5 for the 2 others, 3 of whom are lame. These crews are quite insufficient. I shall therefore leave one of the canoes either tomorrow or the day after. The weather was cloudy with slight showers of rain. We found enough of dry willows to cook with.[9]

[1] White-fronted goose (*Anser albifrons*).

[2] Named by Back for the musk-oxen which his Indian guide, Maufelly, said were commonly found there. See Back, *Narrative of the Arctic Land Expedition*, p. 159.

[3] Named by Back (31 August 1833) because he found it still ice-bound with the river emerging from an ice tunnel. See Back, *Narrative of the Arctic Land Expedition*, p. 158.

[4] Back and his party reached here on 1 July 1834 and took a whole day to portage the rapids. See Back, *Narrative of the Arctic Land Expedition*, p. 303.

[5] Stewart reveals that they also arranged for these Indians to make a cache of meat for them at this location, for their use on their return (PAA 74.1/137).

[6] I.e. the loads were portaged while the canoes ran the rapids empty.

[7] Laferté, who had damaged a testicle. Stewart's assessment was: 'He is sick or rather hurt; perhaps fear has something to do with it' (PAA 74.1/137).

[8] Who had been entrusted with preparing winter quarters at Fort Reliance. See pp. 148–59.

[9] The fireplaces they built were still recognizable thirty-five years later. Warburton Pike, an English sportsman who visited the area in pursuit of musk-oxen in 1890, wrote: 'On the edge of one of these lakes, we stopped for dinner on the spot where Stewart and Anderson separated from their Indian guides before descending the river in 1856 [sic]. The rough stone fireplaces, by which they had economised fuel, were still standing, and Capot Blanc, seated on one of them, gave us a long lecture on the events that had taken place during their expedition, as he had heard the story from his father.' See Warburton Pike, *The Barren Ground of Northern Canada*, London, 1892, p. 170.

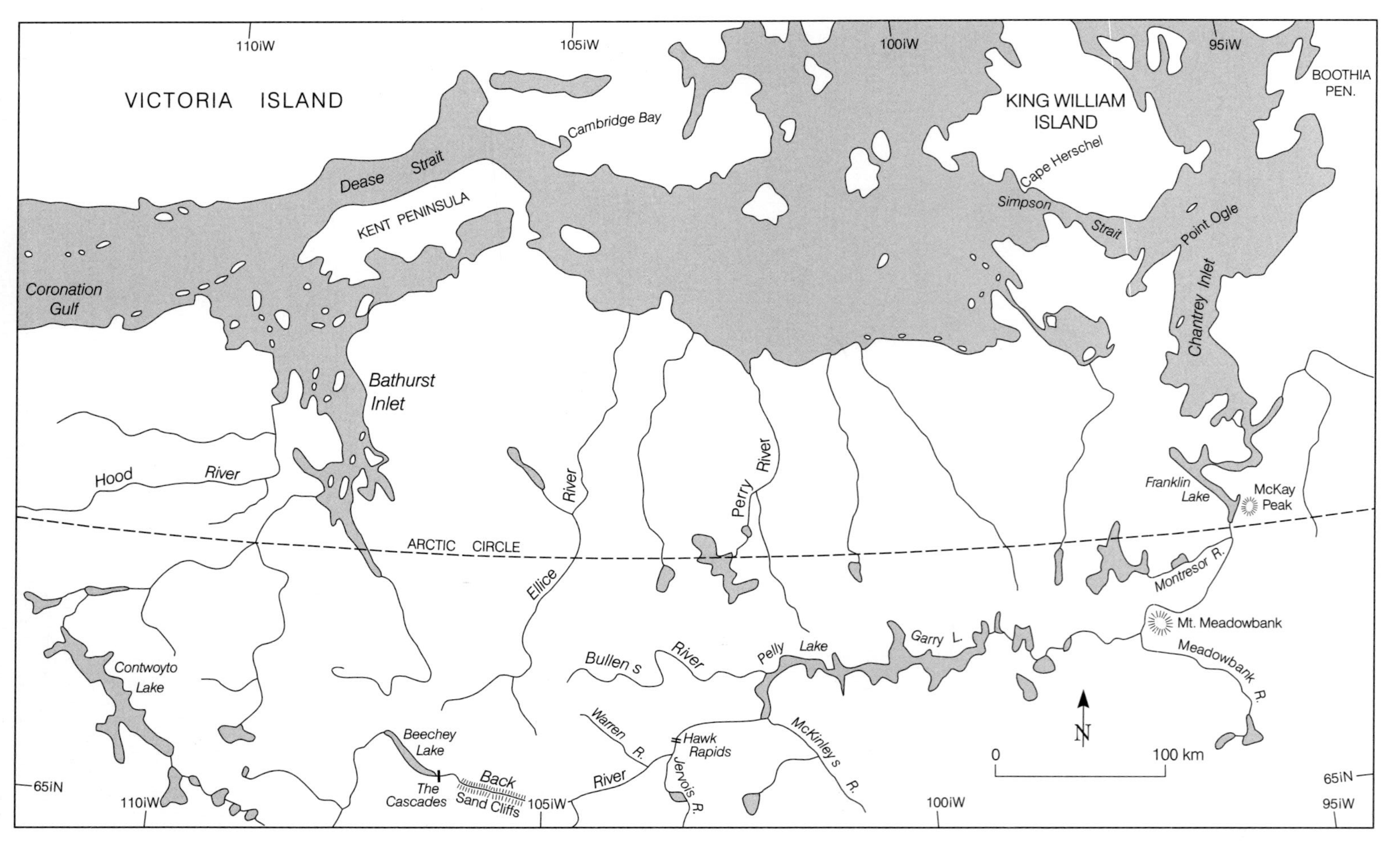

PLATE IX. General map of the expedition's route down the Great Fish (Back) River

In Lake Aylmer we had nothing but heath. Saw a grouse[1] today with its brood. It attacked me bravely. A wolf was also seen as well as a crow,[2] and a few teal,[3] which had long been strangers to us. On arriving opposite the Indian Lodges we found the carcasses of at least 20 deer rotting along the beach. It shows the improvidence of these people.

EXPEDITION

James Anderson	Commanding
J. Green Stewart	2nd Commanding
1. Baptiste Assaminton 2. Joseph Anarin 3. Ignace Montour	Iroquois Boutes[4]
4. Thomas Mustegan 5. Paulet Papanakies	Muskekegon Steer'n
6. John Fidler	Half-breed steer'n
7. Henry Fidler 8. Edward Kipling 9. Donald McLeod 10. Geo. Daniel	Half-breed mid'n
11. Jeremiah Johnson	Muskekegon mid'n
12. Joseph Boucher	Canad'n mid'n
13. Murdoch McLennan	Highland mid'n
14. W. Reid	Orkney mid'n

Sat. 14th. Blowing a N.E. Gale, accompanied by rain and fog, which prevented us from leaving the encampment until 10½ a.m. We were obliged to carry most of the ladings for the remainder of the Rapids, say half way (2 miles), but the canoes and agrets[5] were run with difficulty and rather damaged, particularly one of the Resolution ones, the bark of which is most wretched. Sent back the man I mentioned with the 4 Indians. The ladings were carried at the Rapid where Back nearly lost his boat,[6] but the canoes were merely lifted over a ledge of rock and were run safely with all the agrets. We encamped close to the spot – a little below it – where Capt. Back repaired his boat, and which he left on the 8th at 10 a.m. Two of our present weak crews are so lame that they cannot carry. Encamped at 8½ p.m. Two nets were set, as fish appeared to be running. Two musk oxen[7] were seen at the Rapid of that name.

[1] I.e. a ptarmigan.
[2] Undoubtedly a raven (*Corvus corax*).
[3] Green-winged teal (*Anas carolinensis*).
[4] Bow men.
[5] The canoe accessories and/or the men's personal belongings.
[6] On 4 July 1834; while running the rapids light the boat hit a rock and was partly swamped, although it emerged undamaged. See Back, *Narrative of the Arctic Land Expedition*, p. 312.
[7] *Ovibos moschatus*. Musk-oxen were quite numerous across the entire mainland tundra from Hudson Bay to the Mackenzie Delta at this time, but this mainland population was almost exterminated by commercial hunters (largely trading the hides to the Hudson's Bay Company) by about 1915. The species was given total protection in 1917 and has now recovered to quite a healthy population and has recolonized most of its historic range. See William Barr, *Back from the Brink: The Road to Muskox Conservation in the Northwest Territories*, Calgary, 1991.

Sun. 15th. Left at 4 a.m. The nets produced nothing, though the fish were visibly numerous. This is attributed to the extreme clearness of the water. Ran 10 Rapids with full ladings, except at 2 rapids where Mr. Stewart and myself, 3 men per canoe and 6 pieces, were put ashore.[1] Encamped at 9¼ p.m. at the foot of Malley's Rapids,[2] some distance below Capt. Back's encampment of July 1.[3] I don't find the rapids nearly as bad as I was led to expect by Capt. Back's narrative, and the water is certainly lower than it was when he passed,[4] which renders them in this part of the river worse. Saw some Canada geese. A cache of 1 bag Pemmican was made exactly where Back made his first cache. Wind still N.W. squally, with showers of rain. A little before encamping saw a reindeer,[5] but could not put ashore as we were just entering the rapids. When making this portage a big musk bull was discovered and I had the luck to knock him over. The men are now cutting him up. Query the quality of the meat.[6] We shall sup on a goose shot by Mr. Stewart. The worst canoe was left at the cache. We are now rather deep,[7] but get on well with 7 men per canoe. Some frozen snow was encumbering the shore of a rapids. 5 deer[8] are now running about on the other side of the river. One is a fawn. Slate rocks on the beach at our encampment and 2 or 3 small alders,[9] which we have not seen for some time.

Mon. 16th. Our canoes required so much repairing that we could not leave until 10¾ a.m. All the rapids mentioned by Back were run without difficulty. The water must have been higher and the rapids stronger when he passed. Saw 4 deer and Fidler shot one.[10] Saw 2 bands of Musk oxen, one of 5, the other of 20 animals, besides 5 or 6 solitary bulls, but only one shot was fired at them. 11 grey waveys were also run down.[11] Back's description of the country is in general very correct, but I did not perceive several branches of the river[12] before arriving at L. Beechy,[13] at the

[1] In the fair copy of his journal Anderson remarks: 'I now make it a point to examine every strong Rapid previous to running' (HBC B.200/a/31, f. 4).

[2] So named by Back because his servant, William Malley of the Royal Artillery, went missing for several hours having lost his way on a short cut on the portage. See Back, *Narrative of the Arctic Land Expedition*, pp. 321–2.

[3] Anderson is in error here; Back camped here for the night of 9/10 July 1834. See Back, *Narrative of the Arctic Land Expedition*, p. 322.

[4] Back and his party experienced frequent heavy rains and on one occasion saw the river rise 8 inches (20 cm) overnight (7–8 July). See Back, *Narrative of the Arctic Land Expedition*, p. 36.

[5] Caribou (*Rangifer tarandus groenlandicus*).

[6] In the fair copy of his journal Anderson is a little more explicit: 'The men made a hearty meal off it, but the flavour was so Rank that neither Mr Stewart nor myself could eat it' (HBC B.200/a/31, f. 4^{v}). This is confirmed by Stewart who remarks (with regard to musk-oxen): 'The former we have abjured while deer are to be had as the flavour is none of the best' (PAA 74.1/137).

[7] I.e. the canoes are deeply laden.

[8] I.e. caribou.

[9] *Alnus crispa*.

[10] After eight shots fired by others had missed, according to Stewart (PAA 74.1/137).

[11] White-fronted geese (*Anser albifrons*); these were flightless, moulting birds which were 'destroyed without the assistance of powder' to use Stewart's phrase (PAA 74.1/137).

[12] Anderson is referring to Back's description (13 July 1834): '... and thence the river spread itself into several branches, which not a little puzzled me' (Back, *Narrative of the Arctic Land Expedition*, p. 326).

[13] Lake Beechey, named by Back after his friend Captain Frederick Beechey, who had commanded HMS *Blossom* on her voyage into the Chukchi Sea in 1825–8. See Back, *Narrative of the Arctic Land Expedition*, p. 330.

entrance or head of which we encamped at 9 p.m. Wind dead ahead and strong all day. Weather cloudy and chilly. The rocks at our encampment composed of slate.
Tues. 17th. Left our encampment at 2¾ a.m. and passed Lake Beechy with a fine breeze aft.[1] A complete portage was made at the Cascades.[2] All the rapids below it were safely run, with full cargoes, with the exception of one, where the canoes were lightened of a few pieces and 3 men each. The current carried us on very swiftly, and we encamped at 9¼ p.m. at the 'Sand Cliffs', passed by Back on the afternoon of the 16th. inst. His description of the scenery is most correct; it is beautiful indeed. The mosses which are in full flower and in patches on the cliffs with their green leaves and purple flowers on the cream coloured sand look most beautiful.[3] Back saw immense numbers of reindeer and musk oxen in this part of the river. We saw but 10 of the former and about 40 of the latter; 28 of these were in one drove. They were all sizes; the calves looked like black pigs. Killed 4 Canada geese and 18 grey wavies, which are now moulting. They gave all hands a severe run to catch them. Saw a doe and her fawn cross a narrow part of L. Beechy. 2 wolves were waiting for them. The poor creature was in a sad dilemma, afraid to return on account of us, and to land for the wolves. We shouted and drove the wolves off, and I trust the poor animals escaped their fangs. Observed a great change in the temperature since leaving Lake Beechy: it is much warmer. Capt. Back observed the same thing, and accounted for it by the distance from Bathurst Inlet being increased.[4] Made a cache of a bale of dried meat at our encampment of last night and of one bag Pemmican at the head of the Cascades of Beechy's Lake.
Wed. 18th. Left our encampment at 4¾ a.m. The canoes were lightened at the 2nd cascade[5] and portages made at the first cascade and the 'dalles' previous to arriving at Baillie's River.[6] That stream is now only a few yards in width, though when the water is high it is evidently an imposing stream.[7] Encamped at 9 p.m. about half way between Baillie's and Warren's River.[8] 24 Canada geese were killed. They are all males; no young ones are to be seen. A few musk oxen and deer were seen. The weather was clear and warm. I searched minutely for the Esquimaux marks mentioned by Back, but saw none, either on the banks of the river or on the Gneiss

[1] This lake had given Back much more trouble; it was still partly ice-bound and he had had to wait for a whole day (14 July) for the wind to clear a passage alongshore. See Back, *Narrative of the Arctic Land Expedition*, p. 329.

[2] The first rapid which they did not run, according to Stewart (PAA 74.1/137).

[3] Stewart also found the scenery attractive: 'passed through a small lake at the end of which the scenery changed & became quite picturesque: high sand cliffs covered with verdure & river winding between them in a boiling current' (PAA 74.1/137).

[4] Back remarked (on 16 July 1834): 'As we drew away from the influence of the cold winds coming from Bathurst's Inlet, a proportionate and most agreeable change took place in the weather, and at 2 P.M. of this day the thermometer stood at 68° [F] in the shade and 84° in the sun' (Back, *Narrative of the Arctic Land Expedition*, p. 332).

[5] And lined down in one place, according to Stewart (PAA 74.1/137).

[6] Named by Back after his friend George Baillie, Agent General for the Crown Colonies. See Back, *Narrative of the Arctic Land Expedition*, p. 335.

[7] Back (who saw it after a period of wet weather) described it as: 'a magnificent river, as broad as the Thames at Westminster, joining the Thlew-ee-choh from the eastward' (Back, *Narrative of the Arctic Land Expedition*, p. 334).

[8] Named by Back after Captain Samuel Warren, Superintendent of the Woolwich Dockyard. See Back, *Narrative of the Arctic Land Expedition*, p. 337.

mountains mentioned by Back. Along the bank of the river small stones were often found, placed one on top of the other. But this is evidently done by the washing away of the sand from the stones. Two of Dr. Rae's men say that they do not resemble Esquimaux marks. I saw nothing of the old encampments.[1] 3 kinds of gulls were seen at First Cascade.[2]

Thurs. 19th. Raining and blowing a gale, which prevented us from leaving until 6¾ a.m. About 1 p.m. it began to rain and did not cease until we encamped at 6½ p.m. at the head of the Hawk Rapids.[3] Just before we encamped it rained so heavily and blew so hard that the bowsmen could not distinguish the leads. Saw no musk oxen today, but perhaps a hundred deer. We did not go after them as we have plenty of fresh provisions, having killed 31 large male Canada geese at one run of 10 or 15 minutes. Hundreds of these birds were seen. The so-called Esquimaux marks are seen on the edge of every sandy or gravelly hill, but nowhere else. They point or run in every direction according as the river runs. Blue Lupins[4] are found here in great profusion and several other flowers, among others the dandelion. Warren and Jervois' Rivers were dry.[5]

Fri. 20th. The night turned out fine but cold, and the morning was a lovely one. The rapids were run safely.[6] At this stage of the water, though strong they are not dangerous. Just before reaching McKinley's River[7] we saw fresh Esquimaux caches of deer along the water's edge[8] and crows[9] were seen. Shortly after their tents were seen, 6 men, one of them blind, came down.[10] From signs they made they came down McKinley's River and most probably belonged to the Chesterfield Inlet tribe.[11] Their boots were made of bearskins and muskox soles, and their canoes of deer parchment. Paddles of spruce; spear-heads of iron. One of their women had

[1] Anderson is almost certainly correct in questioning Back's interpretation of various features as being indicative of the presence of Inuit. Anderson's interpretation of the piled stones as being a natural phenomenon seems quite plausible. As to the 'old encampments', here he is alluding to Back's description of 'some trenched divisions of ground, containing the moss-covered stones of circular encampments, evidently the work of the Esquimaux, on whose frontiers we had arrived' (Back, *Narrative of the Arctic Land Expedition*, p. 333. These may well have been sorted circles, a common periglacial phenomenon on the tundra, whereby repeated freezing and thawing produces a sorting of the material to produce a circular arrangement of large stones around a slightly domed area of much finer material, usually 1–2 m in diameter. When well developed they give every appearance of being man-made.

[2] In the fair copy of his journal Anderson reports: 'A Cache of 1 Bag Pemmican and a case containing Ammunition, Tobacco, Tea, etc. was made at the 1st Cascade' (HBC B.200/a/31, f. 4v).

[3] Stewart commented on the lack of fuel here: 'Very little wood to be had; nor any moss or heather; our kettles were boiled and that was all' (PAA 74.1/137).

[4] *Lupinus arcticus*.

[5] Stewart remarks that they had barely enough fuel to boil their kettles (PAA 74.1/137).

[6] Hawk Rapids; Back and his men ran them on 17 August 1834 and named them after: 'three large hawks, which frightened from their aërie were hovering high above the middle of the pass, and gazing fixedly upon the first intruders of their solitude...' (Back, *Narrative of the Arctic Land Expedition*, p. 339).

[7] Named by Back after Rear-Admiral McKinley. See Back, *Narrative of the Arctic Land Expedition*, p. 340.

[8] Back also saw the first indisputable signs of Inuit here: inuksuit (cairns) and hunting blinds. See Back, *Narrative of the Arctic Land Expedition*, p. 341.

[9] I.e. ravens.

[10] In the fair copy of his journal Anderson adds the information: 'They appeared agitated, but attempted nothing hostile. We were soon good friends. The absence of an Interpreter was now felt. Mustegon and McLennan, two of Dr. Rae's men, understood some words' (HBC B.200/a/31, f. 4v).

[11] They were probably Uvaliarlit, or possibly Hanningayarmiut, both identified by Balikci as inland subgroups of the Netsilingmiut. See A. Balikci, 'Netsilik', in D. Damas, ed., *Handbook of North American*

bracelets of round common beads and the oldest man brought down some wolf and white fox skins to trade, which we could not take at present. I gave them all presents of files, knives, needles etc., and the women a mirror and small scissors, gartering and needles. After leaving them we came on two other lodges and 3 men came to visit us, and further on two more, which we did not visit as it was blowing too fresh. The men were short and stout; the women not bad looking, with clean faces, tattooed the same as the female in Capt. Back's book. I regretted much not having an interpreter with us so as to learn the route they take from Chesterfield Inlet (assuming that they come from there). 2 of Dr. Rae's men with me[1] understand and speak a few words. Shortly after leaving the 2nd Esquimaux lodges a gale came on, which shortly after increased to a storm, which nearly swamped us. This was accompanied by showers of hail and clouds of sand, which nearly blinded us. At last I gave up the contest and encamped near Bullen's River[2] at 6 p.m. It was piercingly cold. Capots, cloaks and blankets in general demand. Both yesterday and today we were much incommoded by sand banks. The Esquimaux also made us lose some time. They had evidently not heard of Franklin's party, as we made them understand that white men who had come in ships had died from starvation at the mouth of the River. About 50 or 60 deer were seen today, but neither musk oxen or geese. At the Esquimaux encampments many deer were lying at the water's edge till they get high enough for their taste.[3] They were all does. Several fawns were lying close to the encampments apparently unalarmed. Several deer were also seen.[4]
Sat. 21st. Detained all day by wind and rain.
Sun. 22nd. The gale of yesterday abated a little this morning, but the weather was still miserable when we left our encampment at 2½ a.m. When we reached Pelly's Lake[5] we hoisted sail and carried it most of the day. Encamped at the second Narrows in Lake Garry[6] (Back's Enct. of 20th) at 9 p.m. Saw 2 lodges of Esquimaux at the Rapids between L. Pelly and Garry, but the inhabitants ran away on perceiving us. They evidently have intercourse with the Churchill Esquimaux as there were 2 tin kettles in their lodges, as well as our dags.[7] I left a few articles in each tent and left. A number of young fawns were running about the lodges; I suppose that their dams have been killed. 2 bags of pemmican were cached at our encampment of last night. Very few deer seen. 30 geese were killed.

Indians, Vol. 3, Arctic, Washington, 1984, pp. 415–30. It was probably one of the children from this group who later told Hanbury of this encounter (see p. xi).

[1] Thomas Mustegan (an Ojibway from Norway House) and Murdoch McLellan (from Lewis).

[2] Named by Back after Captain Sir Charles Bullen, Superintendent of Pembroke Dockyard. See Back, *Narrative of the Arctic Land Expedition*, p. 342.

[3] Stewart noted that the expedition members benefited from this situation: 'Fortunately they were recently killed. We got the tongues from them which proved excellent' (PAA 74.1/137).

[4] On this date Stewart makes the first reference to his wife in his journal entry: 'How do you do, dearest Meg, my own dear wife. Hope is supporting me & in ten days we ought to be nearly at our furthest footstep. You are dearest, Meg' (PAA 74.1/137).

[5] Named by Back after Sir John Henry Pelly, Governor of the Hudson's Bay Company. See Back, *Narrative of the Arctic Land Expedition*, p. 344.

[6] Named by Back after Nicholas Garry of the Hudson's Bay Company. See Back, *Narrative of the Arctic Land Expedition*, p. 351.

[7] I.e. daggers of the type traded by the Hudson's Bay Company.

(Course in L. Garry[1]

To 2d Detroit[2] E. by S; mark a small island with gravel etc. shoved up by ice and crowned by square blocks of stone in Situ but in a state of disintegration. Then through a labyrinth of islands and narrow bays to a prominent sand hill; thence to a 3rd Detroit N.E.. Nearby mark a clump of sand hills or cut very picturesquely; thence to Rapid (point) N. by E.; mark a high conical sandhill).

Mon. 23rd. Left at 4½ a.m. Lost most part of the day in finding our road. We were also retarded by cutting through ice 3 feet thick. Encamped at the 3d straits of L. Garry at 10 p.m. (Back's encampment of 21st).[3] Either we are very stupid or the map in Back's work is very incorrect.[4] The day has been the warmest we have had for some time. I shot a deer today,[5] a doe I am ashamed to say, but we had no fresh provisions and the pemmican must be saved. The fawn was half grown and was of course allowed to live. In a bay surrounded by sand hills to the north of the sand hill at the end of the 2nd strait Esquimaux encampments and signs of this spring seen. From a height a chain of lakes leading to the N.E. were seen, by which road I think the Esquimaux come from Lake McDougall.

Tues. 24th. (Encamped 8.30 p.m.). It was near midnight before the men laid down last night. I therefore allowed them to sleep till 5½ a.m. We rounded all the bays by cutting our way through the ice. We were also much retarded by cutting our way through the ice at three points. It was from 2 to 3 feet thick. It is a curious sight to see men working on the ice at this date. We at last reached the rapid at the end of L. Garry, to which we joyfully bid adieu. It falls by three rapids into the river leading to Lake McDougall.[6] This rapid was easily run. At its foot a cache of pemmican (1 bag) was made. The rapids below this, 5 in number, are all strong and dangerous, with the exception of the last one. A little below we camped at 8¼ p.m. 2 decharges were made. At most of these rapids there are several channels. Capt. Back's map (the one affixed to his narrative) is on so small a scale as to be utterly useless in these large bodies of water. 17 geese were killed; no animals were seen with the exception of a young fox. This has been the finest day since we left Slave Lake, clear and very warm.[7] The refraction was very great. Esquimaux ducks[8] seen.

Wed. 25th. Left at 4 a.m. In about 3 hours paddling we reached an easy rapid. This led into an extensive sheet of water[9] where the current became imperceptible. It ran on either hand N. and S. in deep bays. Land was seen in every quarter (Back said no

[1] These 'sailing directions' are inserted into Anderson's text.

[2] I.e. strait.

[3] Back and his party also had great difficulty in negotiating Lake Garry since it was still almost completely ice-bound (19–21 July 1834). See Back, *Narrative of the Arctic Land Expedition*, pp. 346–52.

[4] In the fair copy of his journal Anderson is a little more charitable towards Back: 'Our map (the one attached to Sir George Back's Narrative) is on too small a scale for our purpose' (HBC B.200/a/31, f. 5)

[5] Whether Anderson or Stewart killed it was 'a disputed point' according to Stewart's journal entry of the 24th (PAA 74.1/137).

[6] Named by Back after his friend Lt Colonel Macdougall of the 79th Highlanders. See Back, *Narrative of the Arctic Land Expedition*, p. 357.

[7] And the first clear day since leaving Lake Aylmer, according to Stewart. He was able to determine the latitude as 66°18′N; longitude 98°57′21″; variation 29°11′E (PAA 74.1/137).

[8] Probably king eiders (*Somateria spectabilis*).

[9] Lake MacDougall.

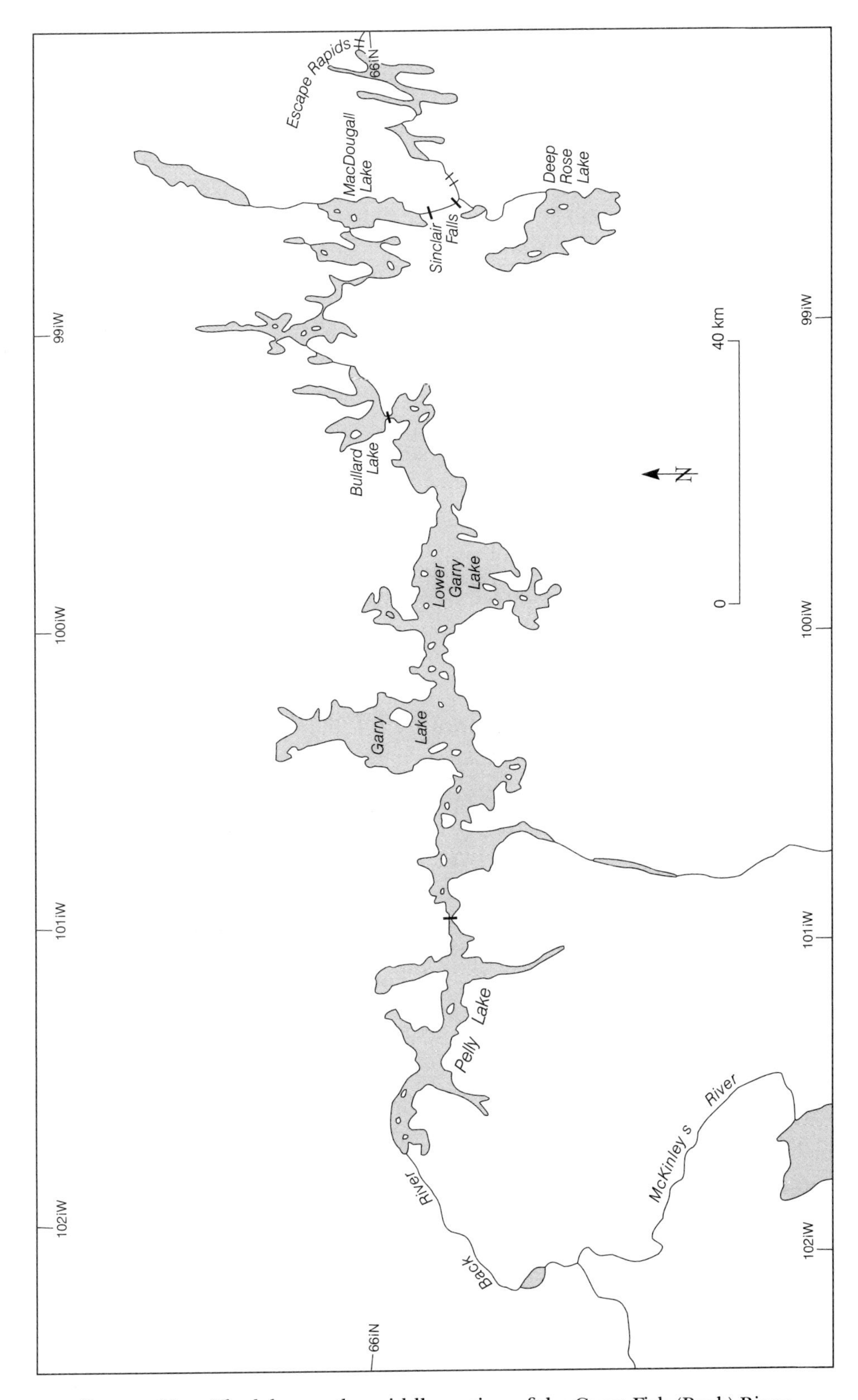

PLATE X. The lakes on the middle section of the Great Fish (Back) River

land to be seen to the N.) though distant. From this we struck due south to the end of Lake McDougall about 10 miles from the rapid. The map is perfectly useless. We ran part of the Rock Rapids (3) but a decharge was made at the last one. We encamped at the foot of the latter (Sinclair's Falls).[1] All these rapids are strong and hazardous. Our Iroquois boutes have had fine opportunities both yesterday and today of exhibiting their matchless skill.[2] Saw 6 or 7 deer and killed 13 male Canada geese. Esquimaux marks were very numerous above the head of Rock rapids and below them to this spot. Made a cache of 1 bag pemmican[3] at the Cascades above this place.

Thurs. 26th. Left at the usual hour. It rained last night slightly. Made a decharge at the Escape Rapid and at 2 of the Sandhill rapids, but ran the others with whole ladings. All of these rapids are strong and long.[4] Two barren does were shot today in the water, one by Mr. Stewart, the other by E. Kippling. 2 or 3 others were seen, and immense numbers of Canada geese; 64 of these were killed in two 'runs' ashore. An ermine[5] and beaver mouse[6] were also killed at Escape Rapid and here we encamped above Wolfe rapids.[7] A cache of 1 bag flour, 1 bag pemmican and a case of tea, etc. at the head of Escape rapid. Some old Esquimaux marks and encampments seen at Escape rapid.[8]

Fri. 27th. ⅓ this day nearly was lost by our mistaking a channel of the river which led us into a deep bay at the bottom of which was a small river.[9] It appears to be frequented by the Esquimaux. The above occurred above Mt. Meadowbank.[10] It was blowing a tempest with rain which prevented the steersman from observing the current in this lake-like expansion of the river. The Wolfe and 9 other rapids were run with whole cargoes; they are all strong, some with whirlpools which must be dangerous in high water.[11] Two large bands of Musk Oxen were seen just before encamping; 2 or 3 deer, 3 wolves, many Canada geese and a hawk.[12] We encamped late about 3 miles below

[1] Named by Back after George Sinclair who served at various times as steersman and bowman of Back's boat. See Back, *Narrative of the Arctic Land Expedition*, p. 364.

[2] Stewart recorded that they negotiated eight rapids or falls during the day, dropping a total of 120–30 feet vertical since leaving Lake MacDougall (PAA 74.1/137).

[3] Together with a paddle and a broken copper kettle, plus an old capot to mark the spot (PAA 74.1/137).

[4] In the fair copy of his journal Anderson adds the further information: 'The Compass is now useless or nearly so' (HBC B.200/a/31, f. 5).

[5] *Mustela erminea*.

[6] Possibly a Northern red-backed vole (*Clethrionomys rutilus*) or one of the lemmings, i.e. Brown lemming (*Lemmus sibiricus*) or Collared lemming (*Dicrostonyx torquatus*) (Banfield 1974).

[7] Named by Back for a pack of nine wolves seen prowling around a musk-ox bull (Back, *Narrative of the Arctic Land Expedition*, p. 368). According to Stewart they camped below these rapids (PAA 74.1/137).

[8] On this date Stewart's entry again ends with a personal note to his wife: 'Dearest Meg, I have been thinking of you all day. How are you today?; perhaps thinking of the Fish River, indulging in the same hopes as I am. Well dearest [illegible], I trust they may be fulfilled for if alive my promise will be kept' (PAA 74.1/137).

[9] Probably Meadowbank River.

[10] Named by Back after Lord Meadowbank. See Back, *Narrative of the Arctic Land Expedition*, p. 370.

[11] As it was they severely strained the canoes (HBC B.200/a/31, f. 5^{v}).

[12] Possibly a Rough-legged hawk (*Buteo lagopus*) or a Peregrine falcon (*Falco peregrinus*). See Godfrey, *The Birds of Canada*, pp. 93–4; 102.

the rapid with whirlpools and Esq. marks. A cache of 1 bag pem'n and 2 nets was made at a bold point at the bend of the river above Mt. Meadowbank.[1]

Sat. 28th. Left at the usual hour. The day was fine, which gave us an opportunity of drying our clothes while breakfasting, only to be wetted again by the spray arising from a strong head wind which retarded us very much. 4 rapids were run, 3 of them very strong. The eddies or whirlpools strain the canoes very much; we cannot keep them tight; they are evidently getting shaky. 2 plovers and immense numbers of Canada geese were seen; 20 were killed. 2 deer were also seen close (does); one of them had a fawn with a leg broken, but the little creature managed to ascend a steep and rugged mountain pretty swiftly on 3 legs. Some good sized willows were gathered. Extensive patches of snow on the right bank of the river. We encamped late a little above Montresor River.[2] (Note: I was nearly upset by the canoe grazing a stone. It was only a shave, the gum only was rubbed off.)

Sun. 29th. Left early. Ran a bad rapid above Montresor River, in which Mr. Stewart's canoe was completely ungummed. We were consequently obliged to put on shore at 6 o'clock to gum, where we breakfasted likewise, and made a cache of 1 bag pem'n. and 1 bag flour. The rapid at McKay's Peak[3] was little more than a strong current. In the rapid below it my canoe was nearly broken though it was an easy one. We had to contend against a strong wind all day. In the evening this was accompanied by a soaking Scotch mist. This compelled me to encamp in case I should miss my road at 7 p.m. near the outlet of Franklin Lake.[4] Esquimaux marks numerous[5] and traces fresh. Saw Esquimaux ducks; no animals were seen, but abundance of Can. geese, of which 53 were killed at one run; they are beginning to fly. Montresor River has a rapid at its mouth; it does not appear a large river at present. 2 small black-headed gulls[6] attacked us at the encampment, even striking at our hats.

Mon. 30th. Left early. The rapids at the outlet of Lake Franklin were partly passed by a portage and partly run. At their foot we saw 3 Esq. lodges, in which were an elderly man, 3 women and a host of children, the others being absent.[7] Large numbers of W. fish and trout[8] were hung out to dry, as well as some deer meat. The

[1] As perceived by Stewart the country was becoming increasingly desolate: 'The right or West bank of the river is more bare of moss or heather than in the upper parts of the river & it is difficult to find enough to boil our kettles. I should fancy that it would require Esquimaux & marksmen to be the permanent inhabitants of this sterile, bleak tract of country' (PAA 74.1/137).

[2] Named by Back after Lt General Sir Thomas Montresor. See Back, *Narrative of the Arctic Land Expedition*, p. 372.

[3] A very conspicuous, black, conical hill which Back's steersman, James McKay, climbed to reconnoitre the route ahead – hence the name. See Back, *Narrative of the Arctic Land Expedition*, p. 373.

[4] Stewart noted that there was still ice on Lake Franklin (PAA 74.1/131).

[5] Stewart noted that one of the Inuit markers was a piece of a kayak (PAA 74.1/137).

[6] Most probably Bonaparte's gulls (*Larus philadelphia*). See Godfrey, *The Birds of Canada*, p. 184.

[7] Back encountered his first Inuit here on his descent of the river. See Back, *Narrative of the Arctic Land Expedition*, p. 379. This is one of two traditional fishing sites, frequented by the Utkuhikalingmiut in summer and autumn for centuries, and indeed still utilized by the people of Gjoa Haven. One of the sites (Akuaq) is located at the rapids by which Lake Franklin overflows; the other, Itimnaarjuk, is located at a second set of rapids (the lowest rapids on the Back River) some 6.5 km farther east. For further details see Jean L. Briggs , *Never in Anger: Portrait of an Eskimo Family*, Cambridge, 1970.

[8] Probably arctic char (*Salvelinus alpinus*).

lodges were made of musk-ox skins dressed with hair inwards.[1] These people made us understand that a party of white men had starved to death (at the sea) after their vessels were destroyed.[2] 2 of Dr. Rae's men understand many words and phrases. In their lodges were copper and tin kettles, both round and of a square form, longer than broad, evidently belonging to cooking stoves. Various pieces of wood, poles and boards of ash, oak, white pine and mahogany were about the lodges, also a brass letter clip,[3] but nothing to identify any person. Some of the boards were painted white. Nothing could be learnt about books or manuscripts.[4] The absence of an interpreter is a sad blow to us. We ran the last falls; they were only an easy rapid at this stage of the water. At some distance below them we saw 2 cyaks, but they turned tail immediately on seeing us and joined 3 others on shore. 2 finally took courage (one an old, the other a young man) to cross to us, but we learnt nothing additional from them. They confirmed the accounts given by the others of the death of the crews of the vessels etc. The weather has been most gloomy and the wind ahead with occasional showers. About 5 it commenced raining in earnest, and increased to such a degree that I gave the order to encamp, but we could find no fit place till 7¼ p.m., when we disembarked thoroughly soaked. No fires could be made so that pemmican and cold water are the order of the day. Some spirits should be provided for an expedition of this kind. The men really require it on such occasions as this.[5] A little before encamping saw a small band of deer in a bay; Canada geese were also numerous. Encamped among the islands about half way between the Fall and Victoria Headland.[6]

Note: On an island below the falls found the head of a blacksmith's tongs, the handles broken off.

Tues. 31st. The rain prevented us from leaving before 5½ a.m. It recommenced just

[1] This group of Inuit, the Utkuhikalingmiut, were unusual in being particularly reliant on musk-oxen for food and skins for tents and clothing. See Barr, *Back from the Brink*, p. 13.

[2] In the fair copy of his journal Anderson is a little more explicit as to this communication: 'They made us understand by pressing the abdomen inwards, pointing to the mouth, and shaking their heads piteously, that these things came from a Kayack, the people belonging to which had died of starvation' (HBC B.200/a/31, f. 6).

In his report to Sir George Simpson Stewart added the information that one survivor had been seen alive:

> One woman in particular, was very explicit about having seen one man on an island at the least extremity. She shewed the way he was sitting on the beach, his head resting on his hands, the hollowness of his cheeks and the general emaciated appearance of the unfortunate person. This she said was four years ago and that, being without provisions themselves, they could not give any assistance and that even if they could have done it would have been too late (PAA 74.1/140).

Anderson later vigorously denied that they had heard any such story from the Inuit (see p. 218).

[3] With the date 1843 (HBC B.200/a/31, f. 6). This was later identified by Lady Franklin as one she had given to her husband. See Woodward, *Portrait*, p. 290.

[4] In the fair copy of his journal Anderson elaborates on this point: 'Printed and manuscript books were shown to the Esquimaux, and we made them understand by signs and words that we would pay handsomely for even a piece of paper: the women were very intelligent, and, I am certain, understood us perfectly; but they said they had none' (HBC B.200/a/31, f. 6).

[5] In the fair copy of his journal Anderson also admits: 'and I myself should have no objection at this moment to a glass of brandy and water' (HBC B.200/a/31, f. 6^{v}).

[6] A very prominent headland, some 200 m high, on the east side of Chantrey Inlet, which Back named after H.R.H. Princess Victoria. See Back, *Narrative of the Arctic Land Expedition*, p. 389.

after embarking, and we had a wretched time of it till we reached Victoria Headland to breakfast at 11 a.m. It then partially cleared up, but we had occasional showers, with fog, till we encamped at 8 p.m. at point Beaufort.[1] Red granite is the prevailing rock at all points on this side of the inlet. Victoria Headland is principally composed of hills of rounded stones, like shingle, though I believe them to be only decomposed rocks. Willows were found at the waterfall at Victoria Headland; fuel of 2 kinds in small quantities,[2] and most of the flowers we saw inland. No animals were seen today, nor any traces of any except a wolf, and two seals,[3] the latter below Victoria Headland. There is no such thing as a deer pass,[4] or any place where even Esquimaux could live.[5] We have seen no marks this afternoon at this encampment. I found all the 'agrets'[6] of an Esquimaux, most of them of deer horn, and a few iron; one had holes evidently drilled by a tradesman. There was also a piece of tin. I suspect they belonged to a dead man. They must have been here some time, as they were in a state of decay. Also Esq. ducks, a loon, and large gulls. Noticed the tide at Victoria Headland.

Wed. Aug. 1st. Detained by wind and rain till 2½ p.m.[7] The wind was from the S.W. and has doubtless cleared away some of the ice. We took the traverse to Montreal Island,[8] and with the aid of the paddle made it in 3 hours. We lost some time among the drift ice, driving very rapidly with wind and tide from Elliot's Bay.[9] We had some narrow escapes, and I was heartily glad to get safe through it. The ice is 6 or 7 feet thick and perfectly sound. We encamped on the north side of a rocky island divided by a channel from Montreal Island. The whole inlet to the north and eastward is

[1] Named by Back after Sir Francis Beaufort, Hydrographer to the Navy. See Back, *Narrative of the Arctic Land Expedition*, p. 393.

As Stewart gleefully points out in his journal entry for this date, the expedition has now 'caught up' with Captain Back, in terms of the time taken for the trip, having crossed the portage to the head of the Back River thirteen days behind him.

Here again, Stewart gives vent to his feelings about the desolation of the area, and the inadvisability of trying to winter: 'The shore here is an extent of rock with scarcely any moss to be seen; such sterility was never seen. Not an animal have we seen but a couple of seals & a band of geese. It is quite madness to think of wintering in such a place. We intend going on 8 days more & then turn back, so that we may have some chance of getting home' (PAA 74.1/137).

[2] In the fair copy of his journal Anderson expressed this more clearly as: 'heath, of two kinds, for fuel, in small quantities' (HBC B.200/a/31, f. 6^{v}). Possibly arctic heather (*Cassiope tetragona*) and Labrador tea (*Ledum groenlandicum* or *L. decumbens*).

[3] Most probaby ringed seals (*Phoca hispida*) or bearded seals (*Erignathus barbatus*). See Banfield, *The Mammals of Canada*, pp. 362–7; 372–5.

[4] A location (e.g. a narrow defile, an isthmus between lakes etc.) where migrating caribou can readily be ambushed.

[5] Stewart, too, was dismayed at the barrenness of this coast: 'The shore here is an extent of rock with scarcely any moss to be seen; such sterility was never seen. Not an animal have we seen but a couple of seals & a band of geese. It is quite madness to think of wintering in such a place' (PAA 74.1/137).

[6] Personal belongings.

[7] Anderson put this period of detention to good use: 'From the top of a high mountain I perceived that the inlet to seaward, beyond Montreal Island, is covered apparently with unbroken ice, except a narrow lane of water leading along the eastern shore. Some drift-ice was also seen towards Elliot's Bay' (HBC B.200/a/31, f. 7).

[8] Named by Back after the city in recognition of the help and hospitality he had received there. See Back, *Narrative of the Arctic Land Expedition*, p. 399.

[9] Named by Back in honour of Captain Elliott of the Admiralty. See Back, *Narrative of the Arctic Land Expedition*, p. 403.

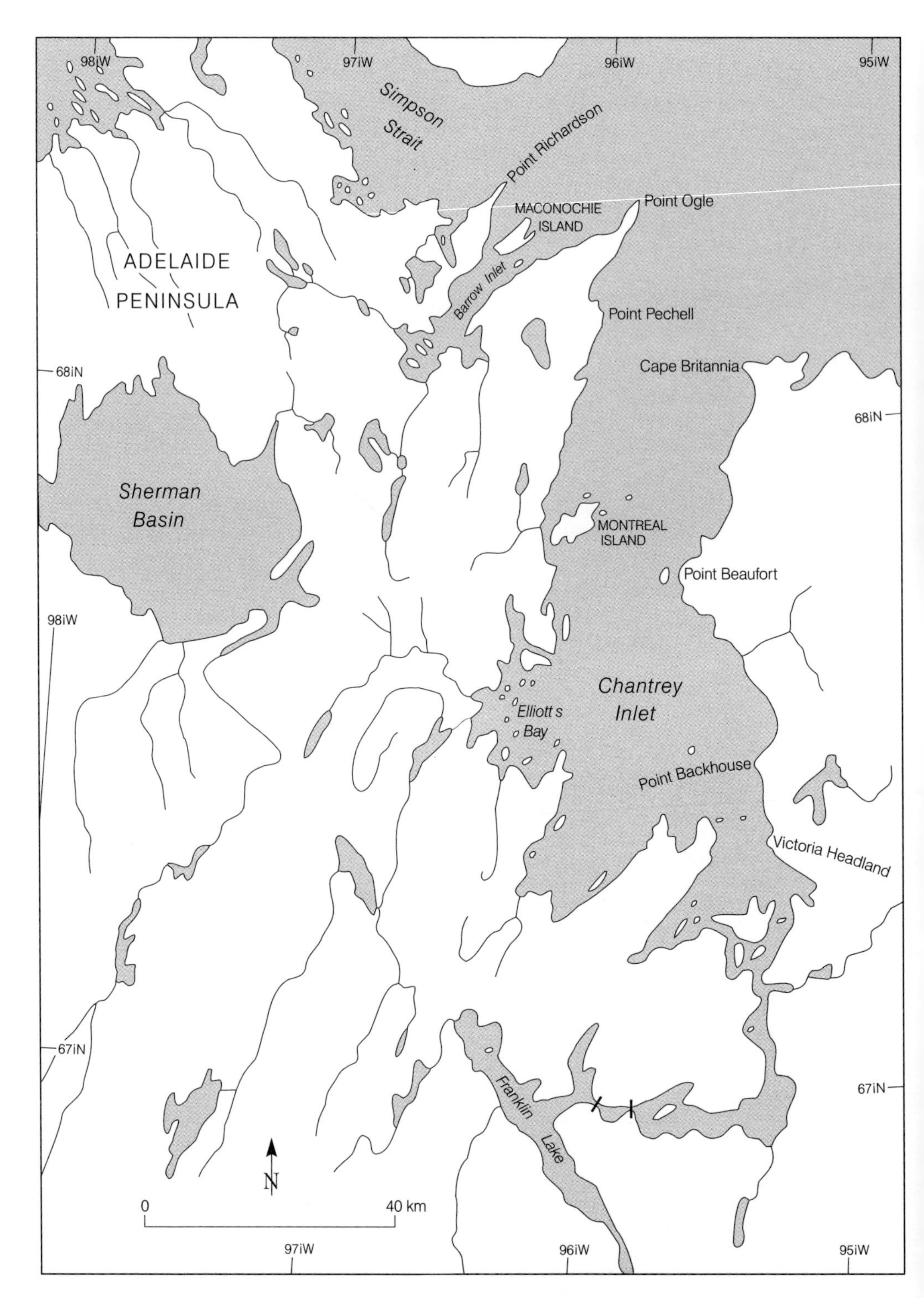

PLATE XI. Chantrey Inlet
showing the farthest point (Maconochie Island) reached by the expedition

blocked with ice. Tomorrow morning the island shall be thoroughly explored for vestiges of the missing party. Saw 2 or 3 seals, some gulls, and many Esq. ducks. A track of a deer was seen on this island.

Thurs. 2nd. The men breakfasted early and left to explore the island.[1] At mid-day we heard shots. I left immediately with Mr. Stewart. We met Bouché and Reid, who showed us sundry articles belonging to a boat, and a chip of wood with 'Erebus' upon it.[2] We then proceeded to the point where these were found, and examined all the Esq. caches,[3] most of which contained blubber and seal oil, but one of them contained a kettle (tin), and others sundry iron works, such as chain, hooks, blacksmith's cold chisel and shovel, and a bar of iron, and the hoops of butts, apparently; a piece of cane, parts of the stands of instruments, a piece of wash rod of a gun, pieces of rope, with the government mark etc., and a piece of wood with 'Mr. Stanley' cut on it (surgeon of the Erebus).[4] The search was continued till late in the

[1] It is clear from the context in both versions of Anderson's journal and from Stewart's journal that Montreal Island is meant, and not the rocky islet on which they had camped.

The search was organized methodically: 'They were divided into two parties – one going to the right, the other to the left. After making the tour of the island they were directed to spread themselves out and cross it' (HBC B.200/a/31, f. 7).

[2] There seems to be some doubt as to the name carved on this piece of wood. While in both versions of Anderson's journal and in Stewart's journal (PAA 74.1/137) it is reported as '*Erebus*', in his report to Sir George Simpson (PAA 74.1/140), Stewart reported it as '*Terror*'.

[3] Stewart deduced that the Inuit had been here shortly before from the freshness of the remains of their fires (PAA 74.1/137).

[4] Anderson expands on this description in the fair copy of his journal:

> ...we were met by two of the men (Reid and Bouché), who informed us that they had discovered the place where the boat was cut up, and confirmed it by showing pieces of plank, etc., and a chip covered partially with black paint, with the name Erebus carved on it. We immediately proceeded to the spot. It is a high rocky ridge on the N.E. extremity of the island. On it were several Esquimaux caches, and among them the spot where it was evident the boat had been cut up. It was strewn with shavings, butts of planks, evidently cut by unskilful hands; small pieces of rope, with the Queen's mark; pieces of bunting, etc. Several of the men having come up, the whole of the caches were opened; in them, besides seal-oil, a variety of blacksmith's tools, a tomahawk, a chain-hook, a piece of a bar of unwrought iron, etc., were discovered, also a bundle of pieces of wood strung together for some purpose; they were of ash, and evidently portions of snow-shoes. On one of them I discovered the name of Mr Stanley carved, the surgeon of the Erebus. Every mound was examined to discover if it were a grave, and the search most zealously carried on till dusk. The only additional things found were some pieces of hoops, parts of instruments, a piece of cane, a piece of the leather of a backgammon board, etc., but not a scrap of paper nor a human bone. The other parties had discovered nothing (HBC B.200/a/31, f. 7v).

Subsequently the fragment of a snowshoe frame with Stanley's name carved in it featured in the official depositions in connection with the Fairholme case. Mr Edward Boyd Roberts, a London furrier, testified that he had supplied Mr Stanley with a pair of snowshoes, made by Mr Rendall of the Hudson's Bay Company. Mr Roberts was shown the fragment recovered from Montreal Island, by John Barrow at the Admiralty and stated that 'I firmly believe that the fragment I have just seen formed part of the frame of one of the snow shoes furnished by me to the officers of Sir John Franklin's expedition'. See *Fairholme vs Fairholme's Trustees, Record, proof and documentary evidence. In Declarator, Court and Reckoning etc., George K. E. Fairholme, Esq. against the Trustees of Adam Fairholme, Esq., Scottish Court of Session, First Division*. (Scottish Record Office), p. 20.

Mr John Rendall, assistant warehouse-keeper at Hudson's Bay House London, testified that he had been employed to make six or eight pairs of snowshoes for the Franklin expedition. He examined the fragment recovered from Montreal Island and stated that he believed it to be part of one of the snowshoes he had made. His grounds for this identification were:

> This piece was of the same sort of wood as those which I made. The frame appeared the same size as

evening, but no traces of the graves were discovered.[1] A band of 10 deer were on the island, of which 5 were killed, 2 by E. Kippling, 1 D. McLeod, 1 J. Johnson and 1 J. Fidler – all fat bucks. Our best hunter, Mustegan, is lame. The day was beautiful, and we had an opportunity of drying everything (which was sadly needed). The whole inlet is full of ice, except to the eastward, where there appears to be some water. Wind light and from the N.E. I promised a reward of £2 to him who found the first traces of the missing party. This was divided by W. Reid and J. Bouché.

Fri. 3rd. Wind moderate, N.E. and N.W.; cold but a fine day. All hands searching for the graves, without success. A few trifling articles belonging to the ships found. Some of the adjoining islands were also examined.[2] Two deer were killed, fat bucks, as were those yesterday, by Mustegan and J. Fidler. The inlet is choked with ice, except along the E. shore.[3]

Sat. 4th. Wind moderate, varying between N.W. and N.E.; clear fine day. As the whole island is completely explored I made an attempt to get over to the Western mainland, but could not succeed. We worked along shore through the ice, along the western end of the island, till we came to nearly the narrowest part of the crossing, the whole inlet appears to be still choked with ice. We can do no more till the ice is driven out. Some Canada geese were seen yesterday; Esq. ducks, loons and plovers are pretty numerous about the island. An Arctic hare[4] was killed by one of the men.

Sun. 5th. We worked through the ice to the western shore,[5] and all hands were employed in exploring the western shore, to the south and north, but no traces of the missing party were found.[6] In the evening we worked our way through the ice[7] opposite to the north-west extremity of Montreal Island. A buck deer was shot just after we put on shore. Mr. Stewart and myself put the first balls in it, and the others afterwards finished the animal.[8] Many deer were seen by the exploring parties, but

---

the frames of those which I made. There was a hole for the cross bar in the piece shewn me, which had been made by a mortice chisel, such as I was in the habit of using. In snow shoes made by Indians this hole would have been made with a knife. The snow shoes I made were of American elm, brought from South America, and which North American natives would not possess. The snow shoes made by them are generally made of birch wood. See *Fairholme vs Fairholme Trustees*, p. 18.

[1] Stewart's interpretation of the finds was as follows:

> It is evident that the natives have been here a short time ago from their fire places being fresh; a great quantity of chips about the same place. These things may have come from the ship but from the rudder irons, the small chip with the vessel's name on it & the surgeon's name we may conclude that a party have been in this neighbourhood & have possibly perished (PAA 74.1/137).

Stewart is thus suggesting that the various finds indicated that a boat party had reached this location.

[2] The Halkett boat was used for this purpose (HBC B.200/a/31, f. 7v).

[3] Although the weather was not very cooperative ('Very cold & rain though the sun was shining now & then through the clouds') Stewart managed to get a sunshot and determined the latitude to be 67°39′21″ (PAA 74.1/137).

[4] *Lepus arcticus*.

[5] As Stewart explains this was quite an exciting crossing: '...crossed to the mainland through an opening that was closed almost as soon as we had come through' (PAA 74.1/137).

[6] This search was organized as follows: 'The party was then divided into two equal portions; one proceeded to search the coast as far as Elliott's Bay, and the other turned to the northwards. Both parties returned without discovering anything' (HBC B.200/a/31, f. 7v).

[7] This ice consisted of 'moderate-sized blocks, which, when the tide is in float, and can be pushed apart' (HBC B.200/a/31, f. 7v).

[8] Stewart claimed that he killed this animal (PAA 74.1/137).

none were killed, as I allowed no guns to be carried as we have plenty of meat. Very little fuel to be found. The shore is low with sand hills inland. Weather in general fine, but foggy in the morning; wind light from N.E. New ice was forming before we put on shore.

Mon. 6th. Near Point Pechell.[1] A beautiful calm day. We have been working through ice the whole day and have reached Pt. Pechell. Either Mr. Stewart or myself, while the other remains with the canoes, have traced the coast from Montreal Island,[2] but not a vestige of the missing party has been discovered.[3] The country in this vicinity is dotted with small ponds of water with ridges of sand and gravel and occasional immense square blocks of grey and red granite; pieces of limestone are also scattered about. Many deer, perhaps 150, were seen. We can find no fuel at our encampment, or within 5 miles of it. Previously we found a scanty supply of the fuel used by Rae.[4] The canoes were much damaged today, and I can shove them on no further; the remainder of my task must be completed on foot.[5] Some Esq. ducks,[6] with their young, loons and laughing geese,[7] with plovers, snowbirds[8] and w. grouse[9] were seen. In the clear sandy bays some whitefish were seen. Many very old Esq. encampments were seen. The entire inlet seems to be choked with ice of great thickness and solidity;[10] notwithstanding the day was warm, new ice formed after 4 p.m.[11]

Tues. 7th. Took an early breakfast and started with Mr. Stewart and all the men, except two of the Iroquois, who were left to arrange the canoes and look after the luggage. We were in light marching order. Five men followed the sinuosities of the coast, while the rest of the party swept the country further inland. For about one-third of the distance the country was intersected by small lakes, the remainder was composed of sand hills, devoid of all vegetation, and between them low valleys,

[1] Named by Back after Sir J. B. Pechell, Bart. See Back, *Narrative of the Arctic Land Expedition*, p. 407.

[2] Stewart's record of how the search was arranged was slightly different: 'Mr A. remained with the canoes & I went along the shore with 4 men as far as Pt. Pechell where Mr A. disembarked, and shortly after we encamped' (PAA 74.1/137).

[3] But they did discover numerous old Inuit camps (HBC B.200/a/31, f. 8).

[4] Probably arctic heather (*Cassiope tetragona*).

[5] Anderson's plans (and thwarted hopes) in this regard were as follows: 'I have, therefore determined on taking the Halkett boat, and to explore the remainder of the peninsula on foot. I shall also, if possible, cross over to Point Richardson and examine it; nothing more can be done. Had we had open water, there would have been no difficulty in reaching Cape Herschel, or even farther' (HBC B.200/a/31, f. 8).

[6] King eiders (*Somateria spectabilis*).

[7] White-fronted geese (*Anser albifrons*). See Back, *Narrative of the Arctic Land Expedition*, pp. 515–16. Anderson thus appears to have used two names ('grey wavies' and 'laughing geese') for the same species, namely the White-fronted goose.

[8] Snow buntings (*Plectrophenax nivalis*).

[9] Rock ptarmigan (*Lagopus mutus*).

[10] In the fair copy of his journal Anderson describes this ice in more detail: 'Not a pool of open water is visible in the inlet; it is as solid as in winter. There is no appearance of decay in the ice, but the rising and falling of the tide has split it into fields which form crevasses not exceeding a foot or two in width' (HBC B.200/a/31, f. 8).

[11] Large numbers of caribou were seen but were left in peace, since there was sufficient meat on hand, as noted by Stewart (PAA 74.1/137).

which are overflown in high tides.[1] In one place the water appears to cross the peninsula, and often nearly cuts through it.[2] If the missing parties died in one of these low spots, their bones must have been either swept away, or buried in the sand. Many very ancient Esq. encampments, but no new ones were seen. Some, perhaps four or five years old, were seen at Point Ogle; among them were found a small piece of cod-line, and a small piece of striped cotton, which were the only vestiges found. We encamped late at the point opposite Maconochie's Island.[3] A very fat buck deer was killed,[4] and a few others were seen. A little beyond Pt. Pechell we crossed a river; it must be a large stream at high water; it ran from the Sd.; I called it Le Mesurier after a relation of Mr. Stewart's. (Ground white with snow this morning.)

Wed. 8th. Early this morning 4 of our best men were ferried across in the Halkett boat and the whole of Maconochie's Is. was minutely examined, without success.[5] The wind drove in the ice so fast into the strait separating the Island from Richardson Point[6] that we were unable to cross over and examine it as I wished. The party killed another fat deer[7] on the Island and returned at 2 p.m. It then began to pour down rain, with a sharp N.E. gale and we were all thoroughly soaked when we reached our encampment about 9 p.m. The last of the party only arrived at 11 p.m. No fuel was to be had, and, of course, no fires could be lighted, so that we passed an uncomfortable night.[8] A little fuel was seen on Maconochie's Is.[9]

[1] Stewart was not greatly impressed by the appearance of Point Ogle: '... which appears to be nothing but sand. Sir Chas. Ogle is not much honoured by such a sterile spot. Found no heather at all' (PAA 74.1/137). Point Ogle was named by Back after Vice-Admiral Sir Charles Ogle. See Back, *Narrative of the Arctic Land Expedition*, p. 409.

[2] Back also noted that the tip of Point Ogle became an island at high tide.

[3] Named by Back after his friend Captain Maconochie, RN. See Back, *Narrative of the Arctic Land Expedition*, p. 423.

[4] Presumably because of shortage of fuel, the men ate most of this caribou raw, but then: 'after much trouble we managed to make a fire of sea-weed, and got a warm supper' (HBC B.200/a/31, f. 8).

[5] There would later be a very intriguing sequel to this visit to Maconochie Island. See pp. 245–8.

[6] Named by Back after Sir John Richardson. See Back, *Narrative of the Arctic Land Expedition*, p. 420.

[7] Killed by Paulet Papanakies and J. Fidler.

[8] After a supper of 'rather ancient pemmican and cold water' (HBC B.200/a/31, f. 8).

[9] Given the circumstances the men were allowed a dram: 'Not being able to make a fire a little Alcohol was given to the men to prevent any bad effects from the cold & wet. Mr A. and myself took a little also, which nearly poisoned us.' One can only hazard a guess as to what was meant by this last remark of Stewart's, who certainly was not a teetotaller. See pp. 217–18.

At this critical turning point of the expedition, Stewart indulged in an assessment of the situation:

> Thus our return is recommenced & though unsuccessful in our search we have followed out our instructions, & have the satisfaction of having done our duty. With light hearts we turn our backs on this cold, sterile region where, with our means it is quite impossible to winter: no fuel or is there a sufficiency of deer to provide a supply for a length of time. Now dearest Meg, I am returning. Keep up a little longer. With God's will we shall embrace once more (PAA 74.1/174).

Anderson assessed the results of their search as follows:

> It may appear strange to any one unacquainted with this desolate region, that not a vestige of the remains of so large a party as are said to have died here, should have been discovered. I can safely say that the whole coast between Elliot Bay and Point Ogle, and the country for some distance inland, has been most carefully searched, as well as the whole of Montreal Island, by as keen-eyed and zealous a set of men as exist; still not a human bone has been discovered. My opinion is, that a party of men, suffering from starvation, would have sought out the lowest and most sheltered spots to haul the boat out and encamp. If they died in such a spot, their bodies have doubtless been torn to pieces and scattered about by wild animals, and their bones covered many feet with sand. There are many such spots all along the west coast and on Montreal Island. Any papers would, of course,

Thurs. 9th. The rain ceased at 7 a.m.,[1] and the canoes were gummed. We started at 9 a.m., and it turned out a beautiful day so that we were able to dry our clothes partially. The ice was even worse than when we were coming.[2] One portage was made, and by dint of shoving the ice aside and cutting it we reached to within 4 miles of our encampment of the 6th at sunset.[3] New ice began to form at 4 p.m. and was thick enough to cut the canoes a little before we reached the encampment.[4]
Fri. 10th. Left our encampment at 3½ a.m. The ice was very close and cemented together with new ice so that we made slow progress and injured the canoes. We therefore breakfasted early and afterwards got on a little better. When we arrived at the strait separating Montreal Island from the West mainland the Halkett boat was launched and a small island examined, on which were some old Esq. encampments. We afterwards proceeded along the south shore of Montreal Island which we found nearly free of ice, and after examining the traverse[5] from a high mountain I determined on risking it, though the eastern land appeared to be lined with ice. We crossed with a fine breeze[6] aided by paddle, and got through the ice easily, there being large openings between the floes. The breeze increased to half a gale and we continued on till 11½ p.m. when we encamped at point Backhouse[7] shortly after which it began to rain at intervals and blow still harder. 2 seals were seen. At this point heather is pretty plentiful, but there are no traces of deer.
Sat. 11th. Unable to move. Blowing very hard between N.W. and N. all day, with squalls of rain. (Note: Ice came on again. Most fortunate we got across yesterday.)
Sun. 12th. Unable to leave the encampment before mid-day. It then lulled a little and we embarked. It was still blowing very fresh from the N.W. with a heavy sea, but we kept on[8] and encamped at sunset above our encampment of the 30th.[9] Showers of rain all day, which turned to snow in the evening. I never experienced such piercing winds as blow on this coast. All of us are in winter rig, but still chilled to the bone. No deer seen today. 3 starving wolves came close to the canoes and stole a piece of pem'n; fortunately for them the guns were wet.[10] (Note: Mountains white this evening with snow.)

---

have been soon destroyed in this climate. Leather-covered books would have been torn to pieces by wolves or foxes. Everything we can do, has been done; and it is evident, from the wretched state of the canoes, that any delay in returning up the river will compromise the safety of the party. There is not the least prospect that the ice in this inlet will break up this season (HBC B.200/a/31, f. 8).

[1] But not before turning to snow, so that the ground was white, according to Stewart (PAA 74.1/137).

[2] The ice had been driven ashore in the interim (HBC B.200/a/31. f. 8).

[3] This was opposite the north end of Montreal Island, according to Stewart (PAA 74.1/137). In the fair copy of his journal Anderson noted that no caribou were seen all day and that all the tracks were leading south (HBC B.200/a/31, f. 8).

[4] Stewart was fully aware of the hazard this posed: 'It is time we were getting home. But more haste less speed answers well through this bay. Were we to break our canoes our lives might be the penalty' (PAA 74.1/137).

[5] To Point Beaufort (PAA 74.1/137).

[6] Having set sails (HBC B.200/a/31, f. 8v).

[7] Immediately north of Victoria Headland, named by Back for his friend John Backhouse, Under-Secretary of State for Foreign Affairs. See Back, *Narrative of the Arctic Land Expedition*, p. 392.

[8] Under sail, despite high seas (PAA 74.1/137).

[9] Well within the mouth of the Back River (PAA 74.1/137).

[10] In the fair copy of his journal Anderson places this event at the previous camp (HBC B.200/a/31, f. 8v). According to Stewart, the men were throwing pieces of pemmican to the wolves. From this he draws

Mon. 13th. Left at 3 a.m. Just after embarking it began to snow and then rain heavily, and this was the case, with a slight interval, all day. Saw the Esquimaux at the rapids leading to Lake Franklin. They now numbered 3 families, consisting of 5 men, 3 women and about 12 lads and children. Endeavoured by all means in our power to find out if they had papers of any description, but they had none.[1] They showed us sundry articles got from the boat, such as tin boilers about 18 in. long by 12 in. broad; an oval frying pan; do., iron; 7 copper boilers and tin soup tureens, a (Ferriers) chisel, a fragment of a handsaw, a piece of the white metal plate of a thermometer and of an ivory rule. Most of their paddles were made out of ash, oak, pieces of mahogany, elm and pine. They made us understand that they had not seen the ships which had been wrecked, but had heard of it from others, and again showed us by signs that the crews of the vessels had died from starvation.[2] We got Esq. boots, etc. for the men and made them presents of a grafting saw each, fish spears, seal spears, knives and bags and sundry trifles for the ladies. We got a little aft wind in L. Franklin. I encamped at the head of the rapid before arriving at McKay's Peak, but Stewart below it, having broken his canoe very badly.[3] No animals whatever seen. (Note: The Esquimaux were just leaving, their fish caches were made. They were leaving for some pass to watch for deer.)
Tues. 14th. Mr. Stewart arrived at 4 a.m. and we then left. The water has fallen so much that we ascended McKay's Peak Rapid with the paddle, and an aft N.E. wind helped us on famously.[4] Encamped late, considerably above Back's encampment of 26th July. It was raining the whole day.[5] Just before encamping a fine rainbow made its appearance. A solitary starving wolf seen today.
Wed. 15th. We were all so wet and stiff that no one awoke until late. We left at 4¾ a.m. The rainbow of last night did not deceive us. The day was beautifully clear and warm, and we carried sail with a fine N.E. breeze for half the day, and made fine progress, having encamped at the Rapids below Wolf's Rapids. This fine day enabled us to dry our clothes and bedding, which were actually getting mouldy. Some of the men began to complain of rheumatism and it is not suprising. I did not take up the pemm. cached on the 27th ult. as it was rather out of the road. We have also enough, and the canoes are rather too heavy. Geese are now flying.[6] Not an

---

the conclusion: 'When wolves starve here at this season what would we do were we fools enough to winter' (PAA 74.1/137).

[1] In the fair copy of his journal Anderson elaborates on this point: 'I exposed the contents of the trading cases, and made them understand they were welcome to the whole if they would give us a book or papers. They understood us perfectly, but said they had none, and to satisfy us, opened up the whole of their caches' (HBC B.200/a/31, f. 8^{v})

[2] Stewart again makes reference to the Inuit having seen a survivor: 'They have evidently seen the boat & one at least of the unfortunate party' (PAA 74.1/137).

[3] Stewart reported this accident as follows: 'In a small rapid below McKay's Peak my canoe ran full swing on a stone & we sunk on reaching shore. Damages were however repaired before bedtime...' (PAA 74.1/137).

[4] They recovered the first cache (left on 29 July near McKay's peak) (PAA 74.1/137).

[5] As Stewart noted the wet weather had been making life miserable: 'We hope to have a little dry weather so that our clothes may lose a little of the humidity contracted since our arrival at the sea & which has continued till now, a sprinkling every day to keep us fresh' (PAA 74.1/137).

[6] I.e. they had completed their moult.

animal has been seen today, but the tracks of deer were seen both yesterday evening and today, all going to the South'd. This accounts for our seeing no deer on Adelaide peninsula on our way back.

Thurs. 16th. This has been a day among the Rapids. The canoes received much damage. In Escape Rapid Mr. Stewart's canoe was broken[1] and mine completely ungummed. Encamped at 6¼ p.m. about 6 miles above Escape Rapid. Mr. Stewart's canoe only arrived at 7½ p.m. We lost also about ¾ hour at breakfast in gumming her. A decharge was made at one strong place in Escape Rapid. Took up our cache in good order.[2] Wind fresh ahead from the S.W. Showery. Yesterday we saw a few sand flies, but today they were in clouds. Neither musk oxen nor deer seen. The geese now fly[3] so that we get no fresh provisions. 3 wolves, a few ermines and several young foxes seen. Last night the aurora was seen for the first time, faint in the South, as well as the Great Bear. Venus we saw some days since.[4]

Fri. 17th. Left at the usual hour. Rained at intervals last night and throughout the day. Mr. Stewart's canoe again broken badly in still water.[5] It was repaired at breakfast time. The river below Sinclair's Falls very shallow. A portage was of course made there[6] and the canoes gummed hastily. The remainder of the Rapids to Lake McDougall were passed safely. We encamped late at the head of the rapid. A doe r.-deer was seen today. I shot 3 white grouse (young ones); they are now ⅔ grown.

Sat 18th. It was blowing such a gale from N.E. this morning that it was impossible to leave before 10 a.m. It had then moderated a little, and though still blowing fresh with a heavy sea, we managed to reach the first rapid in the river (say 10 miles from Rock Rapids) falling into McDougall's Lake. We then hoisted sail and had a fine run for a couple of hours. We got up several small rapids and encamped at 7½ p.m. considerably above our encampment of the 24th ult. The river is now rather shoal, having fallen 10 or 12 feet. Not an animal of any kind was seen.[7] Weather showery.

Sun. 19th. Mounted all the Rapids to L. Garry without accident and encamped at the Narrows at our encampment of the 22nd ult.[8] The first part of the day was clear and calm which enabled us to dry our clothes, only to be again wetted in the evening by heavy rain.[9] Wind variable. We carried sail about 2 hours as far as the 2nd sand hill from nearly the first one. At the last long rapid coming up a decharge was made, it being shallow. Saw swamp berries[10] for the first time coming up. They

[1] Of the damage to his canoe Stewart remarked : '... & no wonder, for now that it is so strongly soaked with water & the bark being so thin the least touch makes a leak in it in spite of Baptiste's exertions' (PAA 74.1/137).

[2] Although an ermine had taken up residence in it (PAA 74.1/137).

[3] The moult had ended.

[4] These comments on the aurora and stars are a reflection of the fact that the night sky was starting to get dark for the first time.

[5] The implied criticism of Stewart in this remark, is borne out by Stewart's report of the same incident: 'Broke my canoe again & I had a row with Mr Anderson about the lading...' (PAA 74.1/137).

[6] Although the canoes were lined up empty, as they were also at Rock Rapid (PAA 74.1/137).

[7] Stewart, however, reported seeing a caribou, though admitted that 'With one yesterday makes two we have had the pleasure of seeing since we left Pt. Ogle' (PAA 74.1/137).

[8] Stewart noted that they have now gained two days on their downstream progress (PAA 74.1/137).

[9] Stewart remarked that this rain was 'to the great detriment of our kettle boiling' (PAA 74.1/137).

[10] Probably cloudberries *Rubus chamaemorus*.

were ⅔ formed. The men chose to compliment me by calling the fine sand hill in the middle of the channel connecting L. Garry with L. McDougall 'Anderson's Hill'. 10 deer were seen this evening. Took up our cache below the rapid at the end of this lake in fine order.

Mon. 20th. Heavy rain and strong gale last night from various points. It was still raining when we embarked at 3¾ a.m. It cleared up partially afterwards. At the rapid between L. Garry and Pelly we saw some Esquimaux, then only women and children when we passed on our way up and they then ran away, but now the men were there and they came to us immediately. They had various articles used by us[1] in the trade which they must have got from the Churchill Esquimaux. There were 3 lodges and 6 men (2 old, 1 middle-aged, and 3 young men); 2 of them we had previously seen at McKinley's River. There were 3 women and 6 children. I think there must be a river falling into the deep bay on the E'd of Lake Pelly. We gave them knives, spears, dags, scissors, etc., and parted famous friends. They gave us some deer meat. Encamped at sunset near the head of L. Pelly. The wind was strong ahead all day. About 25 deer were seen today all going to the Sou'd. The same is the case with the Canada geese.[2]

Tues. 21st. It was miserable weather when we embarked at 3 a.m. It was blowing hard and raining. At Bullen's River[3] we hoisted sail and carried it for about half the day. We were much incommoded by sand banks above Bullen's River. The same was the case in a minor degree when going down. The water in L. Garry and above it does not appear to have fallen so much as below it. Below and at McKinley's River we saw the same Esquimaux as when descending. There were 8 tents; about 10 men were present and 8 women and several children. The women are all of very low stature, good-looking. The young women are only tattooed after they have children. Saw several stone kettles made with 5 slabs sand stone cemented together. These Esquimaux seem a remarkably harmless, honest and clean race. Canoes and tents made of deer skins. Have many of our articles of trade. They made us understand that they came down Mckinley's River, but that it was nearly dry at present. The wind headed us towards the evening and the rain never ceased. We encamped a piece above Mckinley's River. Everything we have is now soaked with rain.[4] We have found plenty of willows since reaching L. Garry.

Wed. 22nd. Left early. 3 Esquimaux came to see us start and accompanied us a short distance.[5] Alders[6] are seen at the Hawk Rapids for the first time. The long line of

[1] I.e. by the Hudson's Bay Company.

[2] Stewart adds the information that despite a headwind and rain all day this was the best day's paddling they had achieved thus far (PAA 74.1/137).

[3] Where they retrieved their cache (PAA 74.1/137).

[4] Making life quite depressing: as Stewart pointed out: 'Everything is wet & were it not the hope of getting home we should be miserable indeed, but dearest Meg I am drawing nearer to you every day so it matters not much what weather comes. The poor men suffer a great deal from this constant wetting; they are notwithstanding in excellent spirits' (PAA 74.1/137).

[5] These Inuit had come for a specific purpose, which quite impressed Stewart: 'When we started in the morning two of them came & showed us where we had left a paddle in going down. So much for unsophisticated honesty. ...I was really ashamed of carrying guns in my belt to guard against such poor innocent creatures as that, though had the camp been larger their demeanour might have been different' (PAA 74.1/137).

[6] *Alnus crispa.*

rapids below and above Hawk Rapids were safely ascended. It then began to rain very hard and continued without cessation till at last I could not endure seeing the men suffering so much and encamped at 5½ p.m., 5 hours above Hawk Rapids, among the sand banks. Mr. Stewart's canoe cannot keep up with mine, and retards us considerably. The fact is, both canoes are now dreadfully leaky and his the worst. Some ripe berries, 'crow berries'[1] were picked; the leaf is red. Several wolves, gulls and crows were below Hawk Rapid, feasting on the drowned deer.[2] But not a deer was seen either today or yesterday. Several bands of Canada geese and grey wavies[3] going to the sou'd. At the rapids between L. Garry and Pelly and below Hawk Rapids appeared to be the only good deer passes we have seen since leaving the coast, tho' there are doubtless others. Heavy rain all night.

Thurs. 23rd. Left at 2½ a.m. 'midst drizzling. It cleared up at breakfast and enabled us to partially dry our clothes etc., but heavy showers soon wetted us again. The sun, however, shone out at intervals. I encamped at 6½ p.m. (to avoid a heavy storm which threatened us) a little below Baillie's River. The wind assisted us a little to-day and the men paddled well. But our progress was much impeded by sand bars which rendered the channel of the river most tortuous. Esquimaux marks as high up as this.

Fri. 24th. Ascended the Cascades etc, above Baillie's River; made 2 decharges. Encamped late at the sand cliff,[4] a little below our encampment of the 17th ult. For a wonder it did not rain till midday, and was positively warm when walking. It then began to rain, and we had occasional showers till evening. 4 deer and a wolf seen. Numerous flights of laughing geese going to the Sou'd. The wind helped us on after midday.

Sat. 25th. Left at 2½ a.m. Wind blowing fresh from the west, with frequent showers of rain and hail. Mr. Stewart's canoe again broken before breakfast, which retarded us a little.[5] We encamped at 9 a.m. at Beechy's Lake at the head of the Cascades. This was, of course, a complete portage. The canoes are now distressingly heavy, particularly mine. No deer seen, but about 20 musk oxen were grazing on the left of the river, below the Cascades.[6] Laughing geese going to the south. It appears that much rain has fallen about here. L. Beechy has only fallen about 6 in. Took up our cache in good order, except a bag containing some meat, which the wolves had got at and devoured. Two of these beasts were seen in the portage. The men gave two of the sand cliffs to Lockhart and Stewart.[7]

Sunday 26th. The canoes required so much gumming, etc. that it was 4¾ a.m. before we left. Strong head winds accompanied by rain and sleet prevailed all day and

[1] *Empetrum nigrum*.

[2] Although excellent swimmers and easily capable of crossing wide lakes and rivers, numbers of caribou are drowned every year when they are swept over rapids and falls while crossing the large rivers of the Barren Grounds.

[3] White-fronted geese (*Anser albifrons*).

[4] At Edward's Butte (PAA 74.1/137).

[5] Stewart noted that 'We got ashore before the canoe filled'. Further delay was caused when 'Our lines broke in Stewart's Rapids. Chafing so much on the rocky bank cuts them in a very short time' (PAA 74.1/137).

[6] The first musk-oxen seen on the trip back up the river (PAA 74.1/137).

[7] I.e. they named them after these individuals.

retarded our progress. Much water was shipped and our crazy canoes bent in with every wave. 20 or 30 musk oxen were seen,[1] but no deer. Encamped at dusk about 5 miles above the Willow Island at the head of L. Beechy. Froze hard at night.
Mon. 27th. Left at 2½ a.m. amidst rain. It cleared up in the middle of the day and we rejoiced to see the sun. Towards evening the rain recommenced. The wind, however, was favourable and helped us along considerably. Encamped when it was nearly pitch dark at the foot of the Long Rapids below where we left our canoe on the 15th ultimo. Mr. Stewart's canoe was again badly broken and he was obliged to encamp below us.[2] With this exception the long line of rapids in this day's march was ascended without accident. Two young laughing geese were killed. Some of our best shots fired at a musk ox bull from the canoe and one ball hit him apparently on the end of the spine and paralyzed his hind quarters; he, however, soon recovered and escaped. A few musk oxen were seen, and at dusk 2 or 3 deer. Many flocks of wavies flying to the Sou'd. Froze hard at night.
Tues. 28th. Detained till 5.40 waiting for Mr. Stewart; this delay is most vexatious. Everything was hard frozen this morning. The tent was as stiff as a board. Found our first cache in good order, and took all the perches, masts & yards of the canoes left there for poles.[3] The canoe was also broken up for firewood.[4] Met with no breakages today. Encamped at dusk above the rapid where Capt. Back repaired his boat and sent back his carpenters. A little snow fell before breakfast, but afterwards the day turned out beautifully fine but very cold, though the wind was fresh from the southward, which retarded us much. 7 deer were seen, but no musk oxen. Some flights of grey wavies were seen. The river is lower than on our way down, but not so low as I had anticipated.
Wed. 29th. Left at the usual hour. Just below Musk Ox Rapid a small band of deer was seen, one of which, a fat buck, was shot by Mustegan. Musk Ox Rapid was very shoal; its ascent by the canoes light and the carriage of the pieces occupied upwards of 6 hours. The canoes were completely ungummed and it took 2 hours to repair them. Encamped at dusk at the head of Musk Ox Lake. It took 1.$\frac{35}{60}$ hours to make the traverse, hard paddling. The day was the first day without rain (and beautifully clear) that we have had since leaving Point Ogle. As the men have behaved so well, and as we have hard work still before them, I have promised them each £5 in addition to their wages, and, moreover, that should their conduct be good, and if they exert themselves on their way to R.R.[5] and Nor. House,[6] that should they arrive before the expiration of their year's time, that they shall receive their whole wages as for the entire year, thus changing their terms from the year, to the trip. H. Fidler and Paulet are to get £5 extra for acting as steersmen. (A cache left by Indians 3 MBr meat).[7]

[1] And fired at unsuccessfully (PAA 74.1/137).
[2] Below Paulet's Rapids, on the opposite side of the river from Anderson and party (PAA 74.1/137).
[3] For poling upstream. Each canoe had only three poles left, the remainder having been lost or broken (HBC B.200/a/31, f. 9^{v}).
[4] This was done only after they had: 'Examined the Canoe to see if it was not better than Mr Stewart's, but found it – bad as that Canoe is – still worse' (HBC B.200/a/31, f. 9^{v}).
[5] Red River.
[6] Norway House.
[7] I.e. caribou meat left by the Indians hired to hunt for the expedition, to the value of 3 Made Beaver.

Thurs. 30th. Another fine day. The men worked splendidly. The river from the lake above Musk Ox Lake is nearly dry, and it was therefore a continuous portage interrupted only by a small pond and lake. We reached the little lake close to Sussex Lake and saw Lake Aylmer close to us. There is still a little ice on the borders of this Lake (the one nearest). Saw 3 deer. One of the men laid up with a sore foot.[1] Saw 3 rock partridges. Numerous flocks of snow geese passing south; partridges etc. Aurora Boralis faint to the Sou'd.[2]

Fri. 31st. Made 2 short portages and passed 2 ponds of water, and then a third portage brought us to lake Aylmer at 6 a.m. The canoes were thoroughly gummed and we embarked at 7. We had to contend with a strong head wind all day, which retarded our progress much. The weather was bad also, foggy in the morning and rain afterwards. Several deer were seen today, perhaps 30, and 2 were shot by Mustegan and J. Fidler, a 2 year old buck and doe. A slight deviation was made, entering a bay running to the S.E. which is not noticed in Back's map. We lost by this about 1½ hours. Encamped at dusk near the Narrows leading to Clinton Colden Lake. We have now lost the willows and are reduced to burn heath.

Sat., September 1st. A beautiful day, most of which was wasted in finding our road, and I am not quite sure if we are, as I supposed, encamped at the Straits leading to Clinton Colden Lake. Back's map makes it appear that the Strait was bounded by the southern shore, whereas a deep bay running E. and S.E. intervenes between the straits and the south shore. The map is utterly useless for such a lake as this.[3] Some very small spruce were seen at the bottom of the bay before mentioned, out of which we got by a short portage to the proper bay. Several deer were seen and 2 shot by E. Kipling and J. Fidler, the former a large fat buck, the latter one of 2 years.

Sun. 2nd. Blowing a gale from the N. and N.E. We were compelled to pull down the tent during the night to prevent its being carried away. The wind abated about 3 p.m., but we did not leave as I had sent off Mustegan to see if this was the right Strait. He returned late with the information that it was. Immense flocks of Canada geese passing all last night and today. The men arranged all their little affairs and dried the contents of their bags (Mustegan killed another buck). Froze sharply.[4]

[1] In the fair copy of his journal Anderson comments: 'This is decidedly the most severe day's work I have ever seen gone through by men since I have been in the Country' (HBC B.200/a/31, f. 10). Considering Anderson's decades of canoe travel, this is a very telling remark, as regards the difficulties of the Back River route.

[2] Stewart cannot refrain from gloating that: 'Tomorrow we hope to be afloat in Lake Aylmer, G.W., having accomplished our trip in far less time than Captn. Back' (PAA 74.1/137).

[3] Stewart noted that Back, too had great difficulty here, although he had a guide (PAA 74.1/137).

[4] This day of inactivity gave Stewart the opportunity for some philosophizing about the trip:

Still on our way home & getting so near to our own kind after a separation which at first was doubtful whether it would ever be concluded, makes delay to be felt more keenly than otherwise. It is now approaching the end of the ninth month since I left my dear wife, uncertain whether I should ever again meet her in this world, doubtful how she would get through an arduous journey to Red River, trusting though in Providence who is so kind, to strengthen her sufficiently for the difficulties of her situation & now that I am approaching the place where we hope to receive letters no wonder that anxiety in their contents should weigh my spirits down. Our journey hitherto has been most [illegible] dangers at every step through a rugged, uninhabited, cheerless country have been passed by us without any serious accident & here we are within [illegible] of our depot, all in good

Mon. 3rd. Reached Clinton Colden Lake at breakfast (7 a.m.). Snow in the morning. Froze sharply. The wind was blowing a gale from the E. and SE., which rendered it impossible to take the traverse to the first point.
Tues. 4th. Detained by the gale till 6 p.m. It then subsided a little and we took the traverse. Shipped much water and our rickety canoes were sadly strained. Encamped at 11 p.m. at the first large island. Ignace killed a buck.
Wed. 5th. Started at 2½ a.m., having just laid down on the beach till we could see. Fog till after breakfast. The whole day has been spent in looking for the river. Back's small map is a snare and a delusion. We are encamped on what we suppose (for the third time) to be the Straits.[1] Many deer seen and some snow geese.
Thurs. 6th. Detained by dense fog till 9 a.m. We then left, tho' the fog was still thick. This compelled us to round several extensive bays. We at length fell on a strait with some current in it, and encamped at dusk near the rapid leading to Artillery Lake. I trust never to be guided by such a map as I have again. Many deer seen, bucks; 3 very fat ones were killed, 2 by J. Fidler and one by Ignace. Traces of Indians were seen.
Fri. 7th. Ran the first two rapids but lowered down part of the 3rd, the water being so low that there was a small fall. Saw very small pines[2] above the 2nd rapid, and afterwards they increased in size, till about half way in Artillery Lake they became a respectable size. A good many deer seen about the rapids, but not many in this lake. The majority of those seen today, does with young. We had strong head wind till about 2 p.m., after which we carried sail with the paddle for about 1½ hours, and it then fell calm. Last night A.B. mod.[3] every where in irregular patches. We encamped at dark opposite what I suppose to be the island called the Rat Lodge.[4] Left a notice for the boats[5] in case we missed them at the Narrows below the Rapids.

---

health, far sooner than anybody ever dreamt of. How thankful ought we to be for God's infinite mercies, his protection which he has vouchsafed to us, from [illegible] many times each day; how often one stroke of a paddle was between us and death, how often have our frail barks been within an ace of being crushed by ice or smashed to atoms on rocks at the mercy of an [illegible] torrent, or precipitated into the various Falls we have passed. The lord of the winds & waves is benificent & merciful to us miserable sinners but we return his benefits by obedience to his will & commandments. May we, dearest Meg, both be improved in mind by this separation & may we remember with gratitude the [illegible] in me during this voyage. May God in his mercy continue his benefits [illegible]; that we may have a speedy meeting is my fondest prayer (PAA 74.1/137).

[1] Stewart found the day equally frustrating, if not more so:

The whole day was spent in pointless searching for the straits. No sun to take observations by & the map is as good as next to nothing. We are going strictly by compass. We are at present in a deep bay about 2 or 3 miles SE of our previous encampment. This delay is much annoying in a place where we expected our troubles were at an end. 'L'homme propose et Dieu dispose' is truly verified in our case. This paddling about in bays without advancing is most harassing to the men who are very easily put out but we hope tomorrow may bring us better fortune (PAA 74.1/137).

[2] In the fair copy of his journal Anderson identifies these trees more accurately as spruce (HBC B.200/a/31, f. 10v).

[3] Aurora borealis moderate.

[4] It would appear that Anderson had confused 'Rat Lodge' and 'Beaver Lodge' (see below). Both are conspicuous, isolated hills on opposite shores of Artillery Lake, Beaver Lodge being the larger, i.e. by appropriate analogy to the relative sizes of a beaver and a 'lodge' muskrat push-up. Beaver Lodge lies exactly on the 63°W parallel, on the east side of the lake, Rat Lodge lies some 14 km to the south-west of Beaver Lodge, on the west side of the lake. For further elaboration, see: Richard Galaburri, 'The Rat Lodge revisited', *Arctic*, 44 (3), 1991, pp. 257–8.

[5] The boats which Lockhart was to bring up from Fort Reliance.

We took the west shore at the 3 first islands marked in the map (4). There is a pretty deep bay running to the W. and S.W. not noticed in the map.[1]

Sat. 8th. Blew a N.W. gale last night, with rain. The waves were beating so high against the shore that we made a portage to a small bay, and after shipping much water, and working our crazy canoes much, we succeeded in crossing to the Beaver Lodge.[2] The wind increased with rain and snow till at last we were compelled to encamp at 7 a.m. at a bay on the west shore where we afterwards discovered Capt. Back built his boats. We were detained here all day.

Sun. 9th. The ground was white with snow this morning and still blowing a heavy N.Wer, with drizzling rain and snow. Took an early breakfast and left at 5 a.m. under double-reefed sail. Kept along the E. shore until we fell in with some Indian lodges, and the rest of the day was spent in trying to find out their road to Fond du Lac des Esclaves.[3] We made a move in the evening to the bottom of a deep, narrow bay. On the E. shore were 10 Indian canoes, several Indian lodges, or at least the poles of lodges, and remains of deer. The Indians had left this about a fortnight since.[4] We are not yet quite sure of our road. It is really too bad that Indians have not been sent to meet us.[5] Numerous flocks of geese and wavies passing; 4 kinds of berries ripe. (Note: Larch[6] found in this bay).

Mon. 10th. Ground white with snow this morning. Very early this morning Mustegan and J. Fidler went to find the straightest road to the height of land. Returned, and after breakfast (8 a.m.) we began the portages, and before night had passed 8 lakes and had made 8 portages[7] – 5 miles, all in a SW direction. The lake we are on is rather larger and the rivers connecting the lakes are more considerable. The last one was navigable. Men much fatigued. Wood increased in size and quantity as we approached Slave Lake. Birch fit for axe handles to be found. Very few deer tracks.

Tues. 11th. Left early. Fell on an Indian track. On entering the river made 2 portages equal to ½ mile and shortly after got sight of Slave Lake. The river was here larger, but shallow and interrupted by several falls and cascades. The pieces were carried straight to the mouth of the river (5 miles) and the canoes were brought down (light) the river partly by water and several bad portages. They did not succeed in reaching the pieces. Mr. Stewart and I crossed the river by wading at a rapid and

[1] Twice during the day (at the breakfast-spot on first reaching Artillery Lake, and again part way down the lake), Anderson ordered fires lit on hilltops in case there were Indians in the vicinity, but with no effect. This was very disappointing since 'we are at a loss which road to take. The Ahtudesy [now Lockhart River] is reported by Captn. Back to be impassable & the other road we know little or nothing about... A pity that our people are not aware we are so close to them; we should soon have guides enough' (PAA 74.1/137).

[2] See p. 146, n. 4.

[3] The head of Great Slave Lake.

[4] This is the north end of the portage route down to Fort Reliance, now known as Pike's Portage.

[5] Stewart expressed his disappointment a little more fully:

> ... the people belonging to [the Fort] appear to have forgotten us entirely. It is rather curious that no marks or Indians are hereabouts to meet us. Rather too bad to be thus deserted by our friends when we expected our troubles were over. My dearest Meg, I can have no patience with them. Anxious are we all to get our letters, to have news of those nearest & dearest to us. We shall, I trust find our way with God's help yet after coming so far (PAA 74.1/137).

[6] Tamarack (*Larix laricina*).

[7] Each portage involved two trips (HBC B.200/a/31, f. 10^{v}).

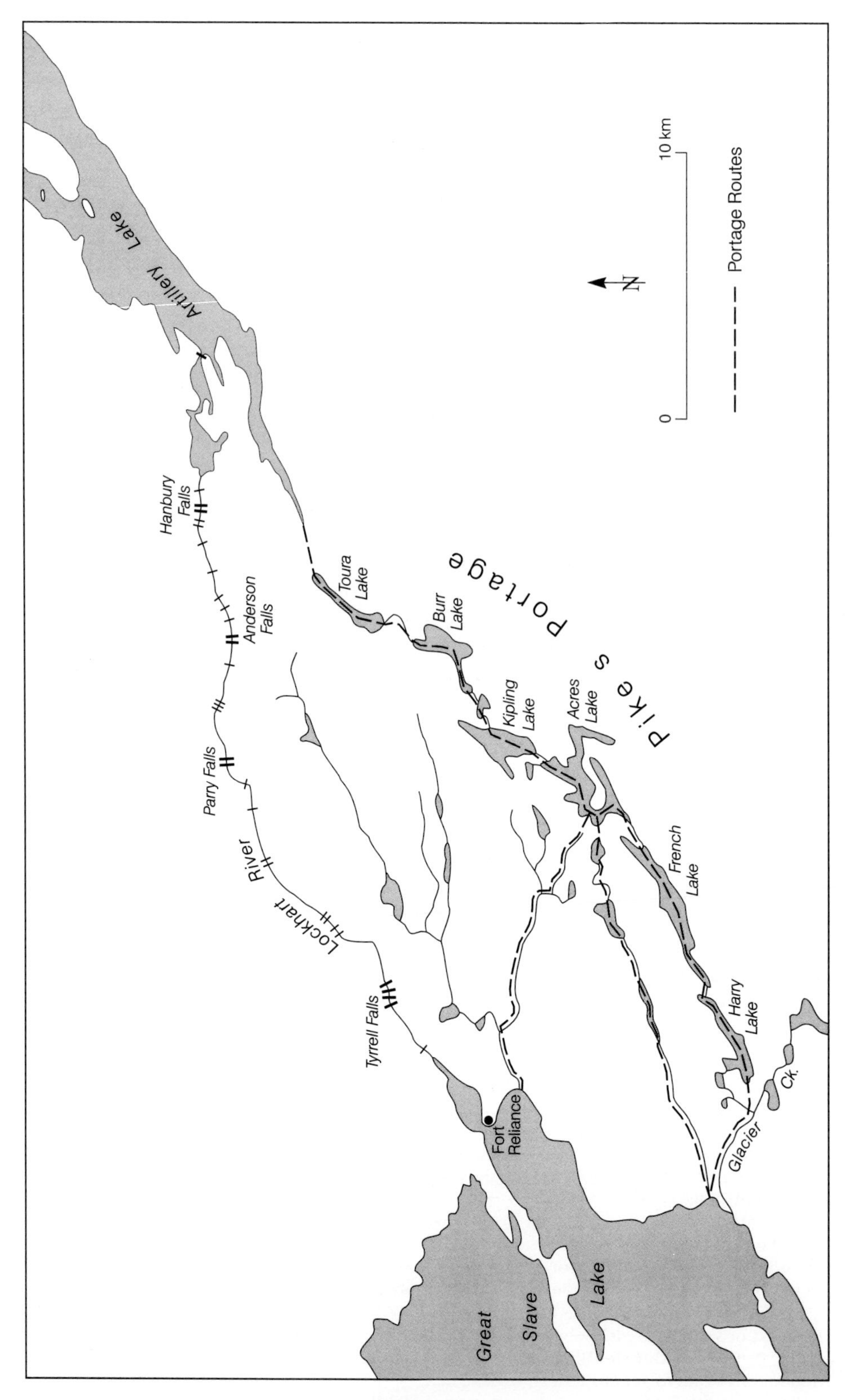

PLATE XII. The Lockhart River, Pike's Portage, Artillery Lake and the site of Fort Reliance

found the Fort. It is built on the old site of Fort Reliance,[1] but on a much smaller scale. We slept here.[2] Mr. Lockhart left yesterday with the 2 boats;[3] he is not far off, as there is a long portage to make, and I sent off immediately to tell him to return.[4] A whiskey jack[5] was seen when we breakfasted, the first for many weeks.

Wed. 12th. The canoes arrived at 10 a.m. After arranging matters I left at 2 p.m. with my canoe,[6] Mr. Stewart remaining to meet Mr. Lockhart. Carried sail to the Point, but afterwards had the wind strong nearly ahead. Encamped late a little beyond Hoar Frost River. A.B.[7] faint.

Thurs. 13th. Left early. Met an Indian about 7 a.m. who gave us some fat meat. We afterwards saw some women and children. Arrived at the Mountain Portage about 3 p.m. Put ashore in the bay beyond it to await the arrival of the 2 canoes from Simpson, which were seen under sail. Took one of these canoes and sent off the other with two crews at 4 p.m. to meet Messrs. Stewart and Lockhart. Gummed and arranged the canoe and proceeded about 6 miles beyond the Mountain Portage at a little river. Wind strong ahead since midday. Shipped much water. These 2 canoes are the proper length, but too narrow and low. The gunwales also are too weak by far.

Fri. 14th. Left early. Met some Indians going to Resolution about 7 a.m. Encamped about sunset at the beginning of the Cut Rocks leading to Pipe Stone Point (opposite the mouth of the large bay). Weather rather cloudy with some light showers. Nearly dead calm and consequently a splendid day for paddling. A great many wavies passing tonight. A.B. superb in streamers and rays of all colours, very active about 11 p.m. Rays apparently descended to within 100 ft. of the water.

Sat. 15th. Left early.[8] Wind rather strong ahead. Encamped at dusk among the Islands where we dined on the 27th June. A.B. very active, same as yesterday. Saw few small poplars before encamping.

Sun. 16th. Left at 2½ a.m. Nearly calm. Men paddled very hard, smoked once and

[1] Built by Alexander McLeod in the autumn of 1833 for George Back's expedition which spent two winters there. For further details see p. 153, n.5.

[2] Having returned to civilization (relatively) Stewart gave vent to his feelings: 'I trust it shall never more be my lot to revisit the Great Fish River. Now for home, dear Meggy. With God's blessing I shall yet be home this winter' (PAA 74.1/137).

[3] The fact that the two parties missed each other can probably be explained by the fact that in its lower (south-western) section, there are two variants of the portage route. The one probably followed by Anderson and Stewart strikes almost straight west from Burr Lake over a low divide to the head of a river (with several lakes) which debouches into Great Slave Lake just over 1 km south of Fort Reliance. Lockhart, meanwhile, had taken the slightly longer, but easier and more heavily used, variant, i.e. from a starting point on the shore of Great Slave Lake about 13 km south of Fort Reliance, up the initial portage (about 5 km) parallel to Glacier Creek to Harry Lake then via French Lake, Acres Lake, Kipling Lake, Burr Lake, Toura Lake and an unnamed lake to Artillery Lake.

[4] Lockhart was to have cached the boats (and the supplies in them) at Sussex Lake for use on a possible continuation of the expedition the following year. Anderson now recalled Lockhart 'as there is nothing to justify me in incurring any further expense' (HBC B.200/a/31, f. 10^{v}).

[5] Canada jay (*Perisoreus canadensis*), an archetypical species of the boreal forest, which the party would certainly not have encountered on the tundra.

[6] To intercept other canoes, on their way from Fort Simpson by prearrangement (HBC B.200/a/31, f. 10^{v}).

[7] Aurora borealis.

[8] In the fair copy of his journal Anderson noted that it was a fine, calm day, and very observantly comments: 'What a difference between the climate of this Lake and Artillery Lake!!' (HBC B.200/a/31, f. 11).

arrived at the last Cariboo Island 9 a.m. Wind a little stronger ahead; took the traverse straight to Stoney Island 3¾ hours (Mustegan never once missed his way through this labyrinth of islands). Supped at 6 and arrived at Resolution at 9 a.m. Men quite fresh.
17th. Resolution.
18th. Res'n.
19th. Res'n.
20th. Res'n.
21st. Res'n. Stewart and Lockhart with one canoe arrived at 1 p.m.[1]
22nd. Equipped men. Remainder expedition men arrived.[2]
23rd. Resn.[3]
Mon 24th. Sent off 2 boats to Simpson. I would now start, but would have to pay off some Indians. Pack up the remainder of the Expedition goods, and, if possible, await the arrival of despatches per 'A' boat[4] which should now be here.
Tues 25th.
Wed. 26th.
Thurs. 27th.[5]

[1] Stewart and Lockhart had eight men with them; they had left Fort Reliance on the 14th after closing up the buildings (HBC B.200/a/31, f. 11).
[2] In the two small boats which Lockhart had taken to Fort Reliance (HBC B.200/a/31, f. 11).
[3] Anderson made the necessary arrangements for Stewart to take his dispatches for Sir George Simpson by express canoe:

> Delivered the Dispatches to Mr Stewart and the party left about 10 a.m. The Canoes have nothing in them but Provisions. The crews are excellent – 9 men in one canoe and 8 in the other – and I fully anticipate that they will reach Isle à la Crosse by open water, and if the autumn be a mild one they may probably even reach Cumberland. A Boat and Canoe laden with Expedition property arrived (HBC B.200/a/31, f. 11).

[4] The Athabasca boat (HBC B.200/a/31, f. 11).
[5] 'The Athabasca Boat arrived' (HBC B.200/a/31, f. 11).

CHAPTER VI

JAMES LOCKHART'S SUPPORT EXPEDITION TO FORT RELIANCE

INTRODUCTION

James Lockhart's mandate was quite a challenging one: he was to erect winter quarters at Fort Reliance (Back's old winter quarters) at the head of the East Arm of Great Slave Lake, in case Anderson decided to prolong his expedition to a second season of searching, and was to take two boats with supplies from Great Slave Lake north to the headwaters of the Back River, where he was to leave them in case Anderson, Stewart and party ran into difficulties and had to retreat up the river in a desperate state. To achieve these ends he was supplied initially with one canoe and four men (three voyageurs and King Beaulieu).

Lockhart and party left Fort Resolution in the early hours of 3 July and reached Fort Reliance on the 11th. Here Lockhart found Back's buildings in a ruinous and rotting state, but decided to build his winter quarters on the same site since the fire-places and chimneys, at least, were still relatively intact and could be re-used. Work on the buildings began immediately.

On 19 July Lockhart set out by canoe to reconnoitre the route north to Artillery Lake via the river draining that lake (now named after Lockhart). The latter consists of a spectacular series of rapids, canyons and falls, the most spectacular of which are Parry and Anderson falls (named by Back). Lockhart stuck to his guns, and although forced to leave his canoe part way, he persevered all the way to Artillery Lake. He sensibly concluded that 'it would be madness to attempt to bring up our little boats to Artillery Lake by that Route'.

Returning to Fort Reliance and obtaining the services of an Indian guide, Lockhart next tried the traditional Indian route via a series of lakes and easy portages from the head of Great Slave Lake to Artillery Lake, now known as Pike's Portage. This time he reached Artillery Lake with ease and was back at Fort Reliance within 31 hours of starting out.

Leaving the men to continue with the building operations at Fort Reliance, on 30 July Lockhart started back west by canoe for Fort Resolution to fetch a boatload of supplies which were to be delivered there by the Portage la Loche brigade from Norway House. He reached Fort Resolution on 7 August and, starting back east in a

PLATE XIII. James Lockhart
(Hudson's Bay Company Archives)

boat which he described as a 'leaky old tub' on the 15th, he reached Fort Reliance again on the 23rd. Here the caribou had arrived on their annual migration, and the main emphasis was now on hunting, in order to lay in a supply of pemmican and dried meat in case Anderson and Stewart and party decided to winter.

The two boats which were to be relayed north to the headwaters of the Back River and were to be cached there, and which Robert Campbell had had made at Fort Liard, reached Fort Reliance on 7 September. On the 10th Lockhart started north with them over the Pike's Portage route to Artillery Lake. The very next day Anderson and Stewart reached Fort Reliance, having somehow missed Lockhart and party, presumably by taking an alternate variant of the portage route. Anderson immediately sent a messenger after Lockhart and by the evening of the 13th the latter was back at Fort Reliance. On the 14th he started back down the lake towards Fort Resolution.

Thus, due to circumstances, all of Lockhart's efforts were to no avail. But he had acquitted himself flawlessly and he certainly merited Anderson's laudatory assessment of his performance.

DOCUMENTS

1. Journal of James Lockhart, 2 July–14 September 1855.[1]

July 2nd, 1855. By the arrival of the McKenzie's River Brigade[2] yesterday we were put in possession of the tools & the material necessary for erecting the required buildings at Fond du Lac. This evening we made a start, encamping about a mile and a half from the Fort.[3] We are however very deeply laden, notwithstanding we left all our clothing and provisions except absolute necessaries. I fear we will have great difficulty in getting along. We are badly manned also, there being only myself and the four Expedition men (E. Kippling, Alexr. Landrie, Ambroise Jobin and King Bolieu[4] Intr.) who were left at Resolution,[5] to work the canoe along.

July 3, Tuesday. Started this morning at 2 o'clock; quite calm. After breakfast, however, the wind rose fair; we immediately hoisted sail and went a considerable distance. About noon the wind veered round to the north, which was right ahead. We paddled away but what could five men do against a strong head wind? so strong that we were unable to get farther than the little Rocky Island. We find evidences of the expedition people[6] having been windbound here also: bird's feathers & fish bones, also some net sinks with deerskin strings still on them. Indians are not so extravagant as to leave strings of deerskin on the stones. We have followed their example and put down a net.

July 4, Wednesday. It blew very hard the whole day. Towards evening it abated a little

[1] HBC B.180/a/1, 1855. The original title is *Journal of the Transactions of the Fort Reliance Party during summer 1855.*

[2] From Portage la Loche.

[3] Fort Resolution.

[4] François Beaulieu's son.

[5] I.e. left by Anderson and Stewart.

[6] Anderson and Stewart had been windbound here on 23 June.

and being heartily tired of our little Island we embarked and set out at 7 p.m. From that time till midnight we battled with wind and waves, when we arrived at Rocky Point. We had intended to make the traverse, but it continued to blow so hard that it would have been madness to attempt it.[1]

July 5th, Thursday. It continued to blow hard till about noon. It then calmed down to such a degree that we were enabled to make a start; crossed over to the Islands. We encamped at the little Straits. We put down two nets.

July 6 Friday. Started at 2 a.m. At 9 we were obliged to put ashore being unable to make any headway against the wind. At 2 p.m. we again set out and went on till past 9. We were very fortunate with our nets last night, catching about twenty jack,[2] two inconnus and one white fish.[3] The afternoon we came suddenly upon a camp of Yellow Knives, who as soon as they saw us coming ran about like lamp lighters, hiding a large quantity of meat & fish which was hanging to dry on poles outside the lodges; the interesting operation of 'caching' was witnessed by us with no very friendly feelings. We saw it was no use putting ashore there so we showered a few not very orthodox blessings on their eyes, skins & buttons and went on our way.

July 7th Saturday. Started this morning at 3 a.m. Arrived at Stone pipe or rather Pipe stone Point at 8.30 p.m. Head wind all day as usual, but not so strong as to make us put ashore altho' very near it. This is the first day since we left Resolution that we have been able to travel all day.

July 8th Sunday. Could not leave our encampment before 7 a.m., the weather was so stormy. Shortly after starting also we were obliged to put ashore again to gum our canoe, she was so confoundedly leaky. So many detentions I declare would weary the patience of Job. Passed Tal-thelleh Straits. On the first point after we had news from our friends in the Expedition, in the shape of a large flat stone on which was marked:

ASE
29th 6/55
JA & JGS

being their 7th day from Fort Resolution. We have encamped on a Rocky Point which might with great propriety be named Blackfly point; they are in myriads.

July 9th, Monday. Left our encampment at 3 a.m. At 10 we came to a small creek where we saw a number of trout playing about and darting past us into the lake. We immediately put ashore and tied the floats and sinks on a net and stretched it across the mouth of the creek from side to side. So much time was occupied with all these preparations that we were afraid that all the fish had escaped but we threw in small stones up above and in the space of about ¼ of an hour we took 14 splendid trout. At the end of the first pipe after dinner I found another stone with writing on it in a very prominent point. The stone had fallen and the rain had nearly obliterated the writing. I could however make out: 'Go to the Mountain Portage in the Bay for .....' the remainder was illegible. Now where was this portage; no one in the canoe could

[1] Anderson and Stewart had also been windbound here on 24–7 June.
[2] Northern pike (*Esox lucius*).
[3] Lake whitefish (*Coregonus clupeaformis*).

tell. The only man who knew anything about this part of the country was the Interpreter.[1] He said we had passed the Portage some distance. I therefore determined to proceed on our journey and trust to luck to reveal the rest. I embarked the stone also and tried the whole afternoon to decipher it without success. The wind rose this afternoon so strong that we with difficulty could hold our own. We encamped late, all thoroughly fatigued.

July 10th. This morning the weather was quite calm. Started at 3 a.m. At mid-day we overtook some Yellow Knives who told us that they had seen a letter at the Mountain Portage which they could not read. Here was the key to the writing on the stone: they have made a Portage and proceeded by some other route, leaving the letter which probably contains instructions which may materially affect our future movements. Thus reasoning I determined to proceed no further till I had seen the letter. We accordingly put ashore and sent King[2] & a small Indian boy who had seen the letter, back for it, directing King to go day & night and not stop till I had the letter in my hand.

The remainder of us went out hunting. We returned with a few ducks & partridges.[3] Landrie saw several Reindeer[4] but killed none. We have just been making grand preparations for a chase after them tomorrow morning.

July 11th, Wednesday. At 6 o'clock a.m. King made his appearance at the encampment with the letter which, having read, we immediately set out. About mid-day the wind became fair for a wonder. We made the Portage and arrived at Old Fort Reliance[5] at 5 p.m. I lost no time but set out at once with Kippling to examine the resources of the place as regard Building materials & firewood. It is a settled point to build on the old site inasmuch as the chimneys are nearly perfect; at least very

[1] King Beaulieu.

[2] King Beaulieu.

[3] Willow ptarmigan (*Lagopus lagopus*).

[4] Caribou (*Rangifer tarandus groenlandicus*).

[5] The original Fort Reliance had been built (at Captain Back's request) by Alexander McLeod at the extreme eastern end of the East Arm of Great Slave Lake in the autumn of 1833. Its location was well described by Back:

> The site of our intended dwelling was a level bank of gravel and sand, covered with reindeer moss, shrubs and trees, and looking more like a park than part of an American forest. It formed the northern extremity of a bay, from twelve to fifteen miles long, and of a breadth varying from three to five miles, named after my friend Mr. McLeod. The Ah-hel-dessy [now Lockhart River] fell into it from the westward, and the small river previously mentioned from the eastward. Granitic hills, or mountains, as the Indians term them, of grey and flesh-coloured felspar, quartz, and in some places large plates of mica, surrounded the bay, and attained an altitude of from five to fifteen hundred feet, which, however, instead of sheltering us, rather acted as a conductor for the wind between E.S.E. and W.S.W. which occasionally blew with great violence. The long sand banks, which ran out between the two rivers, and the snug nooks along the shores, seemed to offer a safe retreat for the white fish during their spawning season, which was now at hand (Back, *Narrative of the Arctic Land Expedition*, p. 190).

Back's description of the original Fort Reliance is as follows:

> ...the house ... like most of those in this country, was constructed of a framework, fitted up with logs let into grooves, and closely plastered with a cement composed of common clay and sand. The roof was formed of a number of single slabs, extending slantingly from the ridge pole to the eaves; and the whole was rendered tolerably tight by a mixture of dry grass, clay and sand, which was beat down between the slabs, and subsequently coated over with a thin layer of mud. The house was fifty feet long and thirty broad; having four separate rooms, with a spacious hall in the centre for

little labour will make them quite good. There is also a large quantity of old wood which will be serviceable and also clay.[1] Raining very hard tonight.

July 12th, Thursday. All hands employed today handling tools,[2] mounting the Grindstone, grinding tools etc., etc. Gave a Yellow Knife 8 meat skins in debt[3] and with difficulty prevailed on him to lend us his canoe to fish for the summer for a shirt. Gave Capot Rouge's son 10 meat measures in debt. He and his father were very obliging to us in the matter of the letter at the Mountain Portage, lending us himself & canoe when the other Indian refused to go. Raining again tonight.

July 13th, Friday. We set to work in earnest early this morning clearing away the clay, stones and Rotten wood from the foundations. Cut four posts ten feet long & squared 8 feet above & 2 under ground, grooved them, morticed & filled the front plate on the four posts. Jobin has got two nets in the water tonight.

July 14th Saturday. We were all at work early and continued to work hard notwithstanding occasional showers of rain throughout the day. In the evening two Yellow Knives arrived bringing some fresh meat which we traded and made an excellent supper. Our net gave us 7 large white fish and 2 Trout.

15th Sunday. Rested from all our labours. Weather changeable with occasional showers of rain.

July 16th, Monday. Were at work before four a.m. but about two hours afterwards it began to rain and has rained almost without intermission all day. Began lacing nets in our lodge[4] being unable to work outside for the rain.

July 17th Tuesday. All hands have worked hard today, having completed the walls of three of our buildings.

July 18th Wednesday. Landrie, Jobin & King employed cutting wood for walls & roofing of mens house. I and Kippling employed cutting roofing in the forenoon; in the afternoon arranging a canoe etc. for a start tomorrow in accordance with instructions received from Mr. Anderson in the letter left at the Mountain Portage. Old Capitaine arrived this evening bringing the wife of Jambe de Bois the expedition guide. In accordance with my instructions I gave her and the wife of King orders to

---

the reception and accommodation of the Indians. Each of the rooms had a fireplace and a rude chimney, which, save that it suffered a fair proportion of the smoke to descend into the room, answered tolerably well. A diminutive apology for a room, neither wind nor water tight, was attached to the hall, and dignified with the name of a kitchen. The men's houses, forming the west side of what was intended to be a square, but which, like many other squares, was never finished, completed our building (Back, *Narrative of the Arctic Land Expedition*, pp. 205–6).

An observatory (12′ by 12′ internal dimensions), built entirely without any iron components, was located about 100 feet from the east end of the house.

[1] Back's party had abandoned Fort Reliance in the spring of 1835; in view of the fact that only twenty years had elapsed by the time Lockhart's party arrived, one might be surprised at the fact that they clearly found little more than ruins. This becomes somewhat more understandable, however when one considers Back's description of the place on his return from his trip down the Back River after an absence of less than four months: "The house was standing, but that was all; for it inclined fearfuly to the west, and the mud used for plastering had been washed away by the rain. The observatory was in little better state…" (Ibid., p. 453).

[2] Fitting handles to tools.

[3] Lockhart gave him ammunition and provisions to the value of 8 Made Beaver, in return for a commitment to bring in a certain quantity of caribou meat.

[4] I.e. tent.

make nets but she would not remain with us at all & King's wife says she cannot. The nets must as a matter of course, stand[1] till some Indian women arrive for the men have no time to lace except when they cannot work outside.

July 19th Thursday. Set the men to work to cover the store. After breakfast I set out with George Kippling and an Indian lad to ascend the River running into Great Slave Lake at Fort Reliance. We toiled & towed up what we soon saw was one continual Rapid for two hours when we made a small portage, after which we proceeded on for a short distance. We then came to what if I mistake not, Back calls Parry's Falls consisting of several splendid falls in immediate succession.[2] We landed on the N.W. side first & succeeded not without extraordinary exertions, in reaching the head of the Falls, but where we could barely walk light, it would be foolishness attempting to carry across our Canoes. We therefore returned and crossed over to the opposite side, where we found after again going up to the head of the falls, that boats might be hauled over if well manned. After attempting to carry our Canoe which, being old and frail and by far too heavy for us to carry so far we hauled her up and covered her up with pine brush. We then had dinner, after which we started to walk up the River, each man carrying his provisions, blankets & Gun etc. We found deer-paths very numerous; they in many places assisted us greatly in our ascent and descent of the mountains. It is quite useless to begin to enumerate all the Rapids & Portages that we saw; for it very soon became evident to us all that no boat could be made by any possibility to ascend this River. We however, out of curiosity, followed it in all its windings & turnings till we arrived in a dreadful Shower of rain at what Back calls Anderson's Fall. We succeeded in forming a shelter of pine trees for the night. The Black flies have nearly driven us mad today.

July 20th Friday. Left our cheerless encampment at 4 a.m. All that we saw yesterday was mere child's play to what we witnessed today. Suffice it to say we followed up all its windings to Artillery Lake where we encamped. We called a council and were unanimous that it would be madness to attempt to bring up our little boats to Artillery Lake by that Route. Another must be found or else follow the track of the Expedition canoes over the Mountain Portage, and there being twenty-five long & short Portages on that route is not a very encouraging prospect. We began to question our young Sageh about what he knew of the Indian route, but he being a fool and did not understand us and we could not understand him. We could arrive at no very satisfactory conclusion. We saw a few fresh deer tracks today but did not eat any of the meat. Rather cold & windy, not bother from flies.

July 21st Saturday. Left our encampment at ½ past 2 a.m. and returned straight across country to our Canoe. The whole face of the country appeared covered with small lakes, which gave us a great deal of trouble at times. We arrived at our Fort again at ½ past 3 p.m.

[1] I.e. making the nets must wait.

[2] In its 30 km course from Artillery Lake to Great Slave Lake the Lockhart River drops some 200 m, in places flowing through rock-cut gorges up to 60 m deep. At Parry Falls, the most spectacular fall on the Lockhart, the river drops over a vertical fall 25 m high into the head of a gorge with a wall of granite 50 m high on one side and 110 m high on the other. In short, this section of the Lockhart presents one of the most spectacular flights of falls and rapids to be found anywhere in Canada.

We have, I find, lost several days by not being able to speak to our Indian at Artillery Lake; he knows a road by a chain of lakes in which the Portages are good and not many of them.[1] George & I (d.v.) will start again on Monday to examine it altho' our feet are dreadfully lacerated from walking among the rocks without socks. Our people have done very little since I left owing to the rain. The Fishing is a total failure, only three having been taken in my absence. The store has been covered and a door put in it in a fashion. Everything else as we left it.

July 22nd Sunday. Rested till the evening when 8 Yellow Knives arrived with meat, tongues & grease which as I must start tomorrow I have traded at once and engaged an Indian Desjarlais to guide me by another and better route to Artillery Lake.[2]

July 23rd Monday. Started this morning from the commencement of the first long portage at 8 a.m., carrying as usual. We arrived within a short distance of Artillery Lake.

July 24th Tuesday. A couple of hours walk brought us to Artillery Lake. This is certainly the road to come in the fall. There are twelve portages, the first the longest and worst. All the others are quite short and trifling. Had breakfast previous to starting for home again. It then began to rain and continued to rain in torrents all day. On we marched through it all and at 3 p.m. arrived at the Fort all well. The covering of the large house has been brought home and a net laced.

July 25th Wednesday. Landrie and Jobin sawing covering. Kippling on the sick list. King and I working away at the roofing. Old Laret-esch arrived this evening with all his band on his way to the hunting grounds, poor as rats, devil a thing to eat and begging as usual. I bought a canoe from one of them for the use of the establishment, 8 M.B.[3] Blackflies desperate today.

July 26th Thursday. Employed all day covering two of the houses, Master's house & kitchen. Raining today again.

July 27th Friday. Got the wood for men's house rafted across and began building but we are unable to get on with our work as we would wish on account of the rain. It has been raining nearly all today again as usual.

July 28th Saturday. This morning Alfred Laferté returned sick from the Expedition. He left the party at Musk Ox Rapid on the Great Fish River,[4] all well with the exception of a few cuts and bruises. Mr. Anderson writes in good spirits. Four Indians arrived with Laferté laden with provisions which I traded of course. The men engaged cutting & sawing logs & roofing today. Raining again.

July 29th Sunday. Making preparations for a start on a voyage to Fort Resolution to meet the Boats from the Portage.[5] I had intended to take Landrie with me, but now that Laferté has been put under my Command, and knowing the gentleman very well, I think it better to keep him under my own eye, leaving Landrie to assist in finishing the houses.

[1] This is the chain of lakes and portages, to the south-east of the Lockhart River, leading from Great Slave Lake to Artillery Lake, now known as Pike's Portage.

[2] The Pike's Portage route.

[3] Made Beaver.

[4] He had left the Anderson party at Musk-Ox Rapids on July 14th.

[5] Portage la Loche.

July 30th Monday. Left Fort Reliance at 7 a.m. Our Indian guide informs me that it is shorter to go by 'the Portage' and as I like to see new country I have determined to pursue that route. We therefore struck straight out from the Fort, passed through the Wolf Straits and proceeded along the S.E. side of the Lake. Had dinner on a beautiful point covered with large flat stones squared neatly as if they had been made for gravestones. The shore wherever we landed absolutely covered with Blueberries of which I discuss no small allowance. Passed the large Cape and encamped at 7 p.m. It rained very heavily for some time today. Tonight we have made a shelter of my bed oil cloth & the canoe.

July 31st Tuesday. It blew and rained very hard last night and this morning. Could not leave our camp till 8 o'clock. The swell was so great as to make me sick. We arrived at the portage at 1 p.m. It is not a mile long; we crossed it and had dinner and started at ½ past 3. This evening we passed one of Capt. Back's Fisheries.

I will here make a spring over to the date of our arrival at Fort Resolution, giving merely a summary (*August 7th*) of the principle [sic] occurrences which took place. On the 2nd our road by which we came at the Great Cape where the Indians take pipe stone. We were 2½ days windbound beyond Rocky Point. When it did calm we set out and did not again camp till we came to the Fort.[1]

August 10th Friday. Messrs Campbell[2] and Ross[3] arrived with one Brigade of boats. Good news as far as I have yet heard. Exchanged letters with them. I gave them letters from our friends in the Expedition[4] and they gave letters from home and elsewhere.

August 11th Saturday. Mr. & Mrs. Clarke & Mrs. Peers arrived this morning with the other Brigade. They all left together except Mr. Ross in the afternoon. Mr. Campbell has not provided me with any men to man my boat as promised. I suppose I must remain for a year or two till the Indians come in.

August 15th Wednesday. Some Indians arrived yesterday so I managed with Mr. Ross' able assistance to make out a crew of six lame, halt & blind Indians. Made a start and got to the mouth of the Slave River; found our boat a leaky old tub. Were obliged to encamp early, being unable to keep afloat. Got the chinks stopt up with chips and grease.

It will be quite useless & uninteresting to describe the remainder of our journey to Fort Reliance. Suffice it to say that we travelled when we could and didn't when we couldn't. We caught a great many fish in the nights in the nets. One night especially, near the pipe Stone Point we caught 35 in one net. My black bitch which Ross gave me deserted in one of the large Islands and was never more seen. In the Strait a short distance from Fort Reliance I shot 2 wolves as we sailed thro; they were swimming across when we caught them. We arrived at Fort Reliance on Wednesday (22nd). Found all well. A large number of Yellow Knives are waiting for me to get ammunition. The Reindeer[5] have arrived. Nothing is to be seen but fresh meat,

[1] Fort Resolution.

[2] Chief Factor Robert Campbell, on his way back to Fort Simpson from the rendezvous at portage la Loche.

[3] Chief Trader Bernard Ross of Fort Resolution.

[4] Anderson and Stewart.

[5] Caribou (*Rangifer tarandus groenlandicus*).

heads, feet, skins etc. of all sizes denoting indiscriminate slaughter. I had almost forgot to mention that we sounded the lake opposite Pipe Stone Point; found 84 fathoms, Bottom gravel & mud (504 feet). Opposite Hoar Frost River we sounded again in 106 fathoms (636 feet).

August 23rd Thursday. Engaged all today trading & giving meat debt to a few. In the evening great cleaning of guns & filling of horns[1] in the Fort. I have promised a day's shooting to all hands except Landrie & Jobin who must work away at the saw[2] while they can lift a finger.

August 24th Friday. It rained very hard all this morning till past 9 a.m. As soon as it cleared up a little four of us went to hunt, King, Jambe de Bois, Laferté and myself. I killed five, King four, Jambe de Bois four and Laferté none altho' he swears he wounded one so severely that it was trailing its entrails behind it!

August 25th Saturday. Sent off a band to bring home the meat killed yesterday. Kippling & I took a walk up the River. I killed three Reindeer, he one. When we returned to the Fort we found a whole squad of Indians had arrived. I hear they have made very little provisions having been frightened out of their wits at some apparition they have seen.

August 27th. Settled with the Indians today. Engaged three Indians to accompany me to the head of the Great Fish River.[3] Kippling, Laferté & some Indians went to carry home the meat killed on Saturday. Kippling afterwards took a walk himself and killed three more Reindeer.

August 28th Tuesday. Laferté arranging the saw which has not been going well for some time past. King, Jobin & Landrie making a scaffold to hang fresh meat upon. Jambe de Bois & another Indian were sent to carry home meat. Kippling and I went to show them where it was after which we had a walk & killed five more (I three & he two).

August 29th Wednesday. Went off with a Brigade of people to bring home meat. On our arrival there we saw a few feeding some distance off. King and I went after them. I killed four & he one. Laferté & Jobin sawing.

August 30th Thursday. King, Jambe de Bois, Kippling & myself took a trip to the Portage which we have to make in leaving the lake for Artillery Lake, in order to examine it more minutely, to be ready for the little boats when they arrive. We killed one deer; fresh deer tracks numerous. We find that we can shorten the portage considerably by sticking into some little lakes to one side.

August 31st Friday. Some Indians arrived this morning, the same which Mr. Anderson saw down the Great Fish River. They brought little or nothing here. Laferté employed making a box for pounded meat. Kippling & Jobin barking[4] and mending nets.

September 1st Saturday. King, Landrie & Jambe de Bois sent off to make a stage & hang the meat killed on the 29th ult. at the place where it was killed. Laferté & myself arranging the stove. Kippling & Jobin same as yesterday.

[1] I.e powder horns.

[2] I.e. whip-sawing lumber in a saw-pit.

[3] To cache the boats and provisions as prearranged, in case Anderson decided to return to the mouth of the Back River the following year.

[4] Tanning.

Sepr. 3rd Monday. All hands occupied today to bring across sawing logs and soles for the flooring.
Sepr. 4 Tuesday. Laferté flooring my room. Kippling the men's house. Landrie & Jobin sawing. King and myself busy salting tongues & smoking & drying damp meat.
Sepr. 5th Wednesday. All the men engaged as usual except King who went with me & Jambe de Bois to hunt. Jambe de Bois killed nothing. I killed four & King four. When we returned to the Fort we found our three Indians had arrived, bringing no meat, however. They report deer numerous.
Sepr. 7th. The men all engaged as yesterday. The two long-looked-for Expedition Boats arrived this evening.[1] I am sorry to say in anything but good order. Nearly a third of the whole pemican is quite rotten & useless. I have ten bags which I will put in, instead of ten rotten ones. I cannot do more. I intend starting early tomorrow. Lamalice the steersman of one boat declares himself unable to proceed further; also all the Indians except Saulteaux refuse to going further for love or money. Fortunately there are plenty Indians about.
Sepr. 8th Saturday. It is impossible to start today as it has rained in torrents the whole day. Engaged some Indians at 1 MB pr diem.
Sepr. 9th Sunday. Still raining tonight. It has rained all day.
Sepr. 10th Monday. Started at daylight. We got all the pieces across the longest and worst portage in the road to Artillery Lake,[2] also one of the Boats half way. Jobin also left with us for the fishing.
Sepr. 11th Tuesday. Crossed both boats before breakfast.[3] Made three portages all together which brings the boats into the first long lake. I walked round the little lakes trying to get some fresh meat to spare our dried provisions. I saw 2 deer but was unfortunate and did not get a shot at them. Passed thro' the lake; made another portage & encamped shortly after. The men saw two more deer here.
Sepr. 12th Wednesday. We had made two portages this morning and passed a very

[1] The two boats which Anderson had ordered Campbell to have built at Fort Simpson, see pp. 96–7. The man in charge of the boats, Norbert Sauvé, delivered the following letter from Robert Campbell, dated at Fort Simpson, 21 August 1855:

> The two Boats & Canoes built for the A.S.E. will leave her in course of the day. Bills of lading of the same & list of the crew will be handed to you by Norbert Sauvé.
>
> Learning that the route is shorter & safer by Resolution side of the Lake they are directed to pass by that way, thence steer for Fort Reliance, on arrival at which they will place themselves under your orders, agreeable to Mr. Anderson's request, who in his instructions to me says that yourself & a party of men would meet them there or elsewhere to assist and guide them to where the Boats & cargoes would be 'cached' on the Banks of the Thlewycho-desse and as he had no doubt left instructions with you regarding these matters I leave you to carry them into effect.
>
> The canoes are I believe intended to bring back our people to those winter quarters and I would feel obliged if you would be so good as send them back with as little delay as possible.
>
> P.S. Should the two Canoes arrive at Reliance along with the Boats, the two Steersmen will be of course at your disposal should you require their services to assist bringing the Boats to their intended place of deposit.
>
> R. Campbell. (HBC B. 200/b/30. ff. 160–61)

[2] It must have been while Lockhart and party were making this portage that they passed Anderson and party coming south by a slightly different route.

[3] I.e. took both boats across the portage.

long lake & were preparing breakfast when Laferté made his appearance bearing a letter from Mr. Anderson who had returned from the expedition, stating that he had fulfilled their instructions and there was no occasion for any further operations and directing me to return with the boats to Fort Reliance. We immediately after breakfast retraced our steps and returned tonight to the long portage. We will see the fort tomorrow.

Sepr. 13th Thursday. All hands early at work carrying pieces and by 5 p.m. pieces, Boats and everything was over. Fair wind; hoisted sail and arrived at Fort Reliance. Find Mr. Stewart alone. Mr. Anderson had already gone before to Fort Resolution. I have got orders to shut up shop and prepare to start tomorrow morning.

Sepr. 14th Friday. Up very early settling with the Indians and packing up. Started from Fort Reliance in a light canoe with Mr. Stewart at 10 a.m. leaving King[1] supplied with ammunition & tobacco to trade more provisions before he leaves. The Boats left at the same time as we did.

[1] King Beaulieu.

CHAPTER VII

THE EXPEDITION DISPERSES

INTRODUCTION

From Fort Resolution the expedition dispersed. Anderson departed by boat for Fort Simpson to resume his duties as Chief Factor in charge of the Mackenzie District. He left Fort Resolution on 28 September, and encountering generally fair winds, made excellent progress across Great Slave Lake and down the Mackenzie, reaching Fort Simpson on 3 October.

Stewart, accompanied by Lockhart, was entrusted with delivering the expedition's dispatches to Sir George Simpson at Lachine. The party left Fort Resolution in two canoes on 22 September and travelled up the Slave River to Fort Chipewyan, then up the Athabasca and Clearwater to Portage la Loche. Here, on 7 October, they found some horses, which greatly facilitated their crossing of this notoriously long and difficult portage. On the other side of the portage they transferred to two boats and continued east via the standard route via the Churchill River, Frog Portage, Sturgeon Weir, and the Saskatchewan River to Grand Rapids, where the latter river debouches into Lake Winnipeg.

Here Stewart made a decision which would raise Anderson's ire: he deviated from the direct route up Lake Winnipeg to Fort Garry and instead headed north-east across the lake to Norway House, where his wife was waiting (see p. 145). He reached Fort Garry (accompanied by his wife) on 3 November; the voyageurs from Red River were paid off here. On 13 November Stewart and his wife, along with the three Iroquois, headed south to link up with the railway system in the United States. They reached St Paul around 11 December and were in Montreal on the 22nd.

DOCUMENTS

1. Journal of James Anderson, Fort Resolution to Fort Simpson.[1]
Fri. 28th. Left Resolution at 10 a.m., the 'A' boat[2] having arrived yesterday

[1] HBC E.37/3.
[2] The Athabasca boat, bound from Portage la Loche to Fort Simpson.

afternoon. Boat very deep.[1] Wind moderate till we came to the last islands to go to Isle aux Morts, where we were compelled to encamp. Drizzling rain.
Sat. 29th. Wind N.E. with a very heavy swell, which compelled us to put on shore at 1 p.m. at Sulphur Springs, where we were wind bound all day. Showers of rain and snow at intervals all day. A.B. faint in the evening. Sharp frost.
Sun. 30th. Left at the first appearance of dawn. Carried sail with a very light breeze, and pulled. Supped at Point des Roches, and reached Big Island about 2 a.m. of
Mon. 31st. Arranging different affairs for Fort Rae etc. Left B. Is. at 1 p.m. Soon after hoisted sail to a light breeze and at the same time pulled. Water extremely high, both in the lake and the river.[2] Supped at 7 p.m. a little below Point de St. restaux.[3] Then hoisted sail to a light air of wind. Day broke when we had entered a little river on the left bank, where we were nailed till the next morning.
Wed. 2nd. Encamped at Spence's River.
Thurs. 3rd. Reached Fort Simpson about 9 p.m.[4]

2, Journal of James Stewart, Fort Resolution to Fort Garry, 22 September–3 November 1855.[5]

Saturday 22nd [September]. After bidding Msrs. Anderson & Ross goodbye, Mr. Lockhart & myself started for Red River in two Canoes. We are now on our way home & with joyful hearts, thankful for having escaped the dangers of the voyage as well as for having been allowed to fulfil our instructions. Met the two boats we had left behind a short distance from the Fort. Beautiful weather. We encamped below the old Fort; the water is still very high in the river and it is difficult to find a place to put ashore comfortably. We have not as yet met with the A. canoe. I hope it has not passed us.
Sunday 23rd. Cloudy & wind N.W. Encamped at the Grand Batture. Saw some ducks but no geese. They must have all gone south by this time. No A. canoe yet.
Monday 24th. Started at daylight. Met the A. boat about 4½ p.m. & shortly after encamped to read our letters. The day was cloudy but fine. I sorry to learn that I had no news from my dear wife.
Tuesday 25th. Started as usual & encamped at Salt River where we got some fish & a little milk. The potatoe crop has failed here as elsewhere. Beaulieu is not here himself.
Wednesday 26th. Remained till 1 p.m. in consequence of wind & rain. At that hour we started & encamped a little below the Noyé.[6] Saw a fire on the N. side of the river but did not see who it was.
Thursday 27th. Started at daybreak & breakfasted at the [illegible] Portage, having passed the Noyé by water. Passed the rest[7] during the day & encamped at L'Ile du Raquet. Cloudy but no rain, mild.

[1] Being laden with unused expedition supplies and equipment (HBC B.200/a/31, f. 11).
[2] The Mackenzie River.
[3] Point Saresto.
[4] '... having been absent 4 Months and a week' (HBC B.200/a/31, f. 11).
[5] PAA 74.1/137.
[6] The Rapids of the Drowned, the most northerly of the Slave River rapids, beside the present community of Fort Smith.
[7] The rest of the Slave River rapids (Mountain, Pelican and Cassette rapids).

Friday 28th. Started at the usual time. The wind assisted us a little & we encamped at the S. end of Duck Isle Portage. Several showers of snow fell during the day.
Saturday 29th. Arrived at A.;[1] just as well. We brought up Pemmican & flour to I.L.C.,[2] for 50 lbs dry meat was all they had.
Sunday 30th. Left the Fort, crossed the Lake & encamped at the entrance of the Riviere Embaras[3] in the mud. I slept in the canoe. Plenty geese & ducks seen flying about yet, so that our chance of reaching the portage[4] by open water is good.
Monday 1st October '55. Passed a camp of Indians, Piché's band who are as much dogs as any other Montagnais. Got 4 geese from the Fort's men & encamped above the old Fort. Fine weather.
Monday 2nd. Squally with showers of rain. Encamped above the Pointe aux Trembles.
Wednesday 3rd. Hard frost last night. We were off early with a very strong NW wind & two reefs in. Passed the Pierre au Calumet & encamped above the Tar.[5] A little snow fell during the day; it cleared at night but continued to blow hard & to freeze.
Thursday 4th. Started early with a gale of wind. Sailed the whole day with two reefs in & encamped beyond the first winter portage in Little A.[6] river. A few flakes of snow fell & the frost had begun before we encamped.
Friday 5th. [Illegible] a fine day. Encamped on the left hand side of the river in sight La Bonne Mountain.
Saturday 6th. A beautiful day. Started at the usual time. Passed the Cascades, La Bonne, Grosse Roche & encamped after carrying everything over the Pine Portage.
Sunday 7th. Started at daylight; breakfasted at the White Mud. Arrived at Portage la Loche at 1 p.m. Took dinner at the top of the hill[7] & encamped at this end of the Portage at 7 pm. Jas. Bouché found some horses at the other end & that enabled us to cross at once. Such a day as this is worthy of note. It is not often it has been accomplished. 7th day from Athabasca. Most beautiful autumn weather.

[1] Athabasca (Fort Chipewyan).
[2] Ile-à-la-Crosse.
[3] Which provides a shortcut into the Athabasca River.
[4] Portage la Loche.
[5] A site on the banks of the Athabasca River with conspicuous surface showings of tar, regularly exploited by the Hudson's Bay Company. On his way upriver in July 1851 Commander W. J. S. Pullen described the site as follows:

> These pitch springs issue from the face of a hill (on the right bank of the Athabasca), and about a couple of hundred yards from the water. There are several others in the river; but it is from these the company get their supplies for the use of boats etc. They occupy a space on the surface of about nine feet square. From the upper or higher part, the pitch issues clear and pure in a narrow stream, mixing up at the bottom or lower part with mud and sand in a thick mass, which is what the kegs are filled with, and purified by boiling (Pullen, *The Pullen Expedition*, p. 165).

These are the surface showings of the Athabasca Tar Sands, in recent years the object of massive exploitation.
[6] Little Athabasca, now Clearwater River.
[7] The portage rises by a brutally steep climb out of the Clearwater River at its north end. For a description of this hill, the view from its summit, the details of the portage and the impressions it made on travellers, see Walter Kupsch, 'A Valley View in Verdant Prose: the Clearwater Valley from Portage la Loche', *The Musk-Ox*, 20, 1977, pp. 28–49.

Monday 8th. After arranging the boats[1] we started. Breakfasted at Montagnais' & encamped at La Cimetiere. Cloudy with appearance of rain. Mr. L. has had his troubles coming this year [illegible] and a high water.
Monday 9th. Breakfasted at the Barriere. 9.30 Gros Ventre's. Fair wind across the lake. Encamped at the Detroit Clear Lake.
Wednesday 10th. Arrived at Ile à la Crosse at 4½ p.m. to the surprise of Mr. Deschambault. I had the pleasure of receiving a visit from Monsr. Taché.[2]
Thursday 11th. Left our kind host after breakfast and encamped below the Little Shaginnah[?]. Appearance of rain.
Friday 12th. Ran several rapids. Passed Lac a Primeau & encamped in Thin? Lake. Cloudy & warm; a few drops of rain fell. At the Prise[?] we broke the boat losing 2 hours thereby. Saw some Indians from [whom] we got a few ducks & fish; the last were encamped at the head of Rapide Couche.
Saturday 13th. Encamped at Pine Lake.
Sunday 14th. Ditto Trout Lake.
Monday 15th. Nearly filled the boat at Trout Fall & before we got it emptied it was time for breakfast. Ran the other rapids. Passed the Grand Diable; ran the Otter & encamped at the portage above the Mountain. A Cree who visited us last night came on with us today. He goes to the Fort. The weather continues cloudy but mild. Yesterday we had ice on the oars but today none was found.
Tuesday 16th. Breakfasted at Mr. I. McKenzie's Fort. Paid a visit to Mr. Hunt, after which we started again. Met Mr. H's boat at the Rapid River portage & encamped below the Island portage. A most beautiful day.
Wednesday 17th. Breakfasted at Frog Portage[3] & encamped abo[ut] ½ way in Lac des Bois. Rain at night but mild.
Thursday 18th. Started before daylight. Passed 3 small portages; hoisted; sailed across Pelican & Monson lakes & encamped below the Bush Portage. A regular snowstorm all day. Mr. L.[4] killed an eagle.
Friday 19th. Encamped at the Sturgeon River.
Saturday 20th. Encamped near Cumberland[5] it being too late to go to the Fort.
Sunday 21st. Breakfasted at the Fort and carried on all night. Very cold; ice forming fast.
Monday 22nd. Passed the Pas Mission[6] before breakfast where we took some pemmican & flour. Still very cold. Encamped at the entrance to Muddy Lake.

[1] The implication is that the canoes had been left at the north end of the portage and that the group had now transferred to York boats.

[2] Bishop (later Archbishop) Alexandre-Antonin Taché had established a mission at Ile-à-la-Crosse in 1846, from which he carried out missionary work over a vast area extending from Fort Chipewyan to Reindeer Lake. After a period in the south, during which he was made bishop, he had just arrived back at Ile-à-la-Crosse earlier in the summer of 1855 to organize missions at Lac la Biche and Great Slave Lake. See Jean Hamelin, 'Alexandre-Antonin Taché', *Dictionary of Canadian Biography*, XII (1891–1900), Toronto, 1990.

[3] A very low portage from the Churchill River into the head of the Sturgeon Weir system. At high water on the Churchill some water overflows this portage.

[4] Lockhart.

[5] Cumberland House on the Saskatchewan River.

[6] At the site of the present town of The Pas, Manitoba.

Tuesday 23rd. Encamped early at Spider islands. Blowing too hard to cross the Lake.
Wednesday 24th. Remained all day windbound. Got the boat cleaned & the men washed their clothes etc, I could not be easy under the circumstances, so near to my dear wife, but a little more patience & we shall I trust bring this separation to a close. Weather quite mild; the southerly wind has broken up all the ice. We were just in time yesterday; the Straits were all frozen. We saw some Indians those who were stopped in that current.
Thursday 25th. Started at daybreak. Sailed across the Lake, the Grand Rapid, supped opposite the island & continued on all night.[1] Arrived at Norway House[2] on Friday 26th in the afternoon, where we found all well & I had the pleasure of once more embracing my dearest wife after our long & dreary separation.
Saturday 27th. After getting ready for a new[?] start we left our kind friends with much regret & encamped at the entrance of the river. Beautiful weather.
Sunday 28th. Got as far as the Shoal[?] Islands where we encamped. The weather looks threatening & it has already begun to snow. I trust that we have got time to reach Red River ere winter sets in. Such weather is not the thing for ladies in an open boat & adds considerably to my anxiety
Monday 29th. Remained all day in camp. Blowing & snowing all day.
Tuesday 30th. Still the same rough weather. Made a start & reached Marshal's Point. Snowing all night.
Wednesday 31st. Sailed all day & encamped at Flour Point, it being too dark to go on further. Snowing all day though mild.
Thursday 1st November.
Friday 2nd.
Saturday 3rd. Arrived at the Stone Fort[3] & after dinner went up to the Upper Fort, leaving dear Meg at her father's.

[1] Across Lake Winnipeg.
[2] Stewart would later be severely reprimanded by Anderson for this unauthorized diversion of express boats to pick up his wife at Norway House. See p. 213.
[3] Lower Fort Garry.

CHAPTER VIII

REPORTS OF THE EXPEDITION'S ACHIEVEMENTS

INTRODUCTION

The results of the expedition's efforts were reported variously in different forms and for different readers by Anderson and Stewart. Even before reaching Fort Resolution, while still en route down Great Slave Lake from Fort Reliance, on 14 September Anderson addressed a preliminary report to George Simpson (pp. 167–70). Then on reaching Fort Resolution he compiled a more detailed report for Sir George, which was forwarded along with a fair copy of his journal.[1] At the same time he sent a full account of the expedition to Lady Franklin. Later he reported further to Sir George from Fort Simpson in a letter dated 29 November 1855; at the same time he dispatched an almost identical letter to Eden Colvile at Fort Garry.

Stewart also wrote a report for George Simpson,[2] which he delivered in person on reaching Montreal in December 1855. He also kept a journal, which is generally

[1] HBC B.200/a/31. *Journal of a Voyage from Fort Resolution, Great Slave Lake to the Arctic Sea, 1855*. This is essentially a condensed version of the full journal reproduced here, but it does include as an appendix an additional interesting set of recommendations for future expeditions:

Remarks

1. Each person employed in such an expedition as the above, should be provided with a Seal Skin Capot with hood, Seal Skin Trowsers, Esquimaux Boots, a Waterproof Blanket 3½ or 4 points and Waterproof Bags.
2. Half a Gallon of spirits should be given for each person, as tho' tea is far preferable to spirits, it is often impossible to light fires and the people are obliged to lie down chilled and wet.
3. Kettles should be of a square form, 6 inches deep and of 4 & 5 Gals. Capacity.
4. One of the Officers should be a Surgeon, as serious accidents are liable to occur.
5. In my opinion the route to the Arctic by the Thlewycho offers incomparably greater risk and difficulty than any other that could be adopted, and the chance of finding the inlet at its mouth free from ice, is a very faint one.
6. From the absence of a deer pass and the scarcity of fuel on the Coast or near the Mouth of the River and the almost utter impossibility of taking down the requisite provisions and supplies, it is out of the question for a party to attempt to winter there.
7. Although the risk is certainly greater, Good Canoes are the best craft for this route, but first rate Bowsmen, Steersmen and Middlemen are indispensable.

[2] See pp. 182–6.

less detailed than Simpson's but includes some interesting additional information and commentary.[1] It has never been published in full.

In terms of the public hearing of the expedition's results, the first news to reach the public anywhere was in the pages of *The Times* and the *Minnesotian* of St Paul, Minnesota for 12 December 1855. These articles are clearly the product of interviews with James Stewart, on his way from Fort Garry to Montreal, and are not based on his written report or journal. As Anderson later indignantly pointed out,[2] they contain serious errors, e.g. that the party had found a boat with the name 'Terror' on it, and that there are 100 rapids on the Back River (Back had counted 94). Shortly afterwards (on 24 December) similar accounts (although perhaps slightly less flawed) also appeared in the *Montreal Herald* and the *Montreal Gazette*, once Stewart had reached that city.

The British public first read of the expedition's achievements in *The Times* on the morning of 11 January 1856, which printed Anderson's letter of 17 September to Sir George Simpson.[3] This was subsequently reprinted in the *Journal of the Royal Geographical Society*[4] and in the Arctic Blue Books.[5]

Stewart's and Anderson's accounts diverge on a number of minor points, and on one major point. In his report to Sir George Simpson, in his interviews with reporters for the Montreal newspapers, and in his deposition to the Scottish Court of Session in connection with the Fairholme case, Stewart consistently maintained that he had been told by one of the Inuit women at the rapids below Lake Franklin, that she had seen one of the survivors of the Franklin expedition alive. This statement was categorically refuted by Anderson,[6] but was probably reported accurately by Stewart. Stewart would have been unlikely to utter a falsehood deliberately to the Scottish Court of Session, and perhaps even less in a report to the formidable Governor-in-Chief. Probably more than anything else this statement of Stewart's underlines the degree to which the expedition was constrained by the lack of a competent interpreter, since clearly, had they been able to converse freely with her, the woman in question would have been able to recount much more of the final fate of the Franklin expedition, and could have directed Anderson and Stewart to the exact spot where she had seen the survivor.

DOCUMENTS

1. Private letter from James Anderson to Sir George Simpson, 14 September 1855, Great Slave Lake, en route back to Fort Resolution.[7]

Your kind letter of 14th June reached me on my arrival at Fort Reliance on the 11th

[1] PAA 74.1/137.
[2] See p. 218.
[3] See pp. 170–7.
[4] J. Anderson, 'Letter from Chief Factor James Anderson to Sir George Simpson, F.R.G.S, Governor in Chief of Rupert Land', *Journal of the Royal Geographical Society*, vol. 26, 1856, pp. 18–25.
[5] Great Britain. Parliament, *Further Papers Relative to the Recent Arctic Expeditions in Search of Sir J. Franklin and the Crews of Her Majesty's Ships 'Erebus' and 'Terror', including the Reports of Dr. Kane and Messrs. Anderson and Stewart, Presented to the House of Commons, 1856*, London, 1856, pp. 25–9.
[6] See p. 218.
[7] BCA AB40 An 32.2, ff. 86–93.

instant; allow me to return my best thanks for the congratulations you offer on my promotion, which I am well aware is in a great measure owing to your support.

I trust that my proceedings in regard to the Expedition will meet with your approbation. I can assure you that nothing more could be done. We had many narrow escapes which I have not noticed in my public letter. It is out of the question for any party to attempt to winter on that horrid coast. I had two of Rae's best men with me[1] and they declared Repulse Bay is a paradise compared to it; there is no place more destitute of resources than this; it would be almost impossible to take down supplies by this route for even a small party; no craft could go down the numerous strong rapids with full ladings. The bodies of our unhappy countrymen are undoubtedly buried deep under the sand and gravel, and as for any paper or book being found after a lapse of 5 years, I doubt whether any that were left unprotected would be legible after an exposure to the weather of 5 days & books with leather coverings would have been torn to pieces by wild animals. If the Esquimaux found any they have them no longer in their possession; they most probably threw them aside as valueless or like most Indians destroyed them for fear that they talked and told that they had taken away the property of the party. I have no doubt that the vessels were crushed at winter quarters in the vicinity of the Magnetic Pole, and that the Boothian Esquimaux have many articles belonging to the party and possibly their journals; the only chance of recovering these would be by sending out an Expedition to winter at Prince Regents Inlet and parties to be sent on the ice in the spring through Bellots Straits down Victoria Straits and to return via the Isthmus and the Gulf of Boothia; these parties should be pretty strong and well armed, as I believe these Esquimaux are very independent and saucy.

The ice beyond Montreal Isld. had not moved nor would it I imagine this year; had there been anything like a clear passage I would certainly have gone beyond Cape Herschel with ease. The bark of our canoes was wretched and the woodwork much too heavy; the 'lisses' were planks from ¼ to ⅛ of an inch thick and put edge to edge instead of overlapping; the ribs were much too weak; the lining with sheeting[2] does not answer; it was with difficulty that 6 men could carry mine – they were however capital sea boats. The river was fully as bad as Back represents it in the upper parts but though bad enough below that the rapids, though very strong, were not quite so terrific as he paints them. We passed probably at a lower stage of water than he did. The Iroquois had their abilities well tested; Ignace[3] was my boute;[4] he ran all the rapids first – one excepted – and was always ahead in the bad ice. Old Baptiste[5] was Stewart's boute; he stood the trip well and is a most wonderful man for his age. I think however that his eyesight is not quite so good as it has been. Stewart and I had much walking; we are both in good health. I, however, should not like to undertake such another trip, as though my right leg did not hurt me, I perceived that the veins are still more swollen than they were. I never yet knocked up and should not

[1] Thomas Mustegan and Murdoch McLellan.

[2] Stewart's experiment at lining the canoes with tarred burlap.

[3] Ignace Montour.

[4] Bowman.

[5] Baptiste Assanayunton.

admire doing so now; moreover were such a thing to occur in a trip like this it would be death, as the party could not carry me.

I have for the first time turned Hydrographer. I will send you by next opportunity a map (such as it is) of our route through the barren grounds. At the same time you will receive my journal. I have had no time to make a fair copy and the original is almost illegible. I had to turn guide from the head of Lake Aylmer. Back's map – the one attached to the narrative – was perfectly useless in the large lakes. I was almost knocked up for want of sleep going down, as I had to keep a sharp look out and moreover to get up early to waken the men, Stewart being such a hard sleeper.

There was one subject which I had almost determined on passing over, but on reflection, tho' it pains me much, I see that it would be a dereliction of duty were I to be silent. You may perhaps have observed that I have been silent regarding Stewart in my public letter and may possibly have thought me ungenerous. The fact is I could not praise him with any regard to truth and therefore said nothing. Regarding his conduct before I arrived at Resolution I have little to say; I am sure that he did his best, but I have received complaints regarding his extravagance in the expenditure of provisions both at Athabasca and Resolution.

During our expedition we were and still are excellent friends, except in two or three instances when I was compelled to speak to him; but he was utterly useless to me; for instance he would sleep the greatest part of each day and this when going through a strange country and when it was very possible that something might occur to me[1] – this did not prevent him from sleeping soundly at night, in fact I had the greatest trouble to get him up in the morning. He never certainly opposed any of my measures, but he never suggested anything or encouraged me in any way to go on. I had imagined that he would have been too rash and venturesome, but I found nothing of this kind. His luggage was at least twice as much as mine, and he is too fond of good living for such work as this. I mention these things to you not by way of complaint, but to acquaint you with his character, in case you might think of again appointing him to conduct an enterprise of a similar nature.

I took down the instruments with me. They were however of no use, but only an encumbrance – of course compasses excepted. Stewart got some instruction this spring from B. Ross[2] regarding taking an altitude and working for time and thought that he could do it correctly, but found he was mistaken. He took 4 or 5 observations, the last one at Montl. Isld. when he could not get the latitude within 2½° of Back's, and the chronometer which was 3 min. fast when we left Resoln. had only gained 3 min. at that place! He did not try it after that.

As you desire it, I shall of course proceed to Simpson and resume the charge of

[1] Anderson is implying that Stewart slept so much in his canoe on the downriver trip that he saw very little of the country. Lakes such as Pelly, McDougall and Garry are extremely complex in shape, with numerous arms, islands and peninsulas, and finding the outlet or inlet can be very difficult. Thus Anderson is suggesting that if he himself had been killed Stewart might have been unable to guide the party back to Great Slave Lake. Of course this is rather an unfair (and probably unintended) vote-of-no-confidence in the abilities of the voyageurs.

[2] Bernard Ross at Fort Resolution.

the District; from what I can learn matters have not gone on very well since my departure.

The Expedition accounts shall be sent up by the winter express. Most of the articles sent up will answer either for the trade or for the sale shop; those which will suit neither must of course, be taken at a reduced rate.

Mr. Lockhart has conducted the duties that devolved on him very well. He is a remarkably clever and active officer. I trust that he will receive some recompense for the extra duty that he has performed. His sudden removal was very expensive to him as he had to buy everything that he required here. I think that his salary for the past year should be £150. I should like to have kept him here, but as we have our full quota of officers I could not presume to do so, but if Mr. Campbell (as he says he will) goes out next year, as also Messrs. Mills and Harrison, we shall require several fresh officers – However I shall address you on this subject by the winter express.

May I beg of you to procure for me an elastic stocking and cap for the knee – right leg – and send it by the Portage Brigade; may I also beg that if my despatch be published in the papers or any thing on the subject to send a copy of the paper to my mother, Mrs. A., Georgina L., Simcow C.W.[1] – to my brother Alexander, my uncle Jas. Anderson Esqre, Brechin, Forfarshire N.B., and to my coz. at E. Main,[2] as I really have no time to write except a few words to my mother and brother. I should like also two or 3 copies of the paper to be sent to me. Forgive me for thus troubling you.

17th – Fort Resolution. I arrived here last night. Stewart was to leave immediately Lockhart arrived, perhaps a day after me. The day we left Artillery Lake the ground was white with snow and piercingly cold; here the weather is delightfully warm.

22nd. Messrs. Stewart and Lockhart arrived yesterday and the boats today; they will leave tomorrow.

2. Official letter from James Anderson, Fort Resolution, to Sir George Simpson, 17 September 1855.[3]

I beg to state for your information that the expedition you were pleased to entrust to my command returned here last night.

After having descended Back's Great Fish River, and explored the mainland and islands as far as Maconochie Island, undoubted traces of the missing party were found at Montreal Island; but, I regret to say, that neither the remains of our unfortunate countrymen, nor any persons were discovered. My previous despatches would have informed you that I arrived at Fort Resolution late on the 20th June; the three canoes, built under the superintendence of Messrs. Stewart and Ross, were all ready; they were of an excellent model, the woodwork very strong, but the bark, though the best that could be procured at so short a notice, was very inferior. They were, of course, very heavy; their ladings amounted to 24 pieces of 90 pounds

[1] Simcoe, Canada West.

[2] His cousin, also James Anderson, usually designated as James Anderson (b), and then serving at Eastmain on the east shore of Hudson Bay.

[3] HBC B.200/b/31; HBC A.8/17, ff. 280–91; Great Britain, *Further Papers Relative*, 1856, pp. 25–9; *Journal of the Royal Geographical Society*, vol. 26, 1856, pp. 18–25; *The Times*, 11 January, 1856, p. 8. Through these latter three sources, this was the version of the expedition's activities which became known to the British public at the time.

each, and consisted chiefly of provisions, with a good supply of ironworks, etc., for the Esquimaux; ammunition, nets, Halkett boat, and the luggage of the party. Double sets of poles, paddles, and lines, were also provided. Fifteen men and an Indian guide to Sussex Lake composed the crews; but I found we were too weakly manned, and added three Yellow-knife Indians, who wished to go to their lands. The only thing wanting was an Esquimaux interpreter.

Late on the evening of the 22nd June the expedition made a move to an island about one mile from the fort. Heavy gales from the N.E. detained us between there and Rocky Point, where the traverse is taken, to the 27th; these, however, cleared our road of ice. On the 28th we encamped at the upper end of Tal-thel-leh Strait, we then fell in with the ice; it was about 2 ft. thick. We were employed till the morning of the 2nd July in making our way through it, by cutting, pushing the pieces apart, and making portages over the points of land. Young ice formed every night, and the further we advanced the sounder it got; the canoes, too, unavoidably received some injury.

At that date we had reached the place called 'the Mountain,' in Back's map, and mentioned in his narrative as being a route to the barren grounds, but only practicable for small Indian canoes. The guide engaged by Mr. Stewart was unacquainted with this route; he proposed taking us by a river, falling into Great Slave Lake on its eastern shore, nearly opposite to Fort Reliance;[1] by means of this river, some lakes, and ten portages, he intended to round the head of Great Slave Lake, and fall eventually at the Beaver Lodge, in Artillery Lake.

Hoar Frost River, which Sir George Back ascended with difficulty in a half-sized light canoe in the autumn, would have been at this season impracticable. To get through the ice to either of these rivers would have occupied a long time; and then Artillery, Clinton Colden, and Aylmer Lakes were still before us;[2] it was clear that if some other route were not adopted the expedition would fail. Here was our only chance. One of the Indians had passed by this route as far as the river falling into the upper end of Aylmer Lake; he gave a disheartening description of the difficulties to be surmounted, and the high range of mountains before me was anything but encouraging. I, however, determined on making the attempt, as the route was so much shorter, and we might also expect to have a long stretch of open water on Lake Aylmer at the mouth of the river.

Immediately after breakfast the portage was begun, and though four trips were made before we encamped, the canoes and ladings were carried over 3½ miles of mountains, and across a small lake; at 10½ P.M. our fine fellows were descending a steep mountain with the canoes, singing 'La Violette'. The next day's work was something similar, the third was a little better; after which the mountains subsided to gentle hills, and the lakes were larger; some of them from 20 to 40 miles in length. In short, after passing through twenty-four lakes and twenty-five portages we reached the river falling into Lake Aylmer (which I have called Outram River,

[1] This is the chain of lakes and portages, bypassing the continual falls and rapids of the Lockhart River, and lying a little to the south-east of the latter, which was later named Pike's Portage.

[2] And would still be ice-bound.

after a gallant relative) late on the 7th July. Two easy rapids, and about 12 miles of river, brought us to Lake Aylmer.

I had now to trust my own guidance. As I had anticipated, we found the lake at its mouth free from ice, and had fine paddling for about 30 miles. When we fell in with the ice the whole lake appeared solid and unbroken: the ice was about 3 ft. thick, and perfectly sound.

The north side of the lake which we followed is indented with deep bays, separated from each other by narrow necks of land; round these, and close to the shore, we had to work our way by cutting, poling, and numerous portages across points, with occasional pieces of open water at the bottom of the bays. There were also some crevasses[1] through which we passed at a great risk of being nipped: we had several hair-breadth escapes; indeed one of the canoes was once only saved by pushing poles under her bottom and allowing her to be lifted on the ice. We finally arrived, with our canoes much injured, at Sandy Hill Bay, on the 11th July.

We had now the advantage of Sir G. Back's Map and Narrative; the former – the one attached to his book – was on far too small a scale for our purpose; but the latter was of great service.

It is needless for me to describe the descent of this dangerous river, after the minute and correct description of our gallant predecesor. Notwithstanding the exquisite skill of our Iroquois Boutes,[2] the canoes were repeatedly broken and much strained in the whirlpools and eddies. The river to the small lake falling into Musk-Ox Lake was nearly dry, and the portage work was most severe.

On the 13th July Musk-Ox Rapid was reached: here we found a few Yellow-Knife Indians.

Our four Indians and one man were left here; the latter, with one of the Indians, was to return to Great Slave Lake to join Mr. Lockhart, and the expedition now consisted of the following individuals:

James Anderson,	1st Command	
J. G. Stewart,	2nd	do.
Bte. Assinijunton	} Iroquois Boutes	
Ignace Montour		
Joseph Anarize		
Thomas Misteagun	} Muskegon Steersmen	
Paulette Papanakies		
John Fidler	half-breed	do
Henry Fidler	"	Midman
Edward Kipling	"	"
George Daniel	"	"
Donald M'Leod	"	"
Jerry Johnston	Muskegon	"

[1] By this term Anderson means what, in sea-ice terminology, would be called a lead, i.e. a narrow channel of open water between ice floes.

[2] Bowmen.

Joseph Boucher	Canadian	"
Murdo' MacLellan	Highlander	"
William Reid	Orkneyman	"

On the 15th, the worst canoe, which was completely worn out, was left, and we now proceeded with heavier ladings but better crews. On the 20th the first Esquimaux were seen at the mouth of, and below, Mackinlay River; there were five lodges, and we visited two of them: we here found the want of an interpreter. We had two of Dr. Rae's men,[1] who understood a few words and phrases, and, with the aid of signs, they made us understand that they came down the Mackinlay River. They were not much alarmed, and we soon got excellent friends.

They had evidently seen whites, or had communication with others of their countrymen (most probably those who resort to Churchill) who had intercourse with them, as they possessed a few of our daggers, beads, files, and tin kettles; and one old man brought down some wolf-skins to barter; they were clothed entirely in deer and musk-ox skins. Their canoes were made of deer-parchment, and not a piece of seal skin was seen among them.

Another small party was seen at the rapid between lakes Pelly and Garry. The men appeared to be all absent, and the women and children fled on seeing us. Some small presents were left in the lodges to show our kindly intentions.

In Lake Garry we had to work through about 15 miles of ice; but although ice was observed in some of the other lakes, we met with no farther obstruction from this cause.

On the 30th, at the Rapids below Lake Franklin, three Esquimaux lodges were seen on the opposite shore, and shortly after, an elderly man crossed to us. After the portage was made we crossed over, and immediately perceived various articles belonging to a boat, such as tent-poles and kayack-paddles made out of ash-oars, pieces of mahogany, elm, oak, and pine; also copper and sheet-iron boilers, tin soup-tureens, pieces of instruments, a letter nip, with the date 1843; a broken hand-saw, chisel, etc. Only one man was at the lodges, but the women, who were very intelligent, made us understand, by words and signs, that these articles came from a boat, and the white men belonging to it had died of starvation.

We, of course, by showing them books and written papers, endeavoured to ascertain if they possessed any papers, offering to give them plenty of the goods we had with us for them; but though they evidently understood us, they said they had none: they did not scruple to show us all their hidden treasures. Besides the man, there were three women and eight children, the remainder of this party. Two men and three lads were seen towards evening.

Point Beaufort was reached on the 31st; we were detained there the next day till 2½ P.M. by a S.W. gale; we then took the traverse to Montreal Island; to seaward the ice appeared perfectly firm and unbroken.

When about 3 miles from the island a large stream of ice was observed coming at a great rate before the wind and tide out of Elliot Bay, and the other deep bays to

[1] Thomas Mustegan and Murdoch McLellan.

the westward. Every sinew was stretched to reach the land, but we were soon surrounded by ice, and for some time were in most imminent danger. The ice was from 6 to 7 ft. thick, perfectly sound, and drifting at the rate of 5 or 6 m. an hour. In 15 minutes after we had passed, the whole channel to Point Beaufort was choked with ice. Had we not succeeded in crossing on this day we should have been detained on the eastern shore till the 10th.

We had thus arrived at the first spot indicated by my instructions, on precisely the same day as our gallant predecessor, Sir G. Back.[1]

The two next days were devoted by the entire party to the examination of the island, and the small islands in its vicinity. On a high ridge of rocks, at the S.E. point of the island, a number of Esquimaux cachés were found, and, besides seal oil, various articles were found belonging to a boat or ship – such as a chain-hook, chisels, blacksmith's shovel, and cold chisel, tin oval boiler, a bar of unwrought iron about 3 ft. long, 1½ broad, and ¼ in. thick; small pieces of rope, bunting, and a number of sticks strung together, on one of which was cut Mr. Stanley, surgeon of 'Erebus'. A little lower down was a large quantity of chips, shavings, and ends of plank of pine, elm, ash, oak and mahogany, evidently sawed by unskilful hands; every chip was turned over, and on one of them was found the word 'Terror' carved. It was evident that this was the spot where the boat was cut up by the Esquimaux. Not even a scrap of paper could be discovered, and though rewards were offered, and the most minute search made over the whole island, not a vestige of the remains of our unfortunate countrymen could be discovered.

On the 5th we succeeded in crossing over to the western mainland, opposite to Montreal Island, and the whole party was employed in making a most minute search, as far as the point of Elliot Bay, and also to the northward. As the whole inlet was full of ice which had not yet moved, but was only split into immense fields by the rising and falling of the tide, we could only proceed close in shore at high tide, when by pushing small blocks apart, finding pieces of open water at the bottom of the bays, and navigating through channels of water on the ice, we reached Point Pechell, late on the 6th. The whole coast between Montreal Island and Point Pechell was searched by a land party, always accompanied by Mr. Stewart or myself; many very old Esquimaux encampments were seen, but not a trace of the party.

By this time our canoes had received so much damage, and were so weak and leaky, that it was evident the safety of the party would be hazarded, were they subjected to more rough usage. The ice too, here, was forced on the shore, and there was no prospect of our being able to get through it; I therefore determined to complete the search of the peninsula on foot.

Early on the 7th the entire party with the exception of two of the Iroquois, who were left to repair the canoes, started in light marching trim, taking the Halkett boat with us. Five men followed all the sinuosities of the coast, while the others were spread at equal distances inland, Mr. Stewart and myself taking the middle space. Shortly after leaving the encampment a river was forded; this must be a large stream at a high stage of water; it was called Lemesurier River, after a relative of Mr.

[1] On his pioneer descent of the river in 1834.

Stewart's. No fuel was found in our encampments, and in two hours we left all signs of vegetation behind. The remainder of the peninsula is composed of high sand-hills, intersected by deep valleys, evidently overflown at spring-tides and during gales.

We encamped late opposite Maconochie's Island, and the only vestige of the missing party found was a small piece of cod-line, and a strip of striped cotton, about two inches long and an inch broad; these were found at Point Ogle, in an Esquimaux encampment, of perhaps three or four years of age.

Next morning a piece of open water enabled us to launch the Halkett boat, and explore Maconochie's Island, but nothing was found. It was impossible to cross over to Point Richardson as I wished, the ice driving through the strait between it and Maconochie's Island at a fearful rate. About three in the afternoon we began to retrace our steps through a tremendous storm of wind and rain. The last of the party did not reach the encampment till past ten at night, and as there was no fuel we were obliged to creep under our blankets thoroughly wet, and with no other supper but a piece of cold and rather ancient pemmican.

It was now evident that all that could be done with our means had been accomplished, and that with our frail craft, any delay in returning, would compromise the safety of the whole party. It may be thought strange that the remains of so large a party could not be discovered. It is my opinion, that a party in a starving condition would have chosen a low spot, where they could haul their boat up and have had some shelter; and that if they perished there, that their bones have been long since covered by sand or gravel, forced up by the ice. Any books or papers left open would be destroyed by the perpetual winds and rain in this quarter, in a very short space of time; for instance, a large book, Raper's Navigation, was left open on a cloak at Montreal Island; it was blown open, and the leaves were pattering about in such a way, that had it not been instantly closed, it would soon have been torn in pieces.

No party could winter on this coast: in the first place, there is not enough fuel; and, secondly, no deer pass. About a hundred deer, mostly bucks, were seen on Adelaide Peninsula, on our way to Point Ogle; but not one on our way back. Their tracks were all seen going to the south. On the eastern coast only five deer were seen. It would also be a matter of immense difficulty to get sufficient supplies down Fish River for even a small party.

On the 10th August a shift of wind enabled us to cross over to Point Beaufort, without injury; and a gale brought us to Point Backhouse, at 10½ P.M.

The Esquimaux were still at the rapids of Lake Franklin. Another attempt was made to see if they possessed papers of any description; the contents of our trading cases were offered for any. They showed us all their caches; but nothing of interest was discovered.

The fishery of fresh-water herring and trout appeared to be over, as those we saw hanging to dry on our way down, were all stowed away securely in caches; and the party were on the eve of departure to hunt deer.

Handsome presents were made to them, for which we got boots for the most of the men. The Upper Esquimaux were also seen, and treated in a similar manner.

The weather during the whole trip was dreadful, blowing continually, with rain, snow and hail; and it froze sharply below and above Lake Beechey: our canoes also were very frail and leaky. There was still less water in the upper part of the river than on our way down; from the lake above Musk Ox Lake to Lake Aylmer was almost one continuous portage. That lake was reached early on the 31st August.

Our progress through the lakes was much retarded by strong head-winds and fogs; some time was also lost in finding the very narrow and hidden outlets of Lake Aylmer, and Clinton Colden; at the latter I was disappointed in not finding Indians.

Early on the 9th, we reached a bay at the end of Artillery Lake, on the east shore, near the head of the Aheldessy.[1] It was impossible to descend that river; and we were employed the remainder of the day in discovering an Indian road to Great Slave Lake, through a series of small lakes and a small river.[2] After passing through eight small lakes, and making as many portages, we reached the river, and soon after got sight of Slave Lake. A portage of five miles was made with the pieces. The canoes were partly brought down by water. Mr. Stewart and I reached old Fort Reliance, where the new Establishment was also erected, about 3 P.M. of the 11th, and the canoes arrived at 10 the next morning.

As there was no prospect of discovering anything more from this quarter, I conceived that I was not justified in incurring further expence or risk, and, according to your instructions, determined on sending out the people this fall.

Mr. Lockhart had left the day previous to my arrival with two small boats which, with their ladings, were to have been put en cache at Sussex Lake, in the event of the Expedition having been continued another season. I sent off two men immediately to recall him. Mr. Stewart remained to pack up everything at the fort, while I left the 12th at 2 P.M. to stop the boat coming from Resolution with supplies.

No privation was sustained by the party for want of provisions. We brought three pemmican back.[3] Sir George Back saw immense numbers of deer and musk oxen on his way down. We only saw a few scattered deer with their fawns, the bucks having all passed to the north, and a few herds of musk oxen. On our way down no time was lost in hunting these, as we got as many Canada geese as we wished by running them down – they were moulting, and were all ganders. On our way up, many tracks were seen going south, but no deer, until we arrived at McDougall's Lake, and then only a few does. At the head of the river, and in Lake Aylmer, and Clinton Colden, they were pretty numerous, and among them many bucks in fine conditions. The following is our game list – 289 geese, 25 deer, 1 musk ox.

The conduct of the men was beyond all praise. They sustained hardships and risks of no ordinary description, not only with cheerfulness but with gaiety. The weather on the voyage up was very severe. Not a day passed without rain, sleet, and hail falling between Point Back House and Musk Ox Lake, after which we had occasional fine days. None of the party were provided with waterproof clothes or bags:

[1] The Lockhart River.

[2] Pike's Portage.

[3] Three pièces of pemmican, i.e. 270 lbs.

the canoes also were very leaky; still not a murmur was heard, though their groans at night evinced that they suffered from pains in their limbs.

Trusting that my proceedings will be approved of by Her Majesty's Government, and the Honourable Company.

P.S. Messrs Stewart and Lockhart, with the remainder of the party arrived here on the 20th instant, and will leave to-morrow morning. They will, I trust, reach Ile à la Crosse by open water.

3. Letter from James Anderson to Lady Franklin, 17 September 1855, Fort Resolution.[1]

Your kind and touching letter of 15th Decr. last with a duplicate of your niece's letter under the same date reached me at Fort Reliance on my return from the Arctic coast on the 11th instant.

My despatches to Sir George Simpson giving details of our operations accompany this and will doubtless be made public; to them I will refer you for an account of our proceedings. I have only time to touch on what will interest you most.

You are perhaps ere this aware that we were obliged to leave without an Esquimaux interpreter; we were therefore compelled to have recourse to signs in our intercourse with these people aided by a few words and phrases which two of Dr Rae's men – who accompanied me – were acquainted with; we had also recourse to Washington's vocabulary but it was of little use. I reached Fish River by quite a different route to that of Capt. Back's; it was dreadfully bad but shorter and indeed had I not adopted it the Expedition would have failed in reaching the coast this season.

After many narrow escapes we reached the Rapids at the outlet of Lake Franklin on the 30th July. We there met with the Esquimaux; they are now reduced to 3 families. We here immediately perceived traces of the missing party. The poles of their tents and cayak paddles were made out of ash oars and poles; many of their implements were constructed of elm, oak, mahogany and pine. They also possessed tin and sheet iron boilers of this form, various carpenter's tools, pieces of instruments, a letter nip with 1843 on it, but not a scrap of paper of any description. We showed them printed as well as manuscript books and letters and offered them everything we had for any; though they evidently understood us, they made us understand that they had none, and that the articles they possessed came from a boat brought by some whites, who had all died from starvation.

On the 1st August – the precise day Sir G. Back reached it on his second season – we reached Montreal Island with a hairbreadth escape from being crushed by drifting ice in the Traverse. Two entire days were devoted by the whole party to the examination of this Island. At its S.E. extremity on a high ridge of rocks were a number of Esquimaux caches which besides seal oil contained various articles belonging to a ship or boat, such as chain hooks, tomahawks, blacksmith's implements, a bar of unwrought iron etc. Here also it was evident that the boat had been cut up. There were butt ends of planks evidently sawed by unskilful hands and a

[1] BCA AB40 An 32.2, ff. 102–9; SPRI MS 248/333.

quantity of chips; on one of them the name 'Terror' was carved. Several pieces of ash strung together for some purpose were found in one of the caches; on one of these 'Mr. Stanley' was carved apparently with a knife (Surgeon of the Erebus). Rewards were offered for the discovery of the graves or remains of the party but not even a single bone could be discovered, though the men were most zealous and as quick sighted beings as exist. The strait between Montl. Isld. and Adelaide Peninsula was then traversed and the coast searched most minutely by a land party from Elliotts Bay to Point Pechell. Beyond Montreal Isld. the ice was firm and unbroken from 6 to 7 feet thick, and new ice formed every night; we could only get our canoes as far as that point by navigating them alongshore at high tides and through channels <u>on</u> the ice; we also made occasional portages across points. By this time the canoes were so much damaged and the ice so firm and grounded that it was impossible to take them further. I therefore determined to explore the remainder of the Peninsula on foot. With the exception of 2 men left to repair the canoes the entire party was employed in this task. Nothing was found except a small strip of striped cotton and a piece of cod line in an Esquimaux encampment of perhaps 3 or 4 years standing at Point Ogle. A small piece of open water enabled us to explore McConochie's Island by means of the Halkett boat. The ice was drifting so fast between it and Pt. Richardson that we could not cross over to examine it as I wished.

There was no probability that the ice would break up and even if it had our canoes were in such a wretched state that I could not have proceeded further without sacrificing the lives of the party – we therefore set out on our return. The weather was dreadful; we had rain, hail and snow the entire voyage up to the head of the River, where we had occasional days of fine weather. L. Aylmer was reached on 31st August. Fogs, head winds, etc., retarded our progress to the end of Artillery Lake. I was disappointed in not finding Indians to guide us, and we had to find our own way through a series of ponds and small lakes to a small river falling into Slave Lake close to Fort Reliance, which I reached on the 11th Septr. The canoes arrived the next morning, one of them utterly worn out and the other but little better. I had made preparations for a second season, but as nothing could be done from this quarter, I did not feel justified in incurring further risk or expense. The party will therefore go out by this opportunity. No party could possibly winter on that horrid coast; it is utterly destitute of resources, and it would be impossible to take down supplies by this most dangerous route. What an opportunity Cap. Collinson lost!

In my former communication I presumed to offer a suggestion regarding a vessel being sent out to Prince Regent's Inlet and a party to proceed on foot in the spring through Bellot's Straits to King William's Land and to return via the Isthmus of Boothia. I am still of the same opinion; the party however should be pretty strong – say 12 men and well armed, as I understand the Esquimaux of that quarter are rather bold and troublesome – none but good walkers and experienced men should be sent.

I myself must decline going on any other expedition. Though I am now 43 years of age I feel no diminution of strength or activity but the veins of my right leg are so much swollen that it is likely it might fail me. I trust that your Ladyship will be satisfied that everything that could be done with our means has been accomplished. I can assure you that our lives were risked freely and we have ample reason to be grateful to a merciful God for protecting us through so many perils.

It may appear strange to a person unacquainted with the way in which gravel is forced up by the ice, and how sand drifts during the perpetual gales in these regions that some of the remains of so large a party as are said to have died on these shores, could not be discovered. It is my opinion that a party weak as they were from want of food, would have sought the lowest and most sheltered spots to haul their boat up and encamp. Many parts of Adelaide Peninsula and Montreal Island are overflowed at high tides and during gales, and I believe that during the 5 years that have elapsed since their reported death, their remains must be covered with many feet of sand and gravel. Any book or document left unprotected would be destroyed by the perpetual rains and wind in this region – wolves would destroy any leather-covered book.

Every exertion was made to discover the graves as it struck one that the journals or despatches might have been placed in one of them to preserve them from the Esquimaux; many Indian tribes imagine that paper speaks to the white men and the Esquimaux might have destroyed the books and papers lest they might divulge that they had taken away the property of the missing parties.

Pray excuse this hurried letter.

P.S. I intended to have taken a furlough and visited England but I have been requested to resume the charge of McKenzie's Rr. Distt.

The wind during the whole summer was generally N.E., sometimes N. & N.W. but very rarely indeed veering to the Souwd. J .A.

4. Letter from James Anderson, Fort Simpson, to Eden Colvile, Fort Garry, 29 November 1855.[1]

I trust that you received my letters of 1st May & 15 June last. You will of course see my despatch of 17th Sept. to Sir G. Simpson. It is a hurried composition but will give you a detail of our doings last summer. I shall therefore not trouble you with repetitions, but mention some incidents which may interest you.

You may have noticed that I hardly mention Mr. Stewart in my despatch and may have attributed it to ungenerous motives, tho' I think few who know me well would. Nothing gives me more pleasure than to contribute to the advancement of true merit and I believe that no officer under my command has had reason to complain of me. But as I really could not – with any regard to truth or honor – say anything in Stewart's favor, I was silent. I solemnly assure you that instead of being of any assistance to me, he was a burden and incumbrance. Would you believe that an officer on such an expedition would sleep the greater part of each day, tho' as I ran every rapid first and was always ahead in the crevasses etc. it was most probable that something might occur to me. He is so heavy a sleeper that I could not even trust him to awaken the men; the consequence was, that as I was the guide and obliged to be constantly on the alert, I was half dead for want of sleep going down. I thought that he would have been rash and imprudent, but I had no cause of complaint on this head. I could mention many other things discreditable to Mr. S. as an officer,

[1] BCA AB40 An 32.2, ff. 110–14.

but as I have no wish to make any complaint against him, but solely to exonerate myself from any suspicion of ungenerosity, I shall say no more.

Mr. Lockhart, who made arrangements for another expedition at Fond du Lac, is a remarkably smart and clever officer.

The new Fort was built precisely on the site of Back's old Fort Reliance, but on a smaller scale. The chimnies of the old Fort were used for the new one.

The conduct of the men was beyond all praise. I suppose so severe a trip particularly in regard to portages was never made. My bowsman, Ignace Montour, the runner, made an extraordinary trip. He came up with the canoes to R. R.[1] in the spring, returned to Montreal, came up a second time via the prairies with Dr. Rowand, returned, and then a third time on this expedition as far as Big Island. Old Bap. Assinienton was Stewart's bowsman; he is an extraordinary old man and had also made the spring trip to No. House[2] and came up to Chipewyan on foot; he still carries his pieces[3] as well as the best.

The absence of an Esquimaux interpreter was severely felt. One thing, however, is certain, the Esquimaux have neither papers nor books. Regarding our not finding the bones of our unhappy countrymen, I had not the least doubt they sought out a low, sheltered spot to encamp – there are many such on Adelaide Peninsula and Montreal Island overflowed at spring tides – and died there. Their bodies would then have been torn to pieces by the wolves and foxes, their bones scattered about and then buried by the sand which drifts there like snow. Any leather-covered books would also have been torn to pieces by wild animals, and papers would soon be destroyed in that stormy region.

No party could winter on that bleak coast. It would be impossible to get sufficient supplies down that dangerous river. There is a scarcity of fuel, and we saw no deer-pass near the coast. Dr. Rae's men represent Repulse Bay as an Eden to that horrid coast.

You can conceive that we did not go down such a dangerous river, cross so many long traverses and pass through so much ice – especially crevasses – without having many hairbreadth escapes; we, however, met with no serious accident. I had only the paltry map attached to Back's narrative as a guide – except on the coast where I had fine maps – it was useless in the large lakes, and I had no assistance from Indians in the Upper Lakes as Back had. When I got sight of Slave Lake from the top of a high mountain, I felt as if a load had been lifted off me. Mr. Stewart, having received some instructions from Mr. Ross, the astronomical instruments were taken with us, but after a few trials Mr. S. had to give it up as a hopeless affair.

The roughing we underwent did me good. The veins in my right leg, however, have increased in size. Though we had a great deal of bad walking I found I could get along as well as the best; still it is probable that such a leg might fail (and if it did in such an expedition it would be certain death). I must decline going on another such expedition.

[1] Red River.

[2] Norway House.

[3] The standard 'pièce' of 90 lbs; on portages each man usually carried two pieces at a time.

I suppose that Messrs. Stewart and Lockhart are now at R.R. and that the despatches will reach England about the New Year. I trust that the Hon. Co.[1] & H.M. Gt.[2] will be satisfied with our exertions.

After the arrival of the Athabasca Express at Resoln. I left for this place, which I reached on the 3d. Ulto. Mr. Campbell left for Liard on the 5th. I found matters in rather a perplexed state. Our crops also failed, and by bad arrangements we only received 6000 instead of 16,000 fish from Big Island. Campbell believes – and I am half of the same opinion – that the curse of Glencoe adheres to him.[3]

My children at R.R. are getting on very well and Mrs. A. and my family here are quite well. Trusting that this will find yourself and Mrs. Colville in good health and with best wishes

P.S. I took the liberty of recommending Messrs. W. Hardisty[4] and B. R. Ross[5] to your notice as valuable, well educated officers worthy of promotion; they are so much out of the way that they are almost unknown, and have only their merit and ability to recommend them. The former is of 14 years, the latter of 13 years standing. I trust that they have not been forgotten this year.

5. Letter from James Anderson, Fort Simpson, to Sir George Simpson, 29 November 1855.[6]

I last had this pleasure from Fort Resoln. I fear that you will not be well pleased with either my letter or despatch. The fact is they were both most hurriedly written. The latter, however, gives the facts and I trusted you for working it up in a presentable form. I had promised myself to write my despatch after arriving at Lake Aylmer, expecting to have met Indians there to guide us, but in this I was mistaken, and my whole attention was required to guide the party through those intricate lakes as well as across the Barren Grounds to Slave Lake, and in fact till we again reached the 'Mountain'. When I first got sight of Slave Lake from the top of a high mountain it was as if a heavy load were lifted off me. After a person has had all his faculties on the stretch without relaxation for so long a period as 4 months, directly the necessity for exertion ceases a kind of lassitude supervenes and the mind requires repose to regain its usual tone; at least such is the case with me.

The fall was a late one in this quarter and I think that Messrs. Stewart and Lockhart should be now near Red River. I requested Mr. Swanston to send off Ignace[7] 2 days

[1] Honourable Company, i.e. the Hudson's Bay Company.

[2] Her Majesty's Government.

[3] A reference to the Massacre of Glencoe of 13 February 1692 whereby many of the McDonalds of Glencoe were massacred by a party of Campbells under the command of Campbell of Glenlyon after enjoying the McDonalds' hospitality for several days – in culpable violation of the Highland tradition of hospitality. In Scotland, at least, a stigma has attached to the name Campbell ever since.

[4] William Hardisty, stationed at Peel's River (now Fort McPherson).

[5] Bernard Ross of Fort Resolution.

[6] BCA AB40 An 32.2, ff. 94–6.

[7] Ignace Montour, who evidently had been selected to carry the dispatches back from Red River to Lachine.

after his arrival, so I trust that you will receive the despatches next month. I shall be anxious till I learn if you and H.M. Gt. are satisfied with my proceedings. I know that everything was risked and no trouble spared to fulfil the objects of the Expedition.

By this opportunity you will receive a copy of my Journal.[1] It is nothing more than a record of occurrences. It was useless for me to repeat Back's descriptions of the country and scenery. I have been so occupied since my arrival that I have found it out of my power to make out the map of my route to L. Aylmer; the original will require to be reduced on a correct scale – a work of some time. I shall send it by the March express.[2] The expedition accounts and transfers cannot be sent out till the summer; an interpreter who remained at Ft. Reliance[3] to gather up the meat debts given out by Mr. Lockhart had not arrived when I left Resolution and there are transfers to arrange at Ft. Chipewyan.

I now send out to No. Ho. various articles belonging to the missing party which would have been too cumbrous for Messrs Stewart and Lockhart to take, as it was probable that they would have to leave their canoes on their way to Red Rr.

We may congratulate ourselves that the expedition was finished this fall. Had it been carried on another season, starvation would have been rife in the district. The expenditure of provisions at this place last summer was enormous. Our crops have failed and by faulty arrangements instead of 16,000 fish only 6000 were rendered here from Big Island.

The Expedition also would have been sorrily provisioned as a considerable portion of the pemmican Campbell sent to Fort Reliance was rotten.

I left Fort Resolution – after the arrival of the Athabasca Express – on the 28th Septr. and arrived here after a cool trip on the 3d October. Mr. Campbell left for Liard on the 5th ...

The roughing I underwent last summer did me good. The veins of my right leg have, however, increased in size and I think I cannot delay taking a furlough – say in two years out 1858. I shall have honestly earned it by that time....

6. Letter from James Stewart, Montreal, to Sir George Simpson, 31 January 1856.[4]

According to the orders received from you last winter I started with the men from Montreal and Red River for Athabaska which place we safely reached in the beginning of March. Here we were employed till the ice gave way in making canoes and other preparations necessary for our voyage to the coast. During our stay we were joined by 3 men from Norway House,[5] 2 of whom had been with Dr. Rae the year

[1] This is the fair copy of the journal, now preserved in the Hudson's Bay Company Archives as HBC B.200/a/31.

[2] Anderson's field sketch maps, and fair copies of his final, large-scale maps are preserved in the Hudson's Bay Company Archives (E.37/16 and E.37/29); a portion of these is reproduced as Plate V.

[3] King Beaulieu.

[4] HBC A.12/8, ff. 32–6; also, with minor differences, PAA 74.1/140.

[5] Thomas Mustegan, Murdoch McLellan and Paulet Papanakies.

before and in consequence of their experience they were a good accession to our crews. On the 22nd day of May we left Fort Chipewyan in two canoes for Fort Resolution, Great Slave Lake where the expedition was finally organized, and Mr. Anderson joined it on the 21st day of June. The following morning we started for the sea in 3 canoes manned by 15 men and some 4 or 5 Yellowknives or Chipewyans who were to act as guides as far as Muskox Rapids. After being windbound at Rocky Point and detained by ice portages we arrived at what is called the Mountain Portage on the morning of the 2 July. This is opposite to Cahoutchella or Rabbit Point. By this route we avoided Hoarfrost River, Artillery and Clinton Colden lakes, falling on Lake Aylmer at the northwestern extremity. Our route consisted of some 25 or 26 portages with small lakes between. Our men were much fatigued before reaching Lake Aylmer; this was accomplished on the 7th. Here we were again arrested by ice which in some places was as firm as it would be in the middle of winter. Making portages, cutting ice and availing ourselves of every lane of open water that presented itself we reached the portage to Fish River after seriously damaging our canoes, on the 11th about noon; and considering that we had at last arrived at the proper starting point we encamped early, after making the first portage from Sussex Lake to Muskox Lake. The river was so shoal that the most of the way was one continued portage. From this lake, though, we found more water and we may say the hardest part of the journey was accomplished. On the 20th immediately after breakfast we came on the first Esquimaux tents where we were received very hospitably. From them we could gather no information regarding the ships as they came from Chesterfield Inlet, so much out of the way they did not appear to have even heard of the circumstances. After making them a few presents we continued our route to the sea. Numerous rapids were run in safety (altogether owing to the skilful management of our Iroquois bowsmen) and about noon on the 22nd came to the strait connecting Lake Pelly with Lake Garry. Here we found two Esquimaux tents; the natives on observing us took to the hills in the greatest consternation, leaving everything in their lodges at our mercy. Nothing was disturbed and after leaving some presents to shew them on their return that we were friends, the route was resumed. Some difficulty was experienced in getting through Lake Garry, owing to the defects in the chart, but this, as well as many others were surmounted with a little loss of time and on the 30th in the morning we came to the fall mentioned by Captain Back as being so dangerous.[1] Here a portage was made part of the way and the rest was run. Mr. Anderson and myself walked to the foot of the rapid and descried some tents on the other side of the river, which were visited as soon as we had breakfasted. Here we found that the men, with the exception of one, were absent, but from him and the women we learnt that they knew of the death of a party of Whites at no great distance to the North and from their information we felt assured that we should find no survivors. One woman in particular, was very explicit about having seen one man on an island at the last extremity. She shewed the way he was sitting on the beach, his head resting on his hands, the hollowness of

[1] The fall at the outlet of Lake Franklin.

his cheeks and the general emaciated appearance of the unfortunate person.[1] This she said was four years ago and that, being without provisions themselves, they could not give any assistance and that even if they could have done so it would have been too late. A great many of their lodge poles were evidently the oars of the boat mentioned by Doctor Rae; they also had the boat kettles as well as preserved meat cans & pieces of mahogany and pine. This information was a fresh incentive to our exertions as it appeared to us corroborative of Doctor Rae's report which formed one of the principal motives for our expedition. And the vague, faint hope of yet finding some of our unfortunate countrymen alone incited us to further exertions to reach Montreal Island, and we started immediately, running the last fall in the Fish River with full cargo, a little below which we fell in with some of the men belonging to the tents we had just left. They confirmed what we had already heard, adding that we would find the wreck of the boat, but that no vestiges of the bodies or papers would be discovered. On the first day of August we reached Montreal Island after crossing from Cape Beaufort. In the traverse we encountered a pack of ice through which, with much danger and risk, we passed unharmed. A few minutes delay would not only have prevented us from reaching the island for several days, but would have put an untimely end to the expedition, as the bay was completely covered with ice in about fifteen or twenty minutes. Early on the 2nd the search was commenced and by 11 am we had the good fortune to hear shots fired as a signal of something having been found. Mr. Anderson and myself were not long in reaching the spot, where two of our men had discovered the wreck of the boat, and on turning over the splinters and chips one of them was found which happened to bear the name of the vessel to which the boat belonged, namely 'Terror'.[2] Another piece of wood was found, which proved to be part of a snow shoe made in England, bearing Mr. Stanley's name cut out with a knife on it; this gentleman was surgeon of 'Erebus', Franklin's ship. Parts of their flag and many other things, evidently belonging to white people, were found near this spot. What would we not have given had it pleased Providence to have permitted us to find but one survivor! No words can fully explain the intensity of our feelings at that moment. The fact of the boat's belonging to the 'Terror' and the snow shoe to an officer of 'Erebus', together with what we gathered from the Esquimaux, confirm, without doubt, Doctor Rae's surmises as to the Fate of Captain Franklin's unfortunate expedition, namely that the ships had been crushed and that the survivors had attempted to reach some of the Hudson Bay forts via the Fish River but had perished in the attempt. After Montreal Island had been thoroughly searched, we proceeded to the mainland, on the western shore, where the search was resumed from the point of Elliott's Bay to McConachie's Island, which was also looked through. The examination was most minute, half the party proceeding on foot while the others brought the canoes along the shore through the ice. Nothing, however, was discovered in this part of the country, except two small pieces of striped cotton (such as is used for common

[1] Anderson nowhere makes any reference to this story of a survivor having been seen by the Inuit and he would later reprimand Stewart severely for relaying what he clearly believed was a fabrication to Sir George Simpson, and to the press.

[2] The ship's name was *Erebus* according to Anderson! See p. 133.

shirts) in an old encampment. We were quite satisfied, on the 8th August, that nothing more could be done; the bodies of our unfortunate countrymen have no doubt been covered by the sand on this low shore and the wolves may have assisted considerably in obliterating all traces of them. No prospect of further success being at all possible in this part of the country, the ice beginning to take, the ground covered already with snow, the utter impossibility of wintering on this part of the coast (where the means of living are not to be obtained with a view to extending our operations in spring) we therefore on the 9th of August commenced our return homeward, perfectly satisfied that our hopes of finding any one alive were futile, as well as feeling sure that the ships had been crushed somewhere in the Victoria Straits, and that the people had taken to the ice to attempt reaching Great Slave Lake, as being the nearest point where they would find assistance, all perishing in the attempt. We reached the mouth of the river on the 13th August after innumerable escapes from being crushed by the ice. We breakfasted again with the Esquimaux and they again told us the same story, which relieved us from any doubt that remained of the unfortunate people having died of starvation on the main land west of Montreal Island, and that one was seen to die upon the island itself. We offered them large rewards for any papers – even pieces or slips of paper – but they said they had nothing of the kind, that they would gladly give them to us had they any such thing, but that they must forego 'the pleasure of trading at such a profit.' This was a great drawback to our hopes of satisfying relatives and friends of the unfortunate people but we were obliged to be contented with what to us is evidence quite sufficient of the dreadful fate of our poor countrymen, who have without doubt perished from want. Let me say, Sir, that had the Government put this affair into your hands some five years ago, the Company's officers would no doubt have had the infinite pleasure of being instrumental in restoring these people to their anxious friends and relatives. Unfortunately we were too late and were we to go 20 times over the same ground I am satisfied nothing more could be discovered. The Esquimaux were particular in shewing us that they had no hand in killing them and by what we could judge from their countenances they were sincere in their protestations. After making them presents our homeward route was resumed and after safely ascending Fish River our canoes were once more launched on Lake Aylmer on the 31st of August. Sir George Back's accounts of this river are by no means exaggerated. Some difficulty was experienced in making our way through Clinton-Colden and Artillery Lakes and from Captain Back's description of the Abhaldessy[1] we did not like to venture our crazy canoes in such a river which obliged us to make numerous portages and follow the Indian route to Great Slave Lake, which was reached on the 12th of September, all in good health and spirits, having successfully accomplished the voyage entrusted to our joint command, proving beyond a doubt that Dr. Rae had judged rightly as to the place where the last of the unfortunate people perished.

At this place we found that Mr. Lockhart (the young gentleman we had left behind us to prepare our winter quarters) had succeeded fully in what had been entrusted to him. Fort Reliance, east end of Great Slave Lake, had been rebuilt and

[1] Lockhart River.

a goodly stock of provisions laid in, quite ample for wintering, had we been obliged to do so. Mr. Lockhart himself was absent, having started with two boats, according to his instructions, to make a depot of provisions and warm clothing at the head of the Fish River for the use of the expedition, in case of accident. Mr. Anderson left immediately for Slave Lake while I remained to get everything together, preparatory to our proceeding to Fort Resolution. Mr. Lockhart was recalled, and on his arrival we all embarked for home, reaching Fort Resolution on the 20th September and Red River on the 3rd of November. Our gallant fellows, who had behaved so well during the whole voyage and gone through so many dangers, were paid off, and I am happy to say were returned to their families in as good health as they had left them. During the whole trip no accidents occurred. There was nothing left for me but to proceed to Montreal with as much speed as possible and accordingly on the 13th with the three Iroquois (to whom we owed our lives so often) left Red River and arrived in Montreal on the 21st December, being much delayed by the bad state of the roads between Red River and St. Paul's. The successful issue of the expedition will show you that the arrangements for the safety of the party were well conceived and properly executed and I trust you will feel satisfied that nothing has been left undone. In truth it would have been difficult for me to have made any mistake, having your explicit instructions for my guide. It is with much pleasure that I address you because I am confident that we did as much as was possible to be done; it was the regret of Mr. Anderson and myself that we could not discover the bodies of the unfortunate people but, Sir, it was our misfortune to have been sent too late to effect that desirable object. Apart from this we have fully succeeded in furthering your views and those of the British Government and in confirming Dr. Rae's statements completely, having brought certain proofs of the fate of the last of Sir John Franklin's expedition, having spoken to two different parties who had seen the person in the last extremity.

7. Letter from Edward Hopkins, Lachine, to William Smith, London, 24 December 1855.[1]

...I have the satisfaction to report, for the information of the Governor & Committee, that Chief Trader Stewart arrived here on the morning of the 22 inst. from the Arctic Coast, via Red River and the United States, accompanied by the three Iroquois servants who were sent from hence in November last year to form part of the Expedition under the joint command of Messrs Anderson & Stewart. The result of that Expedition has been the confirmation of the report conveyed to England by Dr. Rae last year. The coast & islands in the locality where the party of whites are reported to have perished in 1850, were carefully searched, and there, as well as from Esquimaux in the neighbourhood, traces of the party were discovered – but no books or papers nor human remains, although the exact spot was visited at which the natives told Dr. Rae they had seen the bodies, but being a low sandy spit, exposed to the sea, the probability is they were washed off or buried in the sand.

[1] HBC A.12/7, f. 627; Hopkins included the newspaper clippings from the *Montreal Herald* and *Montreal Gazette* which follow.

On the Montreal Island, as stated by the Esquimaux, small remains of a boat were found, having been cut up for the sake of the wood and nails; among the chips & fragments, a piece of wood was discovered with the word 'Terror' branded upon it, & another piece has 'Mr. Stanley' (Surgeon of the Erebus) cut on it with a knife: this last is part of a snow shoe & probably of English make, being of oak, a description of wood never used for the purpose in countries where snowshoes are used. These two relics and a piece of rope with the Government mark in it, the step of a boat mast shod with copper, a letter clip dated 1843, some pieces of bunting and the remains of a thermometer, have been brought hither by Mr. Stewart; the more bulky articles Mr. Anderson has retained to be forwarded by another conveyance, consisting principally, I understand, of preserved meat cans, bar iron, ash oars branded with the broad arrow, & some tools. The plan of the expedition laid down in Sir George Simpson's instructions of 18 November 1854 was carefully carried out by all parties entrusted with its execution and with complete success; the officers, servants, craft, and supplies were assembled at the rendezvous on Great Slave Lake at the appointed date & returned thither, after having accomplished the arduous service, without experiencing any privations, or meeting with a single accident to person or property. For fuller details I beg leave to refer to the accompanying copy of Mr. Anderson's report addressed to Sir George Simpson, dated Fort Resolution 17 Septr. 1855....

8. Extract from coverage in *The Minnesotian* (St Paul).[1]

Mr. Stewart, with a party of 14 men, therefore, started from his post, the Carlton house, in 54°N. latitude, on the 7th day of February, 1855, and proceeded to Fort Chipewyan, at the head of Lake Athabasca, in latitude 58°N., at which point they arrived on the 5th day of March. It had been determined to make the trip to the Arctic Sea by water so far as was practicable, and the party therefore remained at this post until the 26th of May, busily engaged in constructing boats, and making other preparations for their dreary journey. At that date the party left Fort Chipewyan, and journeyed by canoe on the Peace River, which connects Lake Athabasca with Slave Lake, some 350 miles in a north-westerly direction, till, on the 30th day of May, they arrived at Fort Resolution, which is situated on an island in Slave Lake, about latitude 61°.

At Fort Resolution the party was joined by Mr. Anderson, who, with Mr. Stewart, had been appointed to the command of the expedition. Here another delay was made, for the purpose of reorganization, and making the last preparations before attempting to penetrate the interminable frozen north. These arrangements completed, the party started out on the 22nd of June for the head of Great Fish River, or, as it is known on the map, Back River, in latitude about 64° North. Thence they followed the course of the stream to the Arctic Ocean. Mr. Stewart represents the navigation of this river as exceedingly dangerous, being obstructed by over 100 difficult rapids. Over all these, however, with nothing more substantial than birch-bark canoes, they passed in safety, and arrived at its mouth on the 30th of July.

[1] *The Minnesotian* (St Paul), 12 December 1855, quoted in *The Times*, 9 January 1856, p. 8.

Here they met with Esquimaux, who corroborated the reports of Dr. Rae, and directed them to Montreal Island, a short distance from the mouth of Back River, as the spot where, according to their instructions, they were to commence minute exploration. From this time until the 9th of August the party were industriously engaged in searches on the island and on the main land between 67° and 69° north latitude. We cannot recapitulate the perils escaped and privations endured by the brave band while seeking to find traces of their countrymen who had perished on those desolate shores. Three times they providentially escaped being 'nipped', as Mr. Stewart expressed it, or crushed between moving mountains of ice. At last on Montreal Island, where their explorations commenced, they found snow-shoes, known to be of English make, with the name of Dr. Stanley, who was surgeon of Sir John Franklin's ship, the Erebus, cut in them by a knife.

Afterwards they found on the same island a boat belonging to the Franklin Expedition, with the name 'Terror' still distinctly visible. A piece of this boat, containing this name, was brought along with him by Mr. Stewart. Among the Esquimaux were found iron kettles corresponding in shape and size with those furnished to the Franklin Expedition, and bearing the mark of the British Government. Other articles, known to have belonged to the expedition, were obtained from the Esquimaux, and brought by the party for deposit with the British Government. No bodies, however, were found, or traces of any. The report of the Esquimaux was, that one man died on Montreal Island, and that the balance of the party wandered on the beach of the mainland opposite, until, worn out by fatigue and starvation, they, one by one, laid themselves down and died too.

The Esquimaux reported further, that Indians far to the north of them, who had seen the ships of Franklin's party, and visited them, stated that they had both been crushed between the icebergs. Mr. Stewart took especial pains to ascertain whether the party had come to their death by fair means or foul; but to every inquiry the Esquimaux protested that they had died of starvation.

Gathering together the relics found, the party set out on their return on the 9th of August last. The return route did not vary materially from that taken on their way north. Mr. Stewart has occupied the whole time since in reaching our city, having come by way of the Red River country, and having been absent in all about 10 months. Mr. Stewart left St. Paul yesterday en route to the Hudson's bay headquarters at Lachine, Canada, to submit an account of his adventures.

9. Extract from coverage in *St. Paul's Times*.[1]

The expedition reached what is called Montreal Island, where they fell in with some Esquimaux, who informed them where the crew of the Terror (one of Franklin's ships) met their untimely fate. They gathered up the remains of a boat having the name of Sir John Franklin on it, a hammer, kettles, part of a blue flag, and other articles belonging to the unfortunate vessel. We are informed by the Esquimaux that they reached the spot just in time to see the last man die of hunger, who was leaning against some object when discovered. He was too far gone to be saved. The

[1] *St. Paul's Times*, 12 December 1855; quoted in *The Times*, 3 January 1856, p. 8.

wolves were very thick there, and no traces of the bones of the men could be seen, supposed to have been eaten by the wolves. The Esquimaux state that it is four years ago since the crew perished. It was on the coast opposite Montreal Island. Their bones lie buried in the sand within an extent of 12 miles. This is the fifth winter since they perished, and the drifting sands of that barren region, being in lat. 68° north, have piled in successive layers on the bones of these noble and ill-fated men. Mr. Stewart describes the region as dreary in the extreme – not a blade of grass or a stick of timber met the eye. No game of any kind could be found. The Esquimaux, from whom their information was obtained by signs, pressed their fingers into their cheeks, and placing their hands on their stomachs, endeavoured to indicate the manner of their horrible death. They were charged with killing them, but merely answered with their sighs.

10. Coverage in *Montreal Herald*.[1]

Return of the Hudson's Bay Company's
Arctic Expedition in search of Sir John Franklin
and confirmation of Dr. Rae's reports

We have been favored by E. M. Hopkins, Esq., (in the absence of Sir George Simpson) with the following outline of the proceedings of the Arctic Expedition, which, by instructions from Her Majesty's Government, was employed by the Hudson's Bay Company to follow up the clue discovered by Dr. Rae while engaged on another exploring expedition, also fitted out by the Hudson's Bay Company, of the fate of Sir John Franklin's party.

It will be in the recollection of our readers that it is scarcely a year ago that we published to the world the first authentic information which had been received, of the lamentable fate of the gallant Franklin and his brave comrades. The intelligence which was conveyed to Dr. Rae in the winter of 1853–54, by the Esquimaux, and in the accuracy of which, that distinguished arctic traveller placed perfect reliance, was received by the public in England with great hesitation, arising, probably, from an unwillingness to believe the mournful facts. That intelligence was in substance that, in the winter of 1850 the Esquimaux saw a party of whites travelling from the northwards towards the Arctic coast dragging a boat over the ice, intending to use it as soon as they reached open water; that the party, about forty in all, made the land near the mouth of a large river (The Great Fish River of Back) and there perished of starvation, to which were added a number of frightful details of their sufferings, which we will not again inflict on our readers. In proof of the truth of these reports, the Esquimaux exhibited and sold to Dr. Rae a great variety of relics, principally silver forks and spoons, marked with the crests and initials of various officers of the ships Erebus and Terror, (Franklin's), and, amongst other articles, a small order, or star, with Sir John Franklin's name engraved on it. These were the tangible proofs conveyed to England by Dr. Rae in confirmation of the tale he collected from the Esquimaux; but his proceedings and conclusions have frequently been called in question, and, therefore, it will be the more gratifying to him now

[1] *Montreal Herald*, Monday 24 December 1855.

that they are fully corroborated, even to minute details of locality etc., in which he might possibly have been mistaken.

As soon as Dr. Rae had laid his report before Her Majesty's Government, it was decided that an attempt should be made to follow up the trace he had obtained, commencing at the point indicated by the Esquimaux as the scene of the last sufferings of the party of whites seen by them in 1850.

The organization and management of this new expedition were wisely entrusted to the Hudson's Bay Company. On the 27th October, 1854, the instructions of H.M. Government and the Company were forwarded to Sir George Simpson at Lachine, where he received them in the middle of November. His great experience and well known ability in affairs of that nature, enabled him to decide with promptitude on the mode of carrying out the expedition, the men to be employed as leaders and in subordinate capacities, the amount of supplies, craft, and all other requisites for the undertaking; and on the 20th November, last year, his instructions were dispatched by special messenger to the Hudson's Bay Territories, all parts of which were under requisition to furnish *matériel*, the whole to be collected at the rendezvous, Fort Resolution, in Great Slave Lake, by the 1st June following; and so complete were the plans, and so carefully had all contingencies been provided against, that in no point was there a failure in carrying out his arrangements.

The officers selected to lead the party were Mr. Anderson, a Chief Factor of the Company, and Mr. J. G. Stewart, a Chief Trader, both well qualified by experience, courage, physical strength, etc. for the arduous duty. The party consisted of these two officers and fourteen men, and left Fort Resolution, a post of the H.B. Company on Great Slave Lake, on the 22nd of June last in two bark canoes, in which they performed the perilous voyage down Great Fish River – a river known to the world for its dangers and horrors by Sir George Back's narrative. From Mr. Stewart we learn that he doubts that the party ever could have got safely down that stream to the coast had it not been for the wonderful dexterity of the three Iroquois voyageurs whom Sir George Simpson had prudently forwarded from Lachine to join the expedition – the three best men of his own canoe.

The party reached the outlet or estuary of the river on the 30th of July, and skirted along its eastern shore as far as point Beaufort, but found no traces to reward their search. From thence they crossed over to Montreal Island, 12 miles distant, lying near the western shore of the estuary; probably, in that crossing, incurring a great peril as any in the gloomy record of Arctic travels, pushing their bark canoes boldly out into the Arctic Ocean, and forcing their way through drifting masses of Arctic ice seven and eight feet thick. But they were prepared to make any effort to reach the island which, as well as Point Aigle [sic], near it, had been the places Dr. Rae understood the Esquimaux to mean when describing where the white party perished in 1850: and they had the melancholy satisfaction of procuring, on that very spot, the fullest possible confirmation of Dr. Rae's report. They also met Esquimaux in that vicinity who had seen the whites, and gave much valuable information. Suffice it to say, that on the island were discovered the remains of a boat, which had been partially destroyed by the natives for the sake of the wood and the metal fastenings. Although there was sufficient left to identify it as belonging to the Franklin Expedition, one fragment of wood (now, as well as some other small relics, in the possession of the Hudson's Bay Company at Lachine) having the name 'Terror' branded on it, while another piece has the name of Mr. Stanley, (Surgeon of the 'Erebus') cut upon it, this latter being part of a snow-shoe, evidently of English manaufacture, being made of *oak*, a species of wood no man accustomed to use snow-shoes would ever select for the purpose. No papers or books, and no human remains, were found; nor was it likely, as four years had elapsed since this tragedy was enacted upon a low sandy beach, exposed to the storms of four Arctic winters, and there is little doubt that either the sea has washed off, or the sand has buried deep, the unfortunates who perished on this spot. The Esquimaux were very friendly, and freely displayed all their treasures, obtained from the boat, or found near it, and these consisted principally of the oars, used by them as tent poles, the boat kettles, the empty preserved meat cases, etc., etc., but no papers, and the natives stated, with every evidence of sincerity, that none had ever been seen or found.

Everything portable was secured by Messrs. Anderson and Stewart and brought back, and are now on their way to Canada; it would be useless to recount them all, but we may mention bar iron, rope with the Government mark on it, oars branded with the broad-arrow, Pieces of bunting (remains of a flag), a letter holder, a step of a mast, etc. etc., all clearly European and all Government supplies. Is anything more wanted?

The weather is described as having been 'execrable', constant storms, with ice, snow, rain, sleet, hail, thunder, and whatever else can be conceived that is disagreeable. It is a part of the coast the natives even consider uninhabitable, merely visiting it for a short time in summer when the deer pass that way. On the 14th August, when the expedition commenced its retreat from the coast, the ground was covered with fresh fallen snow and the ice was forming; in fact, winter had set in. Few further details of the last moments of the lost party have been collected; we may mention one mournful incident reported by an Esquimaux woman, who saw the last man die; he was large and strong, she said, and sat on the sandy beach, his head resting on his hands and this the last survivor of Franklin's Expedition yielded up his brave spirit. Messrs. Anderson and Stewart retraced their steps to Great Slave Lake, from whence the latter continued his journey onwards to Red River settlement, and thence via the Minnesota territory to Montreal, where he arrived on Friday evening last, direct from the Arctic Sea, after upwards of five thousand miles travel in open craft and through uninhabited regions, without a halt. A few facts, taken at random, may serve to bring home to our appreciation what this North-West expedition accomplished and went through. In thirteen months, to a day, the Iroquois who were sent from Lachine to form part of the expedition returned thither, thus performing in one year the same service that Sir George Back got through in three. For sixty days and nights the party saw no fire, there being no timber on the Great Fish River or Arctic coast; and during those sixty days they travelled incessantly in open craft in a wretched climate, never had dry clothes or slept on dry blankets, and never eat cooked victuals except on rare occasions, when they made a little tea by means of a lamp. This party of sixteen in all travelled in bark canoes down one of the most turbulent rivers known, even to North-West voyagers; ventured among the ice on the Arctic sea; and returned to their starting point without meeting a single accident to person or property – and, withal, performed all that was required of them; and had they gone out four or five years earlier would, no doubt, have been instrumental in saving the lives of a portion of Franklin's party.

We think the foregoing narrative is ample corroboration of the wisdom of the recent outcry, to put 'the right men in the right places'.

One word in conclusion as to the Franklin Expedition. The two vessels Erebus and Terror left England in 1845 – were last heard of in 1845. They probably tried several passages, but were baffled by the ice; and finally in 1848 were crushed, probably in Victoria Straits. Many of the crews perished, but one or more boats got off with the survivors, who took all the stores they could collect and travelled Southward towards the Arctic Coast, in the hope of reaching some of the Hudson's Bay Company's posts. The season of 1849 was probably spent on this dreary journey, and renewed in 1850, where they reached the coast at the mouth of Fish River, but in so exhausted a state that they could merely run their boat on the beach and crawl ashore to die. This seems all that is certain, and all that we can ever know, of the fate of the Franklin Expedition.

11. Coverage in *Montreal Gazette*.[1]

Return of the Hudson's Bay Sir J. Franklin searching expedition.
Further discovery of remains

Our readers will remember the great interest excited last autumn by the publication of the painful intelligence procured by Dr. Rae of the unfortunate fate of Sir John Franklin and his companions. Dr. Rae's narrative gave rise to controversy, and a desire that the mystery partially solved by the information given by the Esquimaux last winter to him, should be more fully investigated. In these circumstances Her Majesty's Government decided that an effort should be made to follow up the clues so unexpectedly obtained; and at the same time rescue the survivors, if any, who were seen near the outlet of Back's River; or at least to procure any records that might have been deposited at the place where they were reported to have perished. Her Majesty's Government decided that this task should be confided to the Hudson's Bay Company. They undertook it, and intrusted the organization of the expedition to Sir George Simpson. It consisted of two leaders and fourteen men. The leaders were J. Anderson Esq., and J. G. Stewart Esq. The men were picked from all parts of the Company's territory, for experience, strength and powers of endurance. The leaders were also selected for their ability, experience, zeal and discretion in their dealings with the Indians. The Company laid much stress in obtaining the 'right men' for undertaking the arduous task, and it is well known they have many able men in their service.

We have now information of the result of the expedition. Mr. Stewart, who is, it may be stated, the son of the Hon. John Stewart of Quebec reached this city on Friday last. Mr. Anderson stopt at Fort Simpson.

[1] *Montreal Gazette*, 24 December 1855.

From Mr. Stewart we learn that Sir George Simpson issued the instructions from the Hudson's Bay House, Lachine, in November last. The expedition was organized at Fort Resolution on Great Slave Lake, and started on the 22nd June last. It proceeded down Fish River, in canoes, and reached the sea on the 31st July, in a little over the space of a month. That river, we are informed, is very difficult to navigate for either boats or canoes. Sir G. Back, however, went down in a boat, and possibly that is the reason he took so long a time. Mr. Stewart says he counted one hundred and seven rapids, and then gave up the task. Had it not been that the canoe-men of the party were very expert they would not have succeeded in getting down in safety. There is not a stick of wood on the banks of the river. The water besides being rapid is what is called very 'strong' – that is very trying to the canoes. Mr. Stewart states that the eloquent description of Sir G. Back of Fish River is remarkably accurate.

Having reached the mouth of the river, the expedition still kept to their canoes, and coasted along the east side of an estuary, untill they came to Cape Beaufort. Hence they crossed to Montreal Island, distant about twelve miles. This island is a pile of rocks and swamps with a sandy beach. On it they found the remains of a boat, broken up; together with pieces of oars, masts, etc. On some of the pieces of wood the word 'Terror' was cut or branded in the same way that boats' names commonly are. Large pieces of mahogany and bars of iron were found; also sea-kettles and preserved meat cans. Pieces of the oars and the kettles had a broad arrow marked on them. A piece of a snow-shoe was found, with the name 'Dr Stanley' cut on it with a pen-knife. Some of these remains have been brought on, and others will be. No human remains were found. These the Esquimaux said had been washed away by the sea. Such might have been the case; or they might have been buried by the washing or drifting of sand on the beach. Mr. Stewart informs us that only a very short time would be required to do this. There are, besides, wolves on the island and the neighboring main land, and it is possible they might have eaten the bodies. But, whatever had become of them, they could not be found, or any part of them.

On, or just before the arrival of the Expedition at Montreal Island, the Esquimaux they met knew its object, and voluntarily gave all the information they could respecting it. They stated that the boats' crew had died of starvation; and one woman, the most intelligent among the whole, gave very particular description of seeing the last man die. There was no suspicion of killing or any foul play – the Esquimaux themselves were short of food at the time, according to their statement. Some of the boat's crew died on the mainland, and some on the island.

It is supposed that the ships had been crushed in Victoria Strait or some other place to the north of where these remains were found; and that one surviving boat's crew had endeavored to reach some of the posts of the Hudson's Bay Co. but had not the necessary strength. It is supposed the boat touched point Ogle, a spit of sand a few miles to the North of where the remains were found, and that the crew had succeeded in dragging it to the spot where part of them perished on Montreal Island.

This island and the main land around are represented to us as being of the most inhospitable and dreary character. There is no wood for six hundred miles in any direction. Men who had been with Dr. Rae stated that Repulse Bay was a paradise compared with this place. Mr. Stewart informs us that he and the Expedition were wet for the space of sixty days with thick fog, rain and sleet, without means of drying themselves. When they were on the 9th of August they found snow on the ground, and the ice just taking for the long winter. The Esquimaux were about to take their departure, which they had delayed longer than usual, in expectation of the arrival of the expedition. The Esquimaux do not live there, but only go for two or three months in the summer to fish, and hunt such animals as are to be found on the coast. The wind was strong and searching during the time the expedition was there, and in the absence of wood for fuel or means of escaping from its effects, remaining at the place was in the last degree uncomfortable.

The expedition retraced its steps, and reached Fort Resolution on the 20th of September last; and Mr. Stewart on Friday, as we have above stated, reached Montreal.

The result of the expedition is of much interest; and strongly corroborative of Dr. Rae's narrative. Very precise particulars of the sufferings of Sir John Franklin and his companions will perhaps never be obtained; but that they suffered unspeakably, and perished miserably, can no longer be considered doubtful.

12. Deposition of James Stewart to the Scottish Court of Session in the case of *Fairholme vs Fairholme Trustees*.[1]

I was in joint command with Mr. James Anderson, of the expedition sent out by the Hudson's Bay Company, in the summer of 1855, to ascertain the fate of Sir John Franklin's expedition, and the truth of the information given by the Esquimaux to Dr. Rae.

We succeeded in reaching the spot indicated by the Esquimaux as being the place where the white men perished, and discovered sufficient articles to corroborate the aforesaid statements... No certain human remains, which we assiduously searched after, were discovered. At the entrance of the great Fish River we met a party of Esquimaux, who stated that they had seen the human remains of a party of white men, with only one survivor, and he at the point of death. They described him as a man of large size, and in the last extremity for the want of food. Their reasons for not assisting him were, they were weak from starvation themselves, and pushing on to a depot of provisions at the entrance of the river; further, that he was too far gone for assistance to be of any use. From the frankness and earnestness of their manner, we had no reason to doubt the sincerity of their statement, and their innocence as to their being the cause of their death. Some of the aforesaid party informed us of having been witnesses to the death of the last survivor; also that it had taken place four years previously, which would make it in the summer of 1850. It is impossible to give the day or month in which the events took place, but to the best of my belief

[1] *Fairholme vs Fairholme's Trustees, Record, Proof and Documentary Evidence. In Declarator, Court and Reckoning etc., George K. E. Fairholme, Esq. against the Trustees of Adam Fairholme, Esq, Scottish Court of Session, First Division,* Scottish Record Office. Stewart made this deposition in person during his visit to Britain in 1856. The case revolved around the matter of whether Lt James Fairholme, one of the officers on board HMS *Erebus* had predeceased his uncle, Adam Fairholme, who had died on 23 May 1853, and in whose will James Fairholme was the prime beneficiary.

it was between May and the end of July 1850. I do not believe it possible that any of the party are still living or retained by the Esquimaux. It is my firm belief that none of the party forming Sir John Franklin's expedition are existing – first, from the length of time since the last accounts we have had from them; secondly, because I believe, ere this, that some one of the party would have reached either M'Kenzie River or Slave Lake; or accounts of their being in the captivity of the Esquimaux, through Indians of the Hudson's Bay posts, who are constantly in communication with them [i.e. the Esquimaux], would have been received by some of their officers; thirdly, from the accounts received by Dr. Rae from the Esquimaux of the vessels having been crushed in the ice, and of their having met with a party who had been saved, who were travelling to the southward in the direction of Great Fish River, and who were then short of provisions – all which statements corroborated to a certain degree by the articles found in their possession, which, beyond a doubt, belonged to the missing party, as can be seen in Greenwich Hospital; lastly, by the directions framed by Dr Rae from his information, which formed the basis of our instructions which we received from the Hudson's Bay Company, and after being carried out by us prove to be correct in all its details, still further strengthened by a similarity of statement made by Esquimaux who had not seen Dr Rae, nor those who had communicated with him (Dr Rae). Various reasons might be given for bodies disappearing in such an inhospitable region, particularly after having been exposed for four arctic years – first, they might have been entirely destroyed by wolves or other animals; secondly, they might have been entirely buried in the sand; lastly, if dying near an encampment of Esquimaux, or where they had been accustomed to encamp, it is probable they might have been thrown into the water to save the trouble of burial, which would destroy all vestiges.

CHAPTER IX

CONTEMPORARY REACTIONS TO THE RESULTS OF THE EXPEDITION

INTRODUCTION

As one might have expected, Sir George Simpson (and, one must assume, the Governor and Committee of the Hudson's Bay Company) was quite satisfied with the outcome of the expedition 'which quite fulfilled all that was expected from it by reasonable people'.[1] Sir George was particularly pleased by the positive light which the entire operation cast upon the Company: 'the Company & their officers generally were raised in public estimation by the prompt & complete & very business like manner in which the whole service was carried out from the first design to its final execution...'.[2] On the other hand Lady Franklin and her supporters were less than satisfied; in their view it was 'necessarily an imperfect and ineffective search', although in proposing a motion in the House of Commons for government support for a further expedition to be dispatched by Lady Franklin, Joseph Napier[3] diverted some of the blame on the Government, in that, distracted by the Crimean War, it had not given the Hudson's Bay Company and Anderson adequate assistance (although one wonders how the Government could possibly have been of any practical help). The Government view of the expedition, as relayed to Anderson by Sir John Richardson, was that all that could humanly be done, had been done, and that it would not support a further search.[4]

The interpretation placed on the evidence presented by Anderson and Stewart by Captain Sherard Osborn[5] is that it revealed that a party of survivors of the Franklin expedition had reached Montreal Island and the mouth of the Back River but that: 'Further than this, all is apocryphal.' Osborn's own interpretation was that, rather than the Inuit having broken up a boat at Montreal island, some Franklin expedition survivors had themselves dismantled a boat there in order to build lighter boats or canoes in which to tackle the ascent of the Back River. He further suggested

[1] See p. 198.
[2] See p. 198.
[3] See p. 199.
[4] See p. 199.
[5] See pp. 200–4.

that further traces of the expedition could be expected in the interior, up the Back River, where Anderson and Stewart may well have missed them. On the basis of his knowledge of sledging, Osborn made a remarkably accurate prediction of where *Erebus* and *Terror* were probably abandoned, namely at or close to Cape Felix at the northern tip of King William Island.

The reaction of the acerbic Dr. Richard King was predictable. He first excoriated the Admiralty for delegating the search to 'a commercial company, notoriously ignorant of all things except rat skins and cat skins'.[1] Then he unfairly chastised Anderson and Stewart for not extracting the full details of the fate of the Franklin expedition from the Inuit 'because they had no interpreter', and idiotically criticized them for not having marked the spot where the Franklin survivors had died 'because they had not provided themselves with a simple monument of granite'.[2]

Yet another of King's repeated criticisms is possibly even more unfair. When he and Back visited Montreal Island in 1834, they had left a cache of supplies, which King referred to as 'King Cache'; it had been located and examined by Dease and Simpson during their visit to the island in 1839,[3] but they had found it only because two of their men had also participated in Back's expedition. There is no reference to this cache in either Back's or King's narratives of their expedition.[4] King maintained that Franklin knew of its existence, but for this we have only King's word. King argued that Franklin (or any surviving officer) would naturally have visited Montreal Island specifically to leave at least a message, or perhaps all the expedition's documents, in this cache. Hence probably his strongest criticism of Anderson and Stewart (and/or the Hudson's Bay Company) is that they failed to locate and check the cache in question, because presumably they were unaware of its existence. How they were supposed to know of its existence, when it is mentioned in neither his own or Back's narratives is not explained. The main object of King's harangue to the Admiralty[5] was to propose that a further expedition should be dispatched down the Back River, led by himself, to investigate the cache. An interesting corollary to King's argument, is that, as far as one is aware, the cache has still not been located or examined.

Much more charitable was the assessment published by Captain Francis McClintock, on his return from the expedition on which, guided by Anderson and Stewart's earlier efforts, he discovered the only document ever found which throws some light on the fate of the Franklin expedition, as well as a liberal scatter of skeletons and abandoned equipment along the coasts of King William Island. He noted that 'Mr Anderson ... and his small party, deserve credit for their perseverance and skill'.[6] Obliquely, he criticized the Company for failing to provide 'the necessary means of accomplishing their mission'; those means, of course, included an Inuktitut interpreter.

[1] See p. 205.
[2] See p. 205.
[3] Thomas Simpson, *Narrative*, 1843.
[4] Back, *Narrative of the Arctic Land Expedition*; King, *Narrative of a Journey*.
[5] See pp. 206–9.
[6] See p. 210.

DOCUMENTS

1. Letter from Sir George Simpson, Lachine, to James Anderson, Fort Simpson, 16 April 1856.[1]

Since I last had this pleasure I have received your letters of 25 & 30 March, 6 May, 22 June and also your reports, official and confidential, on the late Arctic Searching Expedition. I acknowledge these letters from hence, instead of Norway House, as usual, as we generally meet the Portage la Loche brigade there about starting, so that it is difficult to find time to write you, – except on the matters of local detail.

I have to congratulate you on the result of the Expedition, which quite fulfilled all that was expected from it by reasonable people, & I believe the public at large, although it has not, as a matter of course, proved palatiable [sic] to Lady Franklin and her clique, who still refuse to be satisfied with the evidence produced. The Government, however, appears quite decided, to consider the late Expedition the final effort in the cause, and have accordingly determined upon granting the reward offered for finding the first trace of Sir John Franklin's fate, which by public advertisement they notify will be awarded to Dr. Rae, unless other parties can profer [sic] a better claim, which we know is impossible. The Government do not appear to contemplate any further expeditions to the Arctic regions, although Lady Franklin is getting up one under the command of William Kennedy; it seems however, to be very little patronised or noticed by the public.

Stewart & his wife, with the three Iroquois arrived here on the 22 December; an outline of your report was immediately published in the Montreal papers; agreeably to your request, copies sent to all your friends (at least all those you designated) & to other persons likely to be interested. It was very favorably received and much credit was given for the prudence & ability of the leaders of the expedition & for the good conduct of the men, while the Company & their officers generally were raised in public estimation by the prompt & complete & very business like manner in which the whole service was carried out from the first design to its final execution...

I have taken sundry occasions for bringing under the notice of the Governor & Committee (or rather, since Mr. Colvile's death, the Deputy Governor & Committee) the claims of yourself & Stewart to some special remuneration for your services on the part of Government, agreeably to the promise held out in the letter from the Secretary of the Admiralty of Novr. 1854 (of which you have a copy) to the Company. They express much interest in your claim, & state it will be presented after they have seen Stewart. It will afford me much pleasure if, on my departure hence for the interior, I can be the bearer of information that your proceedings have gained for you some handsome pecuniary advantage ...

I am sorry to hear you are suffering from the varicose veins in your leg; the elastic stocking & Knee cap which you ask for are now forwarded and I hope may prove

[1] HBC D.4/51, f. 129.

beneficial; to be so however, they should fit closely & it was difficult to judge what would be a proper size for you without a pattern; you will observe, however, that the seam is bound with silk, which, if they are too slack or tight, may be opened, & the size altered to suit you, care being taken not to injure the web in the operation...

2. Extract from *Hansard*, 144, 24 February 1857, columns 1276–9.

The Franklin Expedition. Motion for an address.

Mr. Napier[1] rose to call the attention of the House to the communications with Her Majesty's Government respecting the Franklin Expedition, and the urgent nature of the claim [from Lady Franklin] for a further and complete search.

(Napier) [Recalls Dr. Rae's discoveries]. But the sympathies of the Government were absorbed by the war which then broke out, and the only thing done was to institute an investigation through the medium of the Hudson's Bay Company, in order to obtain a corroboration of the facts communicated by Dr. Rae. This was necessarily an imperfect and ineffective search, yet it served to show that the Government were not disposed to rest satisfied with what Dr. Rae had done. The result of this expedition was to bring to light other memorials, and to strengthen the hope that perhaps some of the unfortunate individuals belonging to Sir John Franklin's party were still struggling for existence.

... (Napier) He was confident that had it not been for the accident of all our energies and sympathies being enlisted in the carrying on of a great war, at the time Mr. Anderson's expedition sailed, much more assistance would have been given to that gentleman.

3. Extract from letter from Sir John Richardson to James Anderson, 6 April 1857.[2]

On the return of the ships from Hudson's Bay last autumn I had the pleasure of receiving your letter of the 3 July 56 together with a copy of your journal of your voyage down the Great Fish River. Dr. King[3] and some other parties desirous of being employed by Government on another expedition of search for Franklin's remains have bitterly attacked Dr. Rae's reports and also asserted that your voyage did not extend to the district that ought to have been carefully searched, but the Govt. have taken a sounder view and when the subject was brought into Parliament by some personal friends of Lady Franklin, declared that they would countenance no further search. Sir Charles Wood quoted in support of the opinion he had formed, Dr. Rae's and my own evidence given on oath before a law commission in Scotland instituted to enquire into the death and succession of Lieut. Fairholme, one of Franklin's companions. To show how carefully you had searched Montreal Island and the neighboring shores I copied all that portion of your journal and sent

[1] Joseph Napier was Lady Franklin's chief spokesman in the House of Commons; in his motion he pressed the government for the loan of a ship and stores for a further search, and for permission for a serving officer to take command. See Woodward, *Portrait of Jane*, p. 294.

[2] BCA AB40 An 32.2, f. 33.

[3] Dr Richard King, who had accompanied Back on his descent of the Back River in 1834.

it to the Secretary of the Geographical Society requesting its publication in the proceedings of that Society.[1] Whether he has done so or not I do not yet know as the volume has not yet reached me. I understood that your chart of the new route to the Fish River and description of the country was to appear in it.[2] If neither are published I will again draw up a paper from your journal to be read before the Society. Lady Franklin is still determined to send out a vessel and has laid aside £10,000 to defray the expense. The route is not yet fixed but it will probably be by Beerings Straits and if so will sail towards the end of the ensuing summer. I should be extremely glad were the scheme to succeed and some journals recovered giving an account of the causes which led to the destruction of Franklin's party, but I must confess I have little hope of the discovery of any records of the kind and am most unwilling to recommend the risk of another attempt to be made.

4. Extract from Osborn, S., ed., 1857. *The discovery of the North-West Passage by H.M.S. 'Investigator,' Capt. R. M'Clure, 1850, 1851, 1852, 1853, 1854.* London: Longman, Brown, Green, Longmans, & Roberts, pp. 366–81.

The Lords Commissioners of the Admiralty took the opinion of some arctic authorities, upon the subject of what could be done towards still further clearing up the tale brought home by Dr. Rae; for there was much about it that was vague, and calculated to keep alive hopes of the most distressing nature to those deeply interested in the crews of Franklin's ships. A gigantic war was pressing upon the resources of our navy both in ships and men, none of them could then be spared; and to meet the outcry of some effort to be made to ascertain if it really was the mouth of the Great Fish River that Franklin's travellers had reached, the Hudson's Bay Company were again requested to send out a party to that locality.

Dr. Rae having declined to take charge of the party which was equipped for this purpose, though he gave every support and encouragement to it, it was consigned to the care of Mr. James Anderson, a chief factor of the company, an officer of high reputation and much experience as a traveller. Lady Franklin, however, earnestly protested against this expedition; she foretold the improbability of its ever reaching King William's Land, and short of that the result would be as inconclusive as Dr. Rae's report, and a loss of very valuable time.

Labouring under many disadvantages, from the short time given to equip and start, Mr. Anderson commenced his descent, from Fort Resolution to the mouth of the great Fish River, on June 22nd, 1855, with three canoes of wooden framing but birch-bark planking, without an Esquimaux interpreter. On July 30th, at the rapids below Lake Franklin, three Esquimaux lodges were seen, and various articles were found, denoting that some of the unfortunate men they were in search of had been there ...

The party next examined Point Ogle, where only a small piece of cod-line and a strip of cotton was found; and on the 8th August, they began to retrace their steps, having held no communication with, indeed seen, no Esquimaux beyond the one

[1] Published as 'Extracts from Chief-Factor James Anderson's Arctic Journal, Communicated by Sir John Richardson' in the *Journal of the Royal Geographical Society*, 27, 1857, pp. 321–8.

[2] Anderson's sketch map was not included.

man and few women at the rapids below Franklin Lake, and never been able to reach King William's Land. This information reached us early in 1856, and goes to confirm Dr. Rae's supposition, that the Great Fish River was the stream upon which the party, he had heard of, had retreated; but instead of clearing up the mystery of what became of them, the whole story leaves the fate of Franklin, Crozier, and their ships' companies as doubtful as ever.

Taking it for granted that the Esquimaux did see thirty or forty men with a boat, as Dr. Rae asserts, what has become of them? If, when they reached the continent, the unfortunates became desperate with mysery, and committed cannibalism, – the practice is by no means rare in those wild regions, and it would assuredly prolong life, – where are the survivors? Is it likely they sat down there and died one after the other? If they were so lost to their own interests and safety as to remain, would not the survivors have scraped the earth over the bones of those who first perished?

Every arctic traveller knows that the tender and oily bones of the seal – even the brittle ones of birds – are found preserved over the whole extent of the arctic regions visited by us. What, then, has become of the bones of thirty men? Five years after the 'Erebus' and 'Terror' left Beechey Island, in Barrow's Strait, all those who visited the scene of their winter quarters found clothing, scraps of paper, and the thousand signs of Europeans having been there, looking just as fresh as the day they were left, and that in a far worse climate than Montreal Island.

Thirty-one years after Sir Edward Parry had been at Bushnan Cove, Melville Island, a traveller (Lieut. M'Clintock) found a spot where that distinguished navigator had, to use his own words, made 'a sumptuous meal of ptarmigan,' and there lay the bones of those very birds strewed about the old encampment! 'I was astonished,' says Lieut. M'Clintock (vide Parliamentary Blue Book, 1852), 'at the fresh appearance of the bones; they were not decayed, but merely bleached, and snapped like the bones of a bird recently killed.'

Esquimaux were not likely to have used dead men's bones. If they had European clothing in their possession, it is hardly likely that they could have concealed it entirely. There is not a musket, pike or cutlass produced: the party were not likely to have gone there unarmed; indeed the Esquimaux acknowledged having seen both powder, shot and ball. And as to Mr. Anderson's theory of the wind blowing away or covering their journals and papers, because his nautical almanacs suffered, it is purely assuming that the officer who headed Franklin's party was such an idiot as to leave his papers strewed about the surface of Montreal Island, instead of putting them in a cache, where, as arctic discovery proves, papers have been preserved and discovered after longer intervals of time than perhaps any other climate would admit of.

Looking, therefore, at the evidence before us, it amounts simply to this, that

'A party from the "Erebus" and "Terror" did reach the Great Fish River, and have left traces at Montreal Island and at the first rapids in ascending the stream!' Further than this, all is apocryphal. Mr. Anderson very naturally went upon his journey, firmly beleiving every iota of the translated account of Dr. Rae's interpreter; indeed, in the absence of any means of communication with the one old man and few women whom he did see, he had no other resource than to connect the traces which lay before him with the report previously made public. But sailors may be allowed to put a sailor's explanation to what lay before Mr. Anderson; and the following is our version of the tale it told :–

On Montreal Island Mr. Anderson found, he says, 'a quantity of chips and shavings and the ends of plank of pine, elm, ash, oak, and mahogany, evidently sawn by unskilful hands.'

Now, no boat supplied to the 'Erebus' or 'Terror' from Her Majesty's yards, which any party of men could have dragged a hundred miles over ice, would have been constructed of plank of so many descriptions; but it is very certain that a party retreating to the Great Fish River, and knowing the long series of rapids and portages in that stream, would have carried with them materials such as plank, which, with the framing of their large boat, would form rough canoes fit for their purpose.

Mr. Anderson distinctly says, 'chips and shavings.' Now a savage, who had never seen a planing instrument, was not likely to be able to produce shavings. After informing us that the plank was evidently cut by unskilful hands, Mr. Anderson says, 'every chip was turned over, and on one of them was found the word "Terror" carved!'. Surely that ominous word is a mute witness against Esquimaux having been the men who there laboured; yet in the next paragraph we read, –

'It was evident that this was the spot where the boat was cut up by the Esquimaux!'

Surely no such fair inference can be drawn. That the party brought carpenters' tools with them, we have the proof in Mr. Anderson discovering, at the lodges near the rapids, 'a broken hand-saw, chisels, etc.'; and perhaps, if a careful list could be procured of every article seen there or at Repulse Bay, some more interesting evidence might be obtained; for even a straw will show the course of a great stream, so may some insignificant trifle throw sudden light upon this sad subject.

The existence of traces further up the river than Montreal Island is a significant fact; and in support of the idea that on Montreal Island preparations were made to ascend the stream, we have another proof in the ash oars being cut or reduced into paddles, – a very necessary measure for a party about to go up narrow and tortuous rivers, and totally unlikely to have been done by the Esquimaux, who have no kyacks or canoes in that part of America. Some of these paddles were found at the rapids likewise.

It is true the women at this spot made signs that these articles came from a boat whose crew perished of starvation; but they did not give a single proof of the truth of the tale, or point out the grave of one of the unfortunate party.

Dr. Rae, zealous for the character of the Esquimaux, repudiates indignantly all idea of their having been treacherous, nor is it at all desirable to give rise to any bloody suppositions upon the matter; but any one, who will carefully read over the able paper of Captain Maguire, in the Appendix of this work, can, as easily as the most experienced traveller, form a correct idea of the character of the Esquimaux generally; and he will then agree with us in thinking that the savage of the polar regions, though not naturally cruel or treacherous, would, like most others, consult his own interests rather than the dictates of humanity, when such a windfall as a boat's crew of starving, scorbutic men, carrying with them untold wealth in the shape of wood, iron, and canvass, fell into their hands, and when they confessed, as those poor fellows evidently did, their direful necessity.

Some of Franklin's people may, we think, have died of disease or starvation at the place upon the continent spoken of by the natives; but that spot has not been reached by us as yet. Others evidently got to an island; there the Esquimaux say the officer perished, and five men likewise. Whether or no, such an island as Montreal Island was very likely to have been chosen by them whereon to await the opening up of the Great Fish River; they would be in a good position for commencing their canoe voyage, and be less likely, whilst employed constructing canoes or rafts, to be interrupted by natives. Granting, therefore that some starved at each place spoken of by the natives (though, until there is proof, people are justified in saying Englishmen can live where Esquimaux can) – granting even that the remainder did so far forget their manhood as to eat the flesh of their shipmates, is it unreasonable to suppose that, when the river opened, some few of those unfortunates started with what they had constructed, abandoning all their unnecessary gear on the island, and at the first portage?

They might have ascended far, and fallen in detail, and yet never, in such a water-intersected region, have been discovered by Mr. Anderson in his descent – the more especially if they, taking Sir George Back's chart, had followed his old track, a track from which Mr. Anderson departed considerably, and with advantage to himself and his party as far as rapidity of journey was concerned. As to holding out a hope of any straggler surviving amongst Esquimaux or Indians, it is not our desire to do so; but those who, by following up a similar train of argument as ourselves, arrive at a hope of such a pleasing and consolatory nature, ought not to be ridiculed for doing so.

They who have kept alive hope, who have urged on expedition after expedition, in spite of failure, in spite of ridicule, and in spite of uncharitable imputations of mania or interested motives, have now reason to feel happy that such trifles did not check their efforts; it remains yet to be seen whether perseverance will not still lift the curtain of this sad but glorious tragedy.

It is not alone the fate of those forty men that we desire to know – they were but a fraction of the lost expedition; there are still one hundred souls unaccounted for! and two of Her Majesty's ships!

To those who urge the expense of arctic expeditions, or the risk of life, as objections to the completion of a task we are pledged to accomplish, the answer is a brief one ... let one line-of-battle ship the less be kept in commission until the question is settled, or some other retrenchment made, if we are in such a bankrupt condition that England cannot afford to seek her missing sons.

We, who have reduced arctic travelling to a mere arithmetical calculation, know very nearly the distance a body of sailors numbering forty could have come from, dragging a heavy wooden boat over the ice, besides the quantity of articles which have been enumerated elsewhere, and which formed, doubtless, but a small portion of what they had with them. Taking therefore, the weight dragged by the forty men as 200 lbs. per man, and the distance accomplished daily about ten miles, an allowance extremely liberal for debilitated seamen, we have the precedents of Captains Richards, Osborn and Penny (who all have had to carry heavy wooden boats as far as possible over the ice) for saying that a journey of about fifteen days, or 150 miles, would be about the utmost distance they could have come from; the more so that sledge travelling was then but little understood, and that the extent of the sledge journeys made from Beechey Island by Franklin's people, as denoted by their cairns, do not exhibit any marked improvement in that respect, Cape Bowden in Wellington Channel, and Cresswell's Tower in Barrow's Strait, being, as far as we know, the limit of their explorations in that quarter; and neither of them would entail a journey of fifty miles.

That the 'Erebus' and 'Terror' are somewhere within the limits of the unsearched area about King William's Land, everything now denotes. One hundred and fifty or two hundred miles from Montreal Island, northward, carries us into the centre of this space, and where Victoria Strait is split into two by the large island called King William's Land. In and about Cape Felix on that island, or near the magnetic pole in Boothia, they most probably got beset; for had they been on Victoria Land, where natives, game, and fish abound, they would, it is fair to infer, have sent their 'forlorn hope' along it towards the Copper-mine or Mackenzie River. How they reached that supposed point, with their ships, time and a discovery of their journals will alone tell. Whether by rounding the west side of Prince of Wales Land, and passing down a channel which some suppose to exist in a south-east direction between it and Victoria Land, or whether, as appears most natural, they took the fine and promising channel which offered to the southward between Cape Bunny and Cape Walker, now called Peel Sound, and so struck the American continent, we can only surmise. But the absence of all cairns, or signs of their having been detained or having landed on either coast of Prince of Wales Land, as far as it is now known, or of North Somerset, leads to the natural supposition that they are nearer to King William's Land than to any other spot – perhaps in some indentation on its northern coast, into which they ran during a late and stormy season, as M'Clure did in the 'Investigator' and John Ross did in the 'Victory,' never to escape with their ships.

It has been argued against the existence of Franklin's ships in that quarter, that he would assuredly have visited the Fury Beach depôt, in Regent's Inlet. We reply to this, that Franklin, through his ice-master and others in his expedition, knew well how worthless it was for his purpose. He knew that, since it had been formed, Sir John Ross had provisioned the 'Victory' from it, that he had retreated upon it, and lived on it with his crew nearly twelve months, and eventually equipped himself there prior to his escape in 1833. After that some whalers had swept nearly everything off the beach; and, to escape the consequences of an Admiralty prosecution, one of the vessels had thrown into Peterhead Harbour a quantity of provisions she had carried off as plunder from the 'Fury' depôt. All this Franklin knew; and when Lieut. Robinson of the 'Enterprise' reached that supposed depôt in 1849, from Leopold Harbour, he found little there besides a cask or two of flour and a few raisins – showing how wisely Franklin had done in not falling back upon it.

5. Extract from: King, R. 1855. *The Franklin Expedition from first to last*. London: John Churchill, pp. 208–24.

Then comes the Hudson Bay Company, dispatched to bury the bodies and ascertain their sad history, and what becomes of it? The man who had pointed out Montreal Island and Point Ogle as the death-spot of the Franklin Expedition, and was intimately acquainted with the locality [Dr John Rae, W.B.] was not to go and bury the bodies and fetch the little history they had bequeathed to their country, – the last message each had delivered to his nearest and dearest relative or friend.

The nation, with one voice, would most assuredly have awarded to him that honour, but that Sir James Graham,[1] with all haste, knowing well the little bit of active mortality he had to deal with, flung him aside,[2] and with him such men as Osborne, Pim, and M'Cormick, before he had an opportunity to appeal to his nation.

What a sad destiny was Franklin's; it extended even to his very remains. Sir James Graham, upon whom fell the duty of providing for the decent burial of these remains, instead of performing this office, which better blood than himself would have esteemed an honour of no little account, delegated that office to a commercial company, notoriously ignorant[3] of all things except rat skins and cat skins,[4] utterly indifferent as to the mode in which they performed the task.

He had no right to do this. He had no right to hand over the bodies of 138 gallant sailors to a commercial company. He had no right to give them any other funeral than that due to them, as belonging to Her Majesty's Service. He surely should have dispatched an officer of Her Majesty's Service, of known ability, to perform that office, and to place a monument over their grave. He has compromised the nation in having thus neglected his duty...

Now what has come of the Commercial Company's Expedition to bury the remains of Franklin and to learn his sad history? They reach Point Ogle and Montreal Island. They find undoubted evidence of the truth of the Esquimaux accounts, and they are content with collecting a few relics to add to Dr. Rae's relics, and return. They never search King Cache of Montreal Island, – because they had no map, – because they had not read the Narrative of Thomas Simpson, – because they had selected a crew who were utter strangers in the land.

They do not ask of the Esquimaux the particulars of the Franklin tragedy – because they could not speak to them, – because they had no interpreter. They did not mark the spot where forty of their countrymen met their death, – because they had not provided themselves with a simple monument of granite. They do not seek for the history, in writing, of their sad fate in the only spot it was likely to be found, – because they had never heard that such a spot had existence. O tempora! O mores!

[1] First Lord of the Admiralty.

[2] One has to assume that King is referring to himself. By this stage he had made himself *persona non grata* with the Admiralty. See Hugh Wallace, *The Navy, the Company and Richard King: British Exploration in the Canadian Arctic 1829–1860*, Montreal, 1980.

[3] Science and Commerce never yet went hand in hand. [This footnote is in the orginal.]

[4] The sable is sometimes called sable-cat: – and musk-rat is the ordinary name of the musquash or lesser beaver, – the little animal which supplied us with beaver hats before silk hats came into use. King, *Narrative*, vol. 1, p. 115 [This footnote is in the original.]

With these feelings I addressed the humble prayer which concludes this narrative to the Lords Commissioners of the Admiralty, which I now address to my country at large, in whose hands now rests the Fate of the Franklin Expedition.

To the Right Honourable the Lords Commissioners
of the Admiralty.[1]

MY LORDS, – Your Lordships are aware that, in the years 1833–35, I was the Medical Officer attached to the Polar Land Journey in search of Sir John Ross, and that, for a considerable period, I commanded the party.

The knowledge which I acquired in that Journey, joined to an anxious desire for the advancement of Geographical science, led me to investigate the causes of the failure of former expeditions, having for their object the discovery of the North-West Passage, and to entertain views as to the means of solving that problem, which were, at that time, at variance with the opinions held by other Arctic travellers, although their soundness has since been established by the discoveries of Sir Robert M'Clure, Sir Edward Belcher, Mr. Thomas Simpson, and others.

In February 1845, when it had been determined by your Lordships to despatch Sir John Franklin, with the Erebus and Terror, to prosecute the discovery of the 'Passage' from Barrow Strait, I pressed upon Her Majesty's Government, although without success, the expediency of aiding the search by means of a Polar Land Journey down the Coppermine and Great Fish Rivers.

In 1847, after a lapse of two years since tidings had been received of the Erebus and Terror, doubts were entertained as to their safety; and on the 10th of June in that year, I submitted to the Government a statement of the grounds which led me to the conviction that the position of the lost Expedition was on the western land of North Somerset, and I proposed to communicate with and convey succour to them by means of a Land Journey down Great Fish River.

My proposal, however, was not entertained; on the contrary, two Naval expeditions were despatched, one from each end of the Continent, and a party was charged with a Land Journey for the purpose of searching the Coast, not in the locality which I had pointed out, but between the Mackenzie and Coppermine Rivers.

It is unnecessary for me to dilate upon the fruitless result of expeditions. On their return, the sympathies of the whole world were aroused to the fate of the Franklin Expedition, and a fleet of vessels was despatched, partly by the State, and partly by private enterprise, in search of the missing navigators; but most unfortunately the coast near the mouth of Great Fish River was again omitted from the search. For the third time I pressed upon the Government the expediency of a Land Journey, for the purpose of examining this neglected spot; and, in a letter addressed to your Lordships, on the 18th of February, 1850, in which I used the prophetic words, – 'The route of Great Fish River will sooner or later be undertaken in search of Sir John Franklin,' I repeated the offer I had previously made to lead a party in the search.

Your Lordships, however, acting upon the advice of the recently appointed Arctic Council, who, to use the words of one of its members, – 'did not think that, under

[1] This letter was also published in: Great Britain, *Further Papers Relative*, 1856, pp. 31–4.

any circumstances, Franklin would attempt the route of Great Fish River,' ignored my plan, and declined my services, and despatched a further Naval Expedition, the crews of which returned from a fruitless search, after the unparalleled desertion of no less than five vessels. Their journey, however, was not altogether without result, for although they failed to find or save the missing navigators, they discovered the long-sought 'Passage' in the identical position, it may be observed, laid down in an imaginary Chart which I had published some years previously, and had upheld against the opinion of other travellers up to the period of the discovery.

In 1854 Dr. Rae was despatched by the Hudson Bay Company to complete a survey of the West coast of Boothia; and, although he informed the public, in his letter addressed to the 'Times', on the 11th of December, 1852, 'that there was not the slightest hope of finding any traces of the lost navigators in the quarter to which he was going,' yet, strange as it may have appeared to him, he ascertained from the Esquimaux, on arriving in Pelly Bay, that about forty white men had perished four years previously at Montreal Island, and on the banks of Great Fish River; – in the very spot, I may observe, where Dr. Rae and the Arctic Council had come to the conclusion that the lost navigators could by no human possibility be found; and in the identical locality which I had never ceased to urge was the precise point which Franklin would endeavour to reach, and where traces of the expedition would infallibly be found.

At the time of receiving this intelligence Dr. Rae was at a distance of about 100 miles from Point Ogle; and it appears, from his official Report to the Hudson Bay Company, that he subsequently arrived at Castor and Pollux River, which is scarcely forty-five miles distant from that spot, and that, instead of hastening forward to verify or disprove the horrible story of cannibalism and death, related to him by the Esquimaux, he turned aside at a right angle, and travelled not less than double that distance, in a northerly direction, up to Cape Porter!

Without pausing to inquire the reason which induced Dr. Rae to turn aside, when he was within forty-five miles from a spot in which so much horrible interest was centred, and when he must have been well aware that neither the Government nor the people of England would rest satisfied until the locality of the reputed tragedy should have been examined; – without pausing, I say, to advert to this inexplicable proceeding on his part, I hasten to remind your Lordships that the accounts thus brought home by Dr. Rae, at once proved the incontestible accuracy of the views which I had so long and unsuccessfully pressed upon the attention of Her Majesty's Government, respecting the locality in which some traces or tidings of Franklin would be found.

In the following year the soundness of my views was at length tacitly admitted, by the despatch of an expedition, in canoes, down Great Fish River, almost in the precise manner which I had so vainly advocated in 1845, 1847, 1848, and again in 1850; and from the official Report of Mr. Anderson, the leader of that expedition, (published in 'The Times' of the 11th instant), it appears that, on the banks of that river, and on Montreal Island, some slight traces of the missing navigators have been found.

It is useless now to inquire what would have been the result if your Lordships had acceded to my earnest and repeated entreaties, and permitted me, in 1847 or 1848, to lead an expedition to the spot where these sad relics have since been found; no doubt can, I think, exist in the mind of any reasoning being, that, if those entreaties had been acceded to, a portion, at least, of the lost expedition would, at the present moment, be alive, and in England.

It is not with any view to my own aggrandisement, or with any feeling of self-laudation, that I submit this hurried analysis of the recent Arctic Expeditions to your Lordships' consideration. If such were my object, I should point out further instances in which the discoveries of Simpson and others have proved the accuracy of my views respecting the conformation of the Polar Regions. But I think it right to place on record a statement, however, hasty and incomplete, shewing the correctness of the opinion which I so long entertained, as to the position in which traces of Franklin would be found, in order that your Lordships may judge whether the further observations, which I feel it my duty to make upon the subject, are not entitled to more consideration than my former suggestions have received at the hands of Her Majesty's Government.

There is an important question now before your Lordships. Has everything, in the power of the English Government, been done to obtain evidence of the death of the Franklin Expedition? I unhesitatingly answer in the negative.

From the statements of the Esquimaux seen by Dr. Rae, taken in connection with the evidence procured by the last searching party, there seems little doubt that a considerable number of white men died at or near Point Ogle, on the western coast of the embouchure of Great Fish River, and that a smaller party, consisting of an officer and four men, died on Montreal Island, – a spot about half a day's journey to the South of Point Ogle. This last party had a boat with them, which was subsequently sawn up by the Esquimaux, who left a quantity of chips, on one of which was found the word 'Terror.' A number of articles of common use, and even of luxury, belonging to the expedition, have been purchased from the Esquimaux, and brought to England, but the inquiries of the last searching party could find no trace of any papers, records, or other written documents!

Such, then, are the simple facts before us, and, without entering upon the vexed question as to the manner in which our unfortunate countrymen met their death, whether by starvation, or by the hands of the Esquimaux, the chief point for enquiry appears to be; – For what purpose did an officer and four men visit Montreal Island? As the iron coast of an inhospitable little Island is the last place to which an Arctic traveller would resort for provisions, it is evident that the visit must be assigned to some other cause, and this point, which seems at present to be a mystery, it is, I think, in my power to elucidate.

On my visit to Montreal Island in 1834, I constructed a hiding-place, which was known by the name of 'King Cache,' and which was subsequently visited and opened by Simpson in 1839, in the same manner as the Cache made by Parry on Melville Island, called 'Parry Sandstone,' was opened by M'Clure in 1852. The existence of my Cache was known to Franklin, and it is my firm belief that he, or the leading survivor of the Expedition, crossed over from Point Ogle for the purpose of searching this Cache, and of depositing there a record of his visit, and that he and his boat's crew subsequently met their death before they could regain the main land.

By whatever means they perished, I think there can be no doubt that the leader, knowing of the existence of my Cache, and trusting that it would be searched ere long by friends from home, would strain every nerve, before he ceased to live, to deposit in this place of safety, not only the memorial of his visit, which he crossed from the mainland for the purpose of placing there, but also the history, which he would most unquestionably have carried with him, of the endurance and the sufferings of that devoted band, and of the heroic constancy with which the officers had sustained the flagging courage of their men, in the speedy hope of receiving that succour which, by a horrible fatality, had been directed to every point of the polar Seas, except the precise spot on which they stood. And the fact that no papers were found in the hands of the Esquimaux, is in itself a strong presumption that the records of the Expedition had been deposited in a place of safety before the death of our hapless countrymen.

In the official report of the leader of the last searching party, my Cache is not mentioned, and, as he would scarcely have omitted to search it, or have forgotten to refer to it in his report, if he had been aware of its existence, I cannot but conclude that, by some further and unexplained misfortune, he started on his journey without being aware that Montreal Island contained any particular spot in which there would unquestionably be found some traces of the missing Expedition.

From these facts, I can only draw the deduction that, in all human probability, a history of the Franklin Expedition still lies buried in my Cache, beneath the rocky shore of Montreal Island, and that it is within the bounds of possibility that this record may be recovered, and that the discoveries of the ill-fated Expedition may yet be published for the advancement of science, and the narrative of their probably unexampled sufferings be made known to the world.

Under these circumstances, I feel assured that the people of England will not consent that the search for the missing Expedition shall rest in its present position. More than two millions sterling has already been squandered in expeditions, which have brought home no tidings of the lost navigators, beyond a few silver forks and other relics, and an apocryphal story, interpreted from the vague signs of the Esquimaux, too revolting in its details to be worthy of implicit belief.

A further Land Journey down Great Fish River may be performed at a cost of about £1000, and this Journey, if your Lordships will give me the command of a party, I offer, for the fifth time, to undertake, in the confident hope that I may yet, at the eleventh hour, be the means of recovering a record of the Expedition, the recital of whose sufferings will otherwise be buried in everlasting oblivion.

I have the honour to be, my Lords, etc.,

17 Savile Row, 23 Jan. '56 — Richard King

Sir, — Admiralty, 28th Jan. 1856

Having laid before my Lords Commissioners of the Admiralty your letter of the 21st instant, volunteering your services to command an Expedition by Land down the Great Fish River to Montreal Island, to search for traces of the fate of the late Sir

John Franklin and Party, I am commanded by their Lordships to acquaint you that they do not think it advisable to undertake such an Expedition.

I am etc.,

Dr. King, M.D. Thos. Phinn

5. Extracts from King, Richard, *The Franklin Expedition: To His Grace the Duke of Newcastle, Secretary of State for the Colonies, A Letter of Appeal.* London, 1860, pp. 13–14 & 26.

Mr. Anderson, in the summer of '55, descended Great Fish River from the Hudson Bay Settlements, examined Montreal Island and the coast of the continent in the vicinity, and the results may be summed up in a very few words. Despatched by the Hudson Bay Company, with insufficient means and information, he was unable to converse with the Esquimaux because he had no interpreter; he did not know there was a particular hiding place in Montreal Island, called King Cache, that was known to Franklin, and where the leaders of the lost Expedition would probably deposit their records. He had no proper map, and being contented with a cursory examination of Montreal Island and the coast about Point Ogle, he never crossed to King William Island, or made further search in the line of march which we now know to have been taken by the fated one hundred and five. He found a few relics, purchased others from the Esquimaux, and after spending seven or eight days about the mouth of Great Fish River hastened homeward with all speed ...

But such a search [via the Back River, in summer] ought not to be entrusted to the agents of a commercial company. To the insufficiency of the equipment of the party dispatched by the Hudson Bay Company, under their factor, Mr. Anderson, is principally to be attributed the meagre results obtained by that Expedition, and the final search for the materials of the history, yet unwritten, of the discovery of the North-west Passage by the Erebus and Terror, ought to be carried into effect under an officer appointed by Her Majesty's Government.

6. Extract from: McClintock, F. L. 1859. *The Voyage of the 'Fox' in the Arctic seas: A Narrative of the Discovery of the Fate of Sir John Franklin and his Companions.* London: John Murray, pp. 2–3.

The Government caused an exploring party to descend the Fish River in 1855; but although sufficient traces were found to prove that some portion of the crews of the 'Erebus' and 'Terror' had actually landed on the banks of that river, and traces existed of them up to Franklin Rapids, no additional information was obtained either from the discovery of records, or through the Esquimaux. Mr. Anderson, the Hudson's Bay Company's officer in charge, and his small party, deserve credit for their perseverance and skill; but they were not furnished with the necessary means of accomplishing their mission. Mr. Anderson could not obtain an interpreter, and the two frail bark canoes in which his whole party embarked were almost worn out before they reached the locality to be searched. It is not surprising that such an expedition caused very considerable disappointment at home.

CHAPTER X

COMPLAINTS AGAINST STEWART

INTRODUCTION

Probably the most unfortunate aspect of the Anderson/Stewart expedition was that there were serious frictions between the two leaders. From this distance one cannot assess whether these frictions had a serious negative impact on the conduct of the expedition, but they cannot have helped.

At several points in Anderson's journal one finds thinly veiled, implied criticisms of Stewart's competence. Thus in the entry for 17 August 1855 one reads: 'Mr. Stewart's canoe again broken badly in still water ...' And on 22 August: 'Mr. Stewart's canoe cannot keep up with mine and retards us considerably.' There can be no doubt that these remarks are indicative of Anderson's increasing dissatisfaction with Stewart's behaviour. In his journal entry for 17 August Stewart reports: 'Broke my canoe again & I had a row with Mr. Anderson about the lading...'

Anderson refrained from direct criticisms and complaints as to Stewart's performance in his official report to Sir George Simpson[1] but in a private letter[2] he claimed that 'it would be a dereliction of duty were I to be silent' on the topic. Anderson claimed that Stewart 'was utterly useless to me'; worse still he claimed that he slept in his canoe for much of the route, and implied that he (Stewart) had seen so little of the route that if he (Anderson) had been incapacitated, Stewart might not have been able to retrace their route. Anderson also voiced identical complaints in a private letter to Eden Colvile.[3]

But it was an incident during Stewart's trip (with the dispatches concerning the events of the expedition) from Fort Resolution to Red River and, ultimately, Lachine, which brought Anderson's ire to the boiling point. From Portage la Loche Stewart and party travelled in two boats; on reaching Lake Winnipeg at Grand Rapids, Stewart took both boats north-east across the lake to Norway House rather than directly south to Red River. His purpose (a very understandable one) was to pick up his wife, Margaret, who was waiting for him there; they had been married only the previous summer and Stewart's journal entries reveal that he had been missing her badly throughout the expedition.

[1] See pp. 170–7.
[2] See pp. 167–70.
[3] See pp. 179–81.

On hearing of this Anderson wrote letters of complaint to Stewart, John Swanston and Sir George Simpson; in Anderson's view Stewart had committed an unpardonable sin by delaying official dispatches for private ends. He (Anderson) felt that Stewart should have sent one boat directly up the lake to Red River with the dispatches.

In his reply[1] Simpson displayed a degree of sympathy, humanity and understanding which is quite out of keeping with the image of the man which historians have tended to portray. He suggested that Stewart had 'strong inducements' to go to Norway House, that he may well have been low on supplies, and that, in terms of the total length of the journey, the dispatches had not been seriously delayed. Simpson advised Anderson to withdraw his complaint about Stewart and to request that his letter be returned to him. This Anderson appears to have done; the letter of complaint is not among Simpson's correspondence in the Hudson's Bay Company Archives; a copy was, however, retained among the Fort Simpson outward correspondence.

But soon after his letter to Anderson, exonerating Stewart in the matter of his side-trip to Norway House, in a letter to Stewart concerning his appointment to the charge of the Cumberland House District, Simpson revealed a totally different side to his character, the side provoked by possible disruption of the smooth running of the Company's affairs, or worse still, by unnecessary expense or loss of revenue for the Company. Citing four separate incidents at widely separated locations (one of these being at Fort Chipewyan, during the preparations for the expedition) he accused Stewart of habitual drunkenness and, worst of all, of drunken brawling with the 'servants'. Simpson states quite clearly that if there is any recurrence of this type of behaviour, Stewart 'will be expelled from the service with ignominy.'[2] Apart from the fact that this letter reveals that alcohol may have been a contributing factor in Stewart's apparent lack of drive during the expedition, it provides intriguing evidence of Simpson's 'network' at all levels within the Company, whereby he was able to keep abreast of the details of the operations and activities at even the remotest post.

Anderson later found grounds for laying a further complaint against Stewart. Among the batch of mail dispatched from Fort Simpson in November 1856 was a letter from Anderson addressed to W. G. Smith, the Secretary at Hudson's Bay House in London. Anderson had received copies of the newspaper accounts of the results of the expedition from Montreal and St. Paul, based on interviews with Stewart, and which Anderson felt, with some justification 'abound in misstatements'. Some of the statements, e.g. that they had found a boat with the name 'Terror' on it, or that there were over 100 rapids on the Back River, are very probably the result of poor reporting and Stewart really cannot be held responsible for them. But one remark which Anderson categorized as 'pure invention', can legitimately be ascribed to Stewart, since it was included in his report to Sir George Simpson, and in his statement (under oath) to the Scottish Court of Session in connection with the Fairholme case.[3] This is the statement that an Inuk woman had informed him

[1] See pp. 215–16.
[2] See p. 218.
[3] See pp. 194–5.

that she had seen a survivor of the Franklin expedition alive. The basis for Anderson's refutation of this statement was that the only two members of the party with any knowledge of Inuktitut (Mustegan and McLellan) were always in his presence, and that he had not heard them translate any such statement.

This is probably the most significant divergence between the two leaders' accounts of the expedition. One is inclined to believe Stewart's statement; he would have been unlikely to make a false statement under oath in a Scottish court, and perhaps even less likely to make the same statement in his report to Sir George Simpson. Given the known strained state of relations between the two leaders it is quite conceivable that at no point during the expedition did they compare notes as to their respective versions of what they understood the Inuit to have told them.

DOCUMENTS

1. Letter from James Anderson, Fort Simpson, to James Stewart, 24 March 1856.[1]
Your short notes from Athabasca and Red River reached me yesterday. They afforded me little satisfaction, and I learnt your proceedings from other quarters.

I regret to say that I totally disapprove of your proceedings after you reached the Grand Rapid; it was very natural for you to proceed to Norway House to join your wife, but you were not justified in retarding an important dispatch for your private ends. You could have taken one canoe and sent on Mr. Lockhart and the dispatches with the 3 Iroquois and 5 other men direct to Red River; they would probably have reached there on the 29th Octr. and the Dispatches could have been sent off – as I directed in the care of Ignace – on the 31st and would probably have reached St. Paul's without being troubled by snow.

It is the first time that I have heard of an Express being saddled with such an encumbrance as a Lady, or retarded on the account of the private affairs of any one.

You tell me that in direct opposition to my orders, you have given A. Laferté[2] the same wages as the other men, because he guided you down English River; now you had a better guide than him in your own canoe namely Thos. Mustegon.

I did not solicit the command of the Expedition but as I was appointed to it, I am determined that my authority shall be respected. I have directed Mr. Swanston to charge your own and Mrs. Stewart's Expences, as well as the extra wages you paid to A. Laferté on your own authority, to your private A/c. As for the retardment of the Express, you will have to answer for that to the higher powers, and I beg to inform you that I have called the attention of the Council to the subject.

I found your Gun very shortly after you left and sent off Landrie with it via Buffalo River in hopes that he would have reached the River before you arrived. I suppose that it has been forwarded.

[1] HBC B.200/6/31, f. 81.

[2] Laferté had been sent back to Fort Reliance from the upper Back River due to sickness.

A Frock Coat of yours brot. from the Youcon has been sold and the proceeds £2 placed to your a/c.

2. Letter from James Anderson, Fort Simpson, to John Swanston, 24 March 1856.[1]

I regret to state that I totally disapprove of Mr. C. T. Stewart's proceedings after reaching the Grand Rapid.

There was nothing to prevent him from proceeding to Norway House in one of the Canoes, and sending on Mr. Lockhart thence with the Dispatches the 3 Iroquois and 5 other men. They would most probably have reached Red River on the 29 Octr. and the Express might have been sent off in charge of Ignace Montour – as I directed – on the 31st; it would probably have reached St. Paul's without being impeded by snow. This is the first time that I have heard of public dispatches of importance being retarded for private ends, or of an Express being saddled with such an encumbrance as a lady.

I directed that A. Laferté should only be paid his bare wages; the man was most undoubtedly a malingerer and tried to create discontent among the other men, besides behaving disgracefully in other ways. Mr. Stewart informs me that (in direct contravention to this order) he has paid him the same as the others because he guided the party down English River; this pretence can not hold good as Mr. Stewart had a far better guide in his own Canoe, viz. Thos. Mustegon.

I request that Mr. & Mrs. Stewart's expences for this trip, as well as the sum paid to A. Laferté, above what I authorised, be charged to the private account of Mr. Stewart; such charges as these I cannot sustain. If Sir George Simpson or H.M. Govt. choose to allow them, they will acquaint you, as I have informed Sir George by this opportunity of what I have done.

3. Letter from James Anderson to the Governor [Sir George Simpson] and Council of the Northern Department, 26 March 1856.[2]

I beg to call your attention to the misconduct of C. T. Jas. Green Stewart when conveying the dispatches of the Land Arctic Searching Expedition last autumn.

Messrs Stewart and Lockhart left Fort Resolution on the 23rd Septr. in two light canoes manned by 17 Men. The autumn was so fine that they reached the Grand Rapid, L. Winnipeg without any material difficulty. Mr. Stewart instead of sending on the dispatches direct to Red River from that place under the care of Mr. Lockhart in one of the Canoes manned by the 3 Iroquois and 5 other men proceeded to Norway House – where his wife was – with both Canoes; they arrived there on the 26th Octr. and proceeded thence in a boat to Red River, which they reached on the 3rd Novr. Here the Express was again retarded till the 12th when Mr. Stewart left on horseback, taking his wife with him. As snow fell shortly after his departure, it was supposed that the Express would have met with much detention.

Mr. Stewart had previously requested me to sanction his going out with his wife. I

[1] HBC B.200/6/31, f. 80.

[2] Ibid., ff. 82–4.

told him I was not authorised to do so, and that even if I were, I should never sanction such a preposterous arrangement. My directions to C. F. Swanston were, to forward it by the 3 Iroquois giving it in charge to Ignace Montour who is a trustworthy man and well acquainted with the route. Had the Dispatches been brought direct from the Grand Rapid and forwarded by the Iroquois, I have no doubt that they would have reached England by the end of Novr. or beginning of Decr.

This is certainly the first time that I ever heard of Dispatches of great public interest being retarded for the private ends of any one, or of an Express being saddled with such an encumbrance as a Lady!!

As it would save much expence and trouble if the Expedition could reach Red River before the closing of the Navigation, as an encouragement to the men to exert themselves to the utmost, I promised that they should get their wages for the whole year even if they arrived at Red River two months before it elapsed. I however made one exception, viz. Alfred Laferté; this man was decidedly a malingerer, had endeavoured to create disaffection among the men, and had otherwise behaved most disgracefully. I directed Mr. Swanston to give this man only his bare wages. In direct contravention to this order Mr. Stewart gave this man the same terms as the others, because he said that he had guided the party down English River. I have no hesitation in saying that this is a false pretence as one of Mr. Stewart's steersmen – viz. Thos. Mustegon, was a much better guide than Laferté.

I did not solicit the command of this Expedition, but as I was appointed to it, I am determined to uphold my authority, and have therefore directed Mr. Swanston to charge the whole of Mr. & Mrs. Stewart's Expences for their trip, as well as the extra wages paid Laferté to Mr. Stewart's private account, until these expences be sanctioned by Sir G. Simpson or H.M. Govt.

Had Mr. Stewart conveyed the Dispatches rapidly to Red River and conducted himself otherwise with propriety, it was my intention to have passed over his conduct during our voyage to the Arctic Sea without any public notice. I now, however, beg to state that during the entire trip he was entirely useless to me, and nothing more than a mere encumbrance, as may be well ascertained by examining any of the men belonging to the Expedition.

Having now plainly brought Mr. Stewart's conduct before you, I shall leave you to deal with him as you think fit.

4. Private letter from Sir George Simpson, Norway House, to James Anderson, Fort Simpson, 14 June 1856.[1]

Since writing to you by the Portage brigade and after the Council had broken up an express arrived from Ile à la Crosse, in advance of the brigade, with a packet from Mackenzie River etc; by which I received various communications from you under dates 16 & 22 February, 24th (2) & 26 March, with enclosures.

The chief topic in these letters is your complaint against Mr. Stewart for want of his diligence in transmitting the report of the Expedition to England. It appears to me you have greatly exaggerated the evil done in the matter, and that you have

[1] HBC D.4/76A, f. 831.

written on the subject under a strong impulse of personal feeling, under a misapprehension of facts; in a style that you will regret on further consideration. It must be borne in mind that Stewart was united with you in the command of the expedition, & whether he made a good or bad use of his authority, was in a position to use his own discretion, when at a distance from you, in the execution of the duty entrusted to him. For my own part, I think he was perfectly right in taking charge of the dispatches himself, rather than entrusting them to the Iroquois; & as for the difference in time occupied between the route from the Grand Rapid direct to Red River or via Norway House, it is so small as to be of little consequence, while he had strong inducements to come this way, & for all I know, may have been short of provisions, in which case he acted sensibly in the course he adopted. After all he made a very good passage to Red River, & did very well from there to Montreal, where he arrived on the 22 December. Your dispatches were in the hands of the British Government early in January, & so far from the public charging the Company with loss of time or mismanagement of the affair, it appeared to be the only point on which all parties were united [illegible] those who directed the [illegible] of the expedition, the want of success as to the object it had in view etc., that the work had been done with a promptitude and businesslike regularity that was highly creditable to the Company and their officers.

Had the dispatches been sent by the Iroquois, I believe greater delay would have occurred; the man who you praise & consider so trustworthy, I regret to say has latterly fallen into irregular habits, rendering it necessary for me to lower his position in our service – in consequence of his fancy for liquor & a tendency to pilfer. The last time he [illegible], he lost every scrap of his own property, & this is the man you wished to be specially employed as the bearer of your dispatches. As to charging any Expenses to Stewart's account, you of course have no authority to give any instructions on that point, nor would Mr. Swanston have acted on your letter until it had been confirmed by Council or myself; & under any circumstances, it is too late to make the change you point out; as I am glad to say the accounts of the Expedition have been presented, paid and finally disposed of. I have mentioned that your letters were received after the Council had broken up; but had they been laid before them, I am satisfied no action would have been taken upon them; they would not have voted that Stewart was in any way culpable in this instance, although we all of us know, perhaps better than yourself, that Stewart has unfortunate failings, for which he received from myself, at the request of the Council, a severe reprimand this season, accompanied by umistakeable hints of the course that would be adopted if he did not entirely reform his habits. If your letter be made public it might lead to very disagreeable consequences, disputes and recrimination, personal quarrels etc., which would not add to the honor of the service. I shall for the present, therefore, suppress it, alluding to it neither to Stewart, nor any member of Council; but if you still press your complaint, I will lay the letter before the Council next year, on your expressing a desire to that effect. I would, however, recommend, as the wiser & most dignified course, that you write to me stating your wish to withdraw the letter of complaint, begging it to be returned to you.

5. Private letter from Sir George Simpson, Norway House, to James Stewart, Fort Garry, 12 June 1856.[1]

You have been appointed to the charge of Cumberland district for the current outfit, but I think it friendly to inform you that the appointment was made with great doubt and hesitation, in consequence of a general feeling that no dependence could be placed on your regularity & sobriety. Since you came out from McKenzie River 3 years ago, several scenes disgraceful to the Service have occurred, in which you have figured prominently and as no such scenes had occurred for years previous to your advent to this part of the country the inference is drawn that they arose from your convivial habits. The cases which have been more particularly specified and reported by eye witnesses were the following:

1st. Near Norway House – when several of the Company's Officers got quite drunk and disgraced themselves in the presence of a large body of Servants – the jollifications ending as usual in a brawl & a threatened fight between you and one of the Clerks, which fortunately evaporated with the fumes of your carousel after some hours' sleep.

2nd. At Old Norway House – when you and Lane passed the greater part of a night in conviviality and your Companion had to be carried to his boat in the morning dead drunk.

3rd. At Carlton – where there were 3 or 4 dances and drinking frolics ending in fighting among masters and Servants, leading to waste of the Company's property and the desertion of 5 of the people, to the great injury of the business of the post for that year.

4th. At Athabasca – on your way to join the expedition, when the small supply of liquor intended to be husbanded with care for use in case of sickness etc. when on the Arctic Coast, was misapplied and consumed, and when, in the course of the orgies, you alarmed the people present by ripping out the parchment window with a large pocket knife (in the depth of winter) to shew your skill in its use.

These several instances are of public notoriety, as there was no secrecy or attempt at concealment in reference to any of them & they were reported far & near by persons who were concerned & others who were spectators. Such affairs bring disgrace on the whole service & are keenly felt by your colleagues, whose character suffers in public estimation as well as your own or any others who were actually concerned in them. The public at a distance does not particularize individuals when judging the Company's conduct & service, but look at the concern as a whole, and when riot and drunkenness are brought home as a charge on their officers, although only a few may be really the culprits, the whole body is stigmatised as of dissipated habits.

I would fain hope you sincerely regret these unfortunate escapades and that your pride, self-respect, and better judgment, will lead you to make a great moral effort to shake off such evil habits, which must otherwise, whether here or in the civilized world, lead to your ruin. I write you without reserve as a friend and well wisher. I have always felt an interest in you and as you are aware have afforded you my countenance & support at all times; as far as activity and zeal has gone you have earned

[1] HBC D.4/51, ff. 188–90.

a good reputation, & until you left McKenzie's River, you were looked upon with favour as a useful member of the concern; now, however, a different opinion prevails and so strong is it, that if during next winter any repetition of irregularity occur you may rely upon it, as sure as fate, that you will be brought before the Council as a delinquent and expelled from the service with ignominy.

This has been the fate of several officers from time to time, but I trust you will escape it.

You must study what I have here said coolly and carefully, and believe that it is fully as disagreeable for me to write as for you to read such plain spoken truths; but I act under a sense of duty and a sincere regard for your welfare. I hope you may receive and act upon my advice and that you will prove yourself an active, zealous and reliable member of the concern, in that case the past will be forgotten and your future prospects as promising as any man of your rank in the service.

6. Letter to W. G. Smith, Hudson's Bay Company, London, from James Anderson, Fort Simpson, 28 November 1856.[1]

I beg to state to you, for the information of the Governor and Committee. That I have recently received some Minnesota Papers and the Montreal Globe and Gazette of 24th Decr. last; all containing accounts of the Expedition down Fish River, from information furnished by Mr. J. G. Stewart the second in command. These accounts I regret to say, abound in misstatements, some of which may excite remark and throw discredit on the Expedition and on the Service. I therefore consider it my duty to furnish you with a refutation of them.

1st. It is stated that Mr. Stewart counted 107 Rapids in Fish River, and then got tired of the Task. Sir George Back who surveyed the River, to the best of my recollection, found but 84.

2nd. It is stated that we were 60 days without seeing Fire or eating Cooked Food, only occasionally getting some hot tea by means of a Lamp. With very few exceptions indeed, we saw fire and ate cooked food not only every day but every meal. The Lamp was never even used.

3rd. I was perfectly astounded on reading the Circumstantial account of 'The death of the Last Man of the Franklin Expedition', said to be derived from an Esquimaux woman. It is a pure invention. I never heard any of the party even surmise (which was all that could be done by people unacquainted with the Esquimaux Language) that the woman alluded to had said any such thing. Dr. Rae's two men who alone understood a few words of Esquimaux, were constantly at my side, and never hinted anything of the kind. The only unmistakeable information we received, was regarding the death of the party, as described in my despatch to Sir George Simpson dated 14th Septr 1855; for everything in it, and in my journal, I hold myself responsible, but for nothing else.

I beg to add that I have furnished no accounts to the Public Prints regarding the Expedition.

[1] HBC A.11/66, f. 4.

CHAPTER XI

ACCOUNTING, AWARDS AND RELICS

INTRODUCTION

From the very start the Admiralty had assured the Company that 'the British public cheerfully bear the whole expense of the expedition',[1] and in his various directives to the various Company officials involved Sir George Simpson had reminded them to keep accurate records of the costs incurred by the expedition. On 29 February 1856 the Company submitted a bill for £6442. 18/6, covering the bulk of the amalgamated costs of the expedition.[2] This bill was settled on 28 March 1856. A bill for additional expenditures amounting to £217. 1/1 was submitted on 13 February 1858 and was settled in May of that year. Thus the total bill for the expedition amounted to £6659. 19/7. The wages of the expedition members (excluding Anderson and Stewart) and not counting gratuities, amounted to £602. 15/4.

On first approaching the Company the Admiralty had also stressed that the personnel selected for the expedition should be informed that the Government promised to 'pay liberally for the service to be rendered, and to reward specially any acts of signal daring and distinguished merit'.[3] After consulting with the Company the British Government awarded £400 to Anderson and £280 to Stewart in recognition of their services during the expedition. Although it is not specified, one assumes this was in addition to their regular salaries of £600 and £300 respectively. When the expedition had returned safely to the headwaters of the Back River on 29 August, Anderson promised every member of the expedition a bonus of £5; this promise was subsequently honoured. In addition Joseph Bouché (or Boucher) and William Reid, each received £1 for finding the first traces of the Franklin expedition on Montreal Island. Bouché and McLellan also received an additional £3 for acting as personal servants to Stewart and Anderson respectively. The highest-paid of the canoemen were the three Iroquois (at £60 per annum against £50 in the case of the remainder); as Simpson noted, he had had to offer them unusually high wages to persuade them to participate, and to make the long winter journey from Lachine to Fort Resolution.[4]

[1] See p. 32.
[2] See p. 221.
[3] See p. 32.
[4] See p. 59.

In addition to these monetary awards each participant (and those of the Company's earlier expeditions in search of the Franklin expedition, e.g. that led by Drs John Richardson and John Rae in 1847–9, also received the Arctic Medal, 187 copies of which were dispatched to the Company from the Admiralty on 21 February 1859.[1]

The assorted artefacts associated with the Franklin expedition which Anderson and Stewart recovered from the Inuit or retrieved from Montreal Island, were shipped home by the Company and transferred to the Admiralty at various times. A case containing the first of these reached the Admiralty on 6 June 1856. A further consignment arrived via the Company's ships from Hudson Bay in the autumn and reached the Admiralty on 13 November 1856.[2] Finally, the last of these artefacts reached the Admiralty as late as mid-June 1859, i.e. four years after the expedition set out. This was a piece of mahogany which the Inuit had been using as a snow shovel, and from which Anderson had contemplated making a souvenir of some sort, but had ultimately decided to relinquish. All these artefacts were forwarded to Greenwich Hospital.

DOCUMENTS

Accounting

1. Letter from R. M. Bromley, Accountant General for the Admiralty, to Archibald Barclay, Hudson's Bay Company, 27 December 1855.[3]

The sum of £3000 having been voted by Parliament in the Navy Estimates for the present financial year, 1855/56, to defray the expenses of an Expedition to search for the relics of Sir John Franklin and the crews of Her Majesty's Ships 'Erebus' and 'Terror', I have to request, as the vote in question will lapse on the 31st of March next, that any claims which the Hudson's Bay Company may have on this Department in connection with the Expedition referred to, may be presented for payment as soon as possible.

I have also to request that you will favour me with an immediate answer hereto, in case it should be necessary in the event of the claims not being presented in time for settlement in the present financial year, to renew any portion of the above vote in the Estimates for the ensuing year, which are now in course of preparation.

2. Letter from W. G. Smith, Hudson's Bay Company, to R. M. Bromley, Accountant General, Admiralty, 2 January 1856.[4]

I am directed by the Governor and Committee of the Hudson's Bay Company to acknowledge receipt of your letter of 27th December, requesting the Company to

[1] See p. 229.
[2] See p. 231.
[3] HBC A.8/17, f. 277.
[4] Ibid.

present any claims that they may have on the Admiralty in connexion with the Expedition that has been organised to search for the relics of Sir John Franklin, and the Crews of Her Majesty's Ships 'Erebus' and 'Terror'.

I am to state in reply that no accounts in connexion with this Expedition have yet been received, but the Governor and Committee have reason to believe that the expenditure during 1854/55 amounts to from £2000 to £3000, the detailed accounts of which are expected to arrive in this country in about a fortnight, and will be presented with as little delay as possible.

I am further to state that the expences for the year 1855/56 will amount to at least an equal sum, as the Accounts about to arrive will not extend beyond the 31st May last.

3. Letter from W. G. Smith, Hudson's Bay Company, to Thomas Phinn, Admiralty, 29 February 1856.[1]

I am directed by the Deputy Governor and Committee of the Hudson's Bay Company to transmit the accompanying Extract of a letter from Sir George Simpson, dated Lachine 28th January 1856, together with the accounts against Her Majesty's Government in reference to the Arctic Searching Expedition, amounting together to £6442.18.6; and I am to request that you will lay the same before the Lords Commissioners of the Admiralty.

Extract of a letter from Sir George Simpson to W. G. Smith Esqr, dated Lachine 28th January 1856.

Herewith are transmitted two accounts against H.M. Government in reference to the Arctic Searching Expedition, amounting respectively to £1852.8.7 and £4589.19.1 Stg. The former is for expences incurred in the summer of 1854 in forwarding from Red River to the Company's posts in McKenzie River district supplies of clothing and provisions to be held in depôt for the use of any parties who might find their way to those posts from vessels then supposed to be set fast in the Arctic Sea. When the supplies were no longer required for that object a portion was made available for the Expedition of last summer, and the remainder, or as much as may be in any way serviceable, will be assumed from time to time for the Hudson's Bay Company, at prices to be fixed on the spot, for which credit will be given hereafter, agreeably to the arrangement made to that end between the Company and the Admiralty. The second account is exclusively connected with the Expedition of last summer under Messrs. Anderson & Stewart, embracing all outlay so far as it has been advised up to the present time, although some supplemental charges may be rendered next season, which however can only be of small amount. I need hardly point out the difficulty of preparing an account of this description including transactions in several of the most remote parts of the Company's territories, but the data furnished to this office have been very carefully collated and reduced to form, the prices being regulated by the tariffs established in 1847 between myself and Sir John Richardson, and followed in all subsequent transactions with Her Majesty's

[1] Ibid., ff. 292–3.

Government. The amount of the account is large, but it will be noticed a considerable portion is for actual cash outlay by the Company (including wages); it is therefore to be hoped that no unnecessary delay may occur in its settlement, as neither interest nor commission have been charged to remunerate the Company for making those advances. The nature of the service prevented anything in the shape of vouchers being obtained for the various items forming the account, but I have no doubt that, under the circumstances, the Admiralty will neither expect nor call for them.

4. Letter from Thomas Phinn, Admiralty, to Andrew Colvile, Deputy Governor, Hudson's Bay Company, 7 March 1856.[1]

In reply to Mr. Smith's letter of the 29th ultimo, transmitting accounts against Her Majesty's Government, in reference to the Arctic Searching Expedition, amounting to the sum of £6442.18.6, I am commanded by my Lords Commissioners of the Admiralty to acquaint you that the Accountant General of the Navy has been directed to cause the above amount to be paid to you.

5. Letter from E. Gandy, Accountant General, Admiralty, to W. G. Smith, Hudson's Bay Company, 28 March 1856.[2]

In pursuance of directions from the Lords Commissioners of the Admiralty I enclose herewith an order on Her Majesty's Paymaster General in favour of Andrew Colvile Esqr., Governor of the Hudson's Bay Company, [marginal note: £6442.18.6] for the sum of six thousand four hundred, and forty two pounds, eighteen shillings and sixpence, being the amount due to the Company for provisions, clothing, etc. supplied to the Arctic Searching Expedition.

6. Letter from W. G. Smith, Hudson's Bay Company, to E. Gandy, Accountant General, Admiralty, 29 March 1856.[3]

I am directed by the Deputy Governor and Committee of the Hudson's Bay Company to acknowledge receipt of Mr Gandy's letter of the 28th Instant, enclosing an order on Her Majesty's Paymaster General in favour of Andrew Colvile Esqr. of the Hudson's Bay Company, for the sum of £6442.18.6 being the amount due to the Company for provisions, clothing, etc. supplied to the Arctic Searching Expedition, and to request that in consequence of the death of Mr. Colvile instructions may be given for the payment of the order to John Shepherd Esqr., the Deputy Governor of the Company.

7. Letter from W. G. Smith, Hudson's Bay Company, to H. Corrie, Admiralty, 13 February 1858.[4]

I am directed by the Governor and Committee of the Hudson's Bay Company to

[1] HBC A.8/17, ff. 293–4.
[2] HBC A.8/17, f. 294.
[3] Ibid., ff. 294–5.
[4] HBC A.8/18, f. 2.

transmit the accompanying Supplementary Account against Her Majesty's Government for expenses incurred by the Arctic Searching Expedition under Messrs. Anderson & Stewart amounting to £217.1.1. and to request that you will be pleased to place the same before the Lords Commissioners of the Admiralty with a view to its payment.

8. Letter from H. Corrie, Admiralty, to W. G. Smith, Hudson's Bay Company, 17 May 1858.[1]

Having laid before my Lords Commissoners of the Admiralty your letter of the 18th February last transmitting a supplementary Account of the Expenses of the Arctic Searching Expedition under Messrs. Anderson & Stewart amounting to £217.1.1, I am commanded by my Lords to acquaint you for the information of the Governor and Committee of the Hudson's Bay Company that they have instructed the Accountant General of the Navy to pay you that amount, and I am at the same time to request that you will send to this Office the remainder of the Articles found belonging to the lost boat of Sir John Franklin's expedition, as Kettles etc.

9. Letter from W. G. Smith, Hudson's Bay Company, to H. Corrie, Admiralty, 19 May 1858.[2]

I am directed by the Governor and Committee of the Hudson's Bay Company to acknowledge the receipt of your letter of the 17th Inst. in which you inform me that the Accountant General of the Navy has been instructed to pay to the Company the Supplementary Account due to them for expenses incurred in connection with the Arctic Searching Expedition under Messrs Anderson and Stewart, and requesting that the remainder of the articles of the lost Boat party of Sir John Franklin's Expedition be sent to the Admiralty.

In thanking you for this communication I am instructed to state that all the articles belonging to the Expedition which have been forwarded to England have already been placed at the disposal of the Lords Commissioners of the Admiralty; and that directions have been given that such others as were brought away by Messrs Anderson & Stewart, and remain in the Indian country be transmitted to England by the earliest opportunity.

10. Wages paid to expedition members.[3]

	Months	Rate p/a	Amount	
Lockhart James	5	75	31/5/	Wages till 1st November
Anarize Joseph	7	60	35/ /	Wages from 1 June to 31 Dec
Assuneyunton Jean Bt.	7	60	35/ /	Do.

[1] Ibid.
[2] Ibid., ff. 2–3.
[3] HBC E.15/10, f. 15.

Bouché Joseph	8	50	33/6/8	
Daniel George	6⅚	50	28/9/5	
Fidler John	6⅚	50	28/9/5	
Fidler Henry	6⅚	50	28/9/5	
Jobin Ambroise	6⅚	50	28/9/5	These men have been
Johnstone Jerry	6⅚	50	28/9/5	credited in all one full year's
Kippling Edward	6⅚	50	28/9/5	wages on account of Arctic
Kippling George	6⅚	60	34/3/4	Searching Exp. per orders of
Laferté Alfred	6⅚	50	28/9/5	Chief Factor James Anderson
Landrie Alexander	6⅚	50	28/9/5	
McLellan Murdoch	8	50	33/6/8	
McLeod Donald	6⅚	50	28/9/5	
Misteagun Thomas	8	50	33/6/8	
Montour Ignace	7	60	35/ /	Wages, 1 June to 31 Dec
Oman William	5⅔	20	9/8/11	Wages till return to Churchill 20 Novr.
Paupaunekis Paulet	12	50	50/ /	
Reid William	4	50	16/13/4	Wages from 1st June to 30 Septr '55
			602/15/4	

E.E.
York Factory
August 1856.

AWARDS

11. Letter from W. G. Smith, Hudson's Bay Company, to Secretary of the Admiralty, 7 April 1856.[1]

I am directed by the Deputy Governor and Committee of the Hudson's Bay Company to intimate to you, for the information of the Lords Commissioners of the Admiralty, that Mr. J. G. Stewart. who was associated with Mr. James Anderson as leader of the Expedition sent, by the desire of their Lordships, to confirm the report of the Esquimaux made to Dr. Rae of the fate of Sir John Franklin's Expedition, has arrived in this country, and to request that you will bring under the notice of their Lordships the following Extract of a letter, under date 27th October 1854, received from the Admiralty: 'The servants of the Company who may be selected to serve on these two Expeditions may be assured that it is the wish of the British Government, and its declared intention, to pay liberally for the service to be rendered, and to reward specially any acts of signal daring and distinguished merit.'

The Deputy Governor and Committee are of opinion that Messrs. Anderson and Stewart have evinced great zeal and judgment in carrying out the wishes of the Government, and confidently leave the question of reward for their services in the hands of the Lords Commissioners of the Admiralty.

[1] HBC A.8/17, f. 295.

12. Private letter from Captain John Washington, Admiralty, to John Shepherd, Deputy Governor, Hudson's Bay Company, 29 April 1856.[1]

The Hudson's Bay Company have written to the Admiralty asking for remuneration for Mr. Stewart and, I suppose (but the letter does not say so) for Mr. Anderson, for their services in the Arctic Regions in search of Sir John Franklin's remains.

As we have no clue to guide us at all in arriving at a decision on the case, will you be kind enough to inform me how long these gentlemen were employed on the above service; at what rate you consider it would be just to pay them, and whether the whole expences connected with the Expedition have come home, as it will be much more convenient to settle all together.

13. Letter from John Shepherd, Hudson's Bay Company, to Captain John Washington, Admiralty, 1 May 1856.[2]

I have been making the necessary enquiries to enable me to reply to your note of the 29th ultimo. I do not think I am warranted in saying that the accounts sent home include the whole expenses connected with the Expedition, yet, as Mr. Stewart is in this country, it would of course be very convenient for him to receive now any reward or gratuity which the Lords of the Admiralty may be pleased to award him.

I learn that our practice has been to allow our officers engaged on such service double pay during the period it continues, and I presume that their Lordships will think that measure of reward the smallest they could propose. The period occupied in the search by Messrs Anderson & Stewart appears to have been six months, and their pay in the Hudson's Bay Company's service is for Anderson £600 per annum, and for Stewart £300 per annum.

You will observe that in our letter of the 7th ultimo the names of Anderson and Stewart were both mentioned in the last paragraph.

14. Letter from Thomas Phinn, Admiralty, to W. G. Smith, Hudson's Bay Company, 17 May 1856.[3]

In reply to your letter of the 7th April last, reporting the return to this country of Mr. James Anderson and Mr. J. G. Stewart (who were sent to confirm the report of the Esquimaux as to the fate of Sir John Franklin's Expedition) I am commanded by my Lords Commissioners of the Admiralty to acquaint you that they have been pleased to award the sum of Four hundred pounds to the former, and Two hundred and eighty pounds to the latter gentleman, as an acknowledgment of their services; and I am to request that you will at the same time express to the Deputy Governor of the Company their Lordships' thanks for the interest & trouble he has taken on the subject.

P.S. The Accountant General of the Navy has been directed to make the abovementioned payments upon application to him at Somerset House.

[1] Ibid., f. 296.
[2] Ibid., ff. 296–7.
[3] Ibid., f. 297.

15. Extract from letter from W. G. Smith, Hudson's Bay Company, to Thomas Phinn, Admiralty, 5 June 1856.[1]

... I am at the same time directed by the Governor and Committee to acknowledge receipt of your letter of the 17th ultimo, stating that their Lordships had been pleased to award the respective sums of £400 and £280 to Messrs Anderson & Stewart in acknowledgement of their services whilst in command of the Expedition, which was duly communicated to those gentlemen.

16. Letter from W. G. Smith, Hudson's Bay Company, to J. G. Stewart, 23 May 1856.[2]

I am directed by the Governor & Committee to inform you that the Lords Commissioners of the Admiralty have awarded you the sum of Two hundred & Eighty Pounds in acknowledgement of your Services on the recent Arctic Expedition to confirm the report of the Esquimaux as to the fate of Sir John Franklin and his party, and that the amount, when received, will be carried to your credit with the Hudson's Bay Company.

I am at the same time to add that the energy and judgment evinced by Chief Factor Anderson & yourself in effecting your arduous undertaking, are duly appreciated by the Governor & Committee.

17. Letter from W. G. Smith, Hudson's Bay Company, to James Anderson, 23 May 1856.[3]

I am directed by the Governor and Committee to inform you that the Lords Commissioners of the Admiralty have awarded you the sum of Four Hundred Pounds in acknowledgment of your services on the recent Arctic Expedition to confirm the report of the Esquimaux as to the fate of Sir John Franklin and his party, and that the amount, when received will be carried to your credit with the Hudson's Bay Company.

I am at the same time to add that the energy and judgment evinced by yourself and Chief Trader Stewart, in effecting your arduous undertaking are fully appreciated by the Governor and Committee.

18. Letter from W. G. Smith, Hudson's Bay Company, to Thomas Phinn, Admiralty, 2 July 1856.[4]

With reference to your letter of the 17th May announcing the awards made by the Lords Commissioners of the Admiralty to Mr. James Anderson and Mr. J. G. Stewart, in consideration of their services whilst employed on the Expedition sent to confirm the report of the Esquimaux as to the fate of Sir John Franklin's party, I am directed by the Governor and Committee of the Hudson's Bay Company to state that as both those gentlemen are now in the Indian country, and not likely to visit England for some years, they will be unable to make personal application for the

[1] Ibid., ff. 297–8.
[2] P.A.A. 74.1/141.
[3] HBC E.37/11, f. 23; BCA AB40 An 32.2, f. 57.
[4] HBC A.8/17, f. 301.

money, and to suggest, should it meet the approval of the Lords Commissioners, that payment be made to the Hudson's Bay Company on their behalf.

I am to add that Mr. Stewart had left England previous to the receipt of your letter of the 17th May.

19. Letter from B. Osborne, Admiralty, to W. G. Smith, Hudson's Bay Company, 4 July 1856.[1]

Referring to my letter to you of the 17th May last, and to your letter of the 2nd Inst., I am commanded by my Lords Commissioners of the Admiralty to acquaint you that under the circumstances stated by you the Accountant General of the Navy has been directed to pay into the hands of the Governor and Committee of the Hudson's Bay Company the sum of £400 for Mr. Anderson, and £280 for Mr. Stewart, in consideration of their services whilst employed on the Expedition sent to confirm the report of the Eskimos as to the fate of Sir John Franklin and his party. I am to request that those gentlemen's acknowledgements of having received this money be transmitted to this office.

20. Letter from R. M. Bromley, Accountant General, Admiralty, to W. G. Smith, Hudson's Bay Company, 10 July, 1856.[2]

... Their Lordships having by their further order of the 4th Instant, in consequence of the protracted absence of Messrs. Anderson & Stewart, directed the sum of [marginal note: £680] Six Hundred and Eighty Pounds to be paid to the Hudson's Bay Company, being £400 for the former, and £280 for the latter, in consideration of their services whilst employed on the Expedition sent to confirm the report of the Esquimaux as to Sir John Franklin's fate, I beg leave also to enclose a Bill for that amount.

21. Letter from W. G. Smith, Hudson's Bay Company, to Secretary of the Admiralty, 15 July 1856.[3]

I am directed by the Governor and Committee of the Hudson's Bay Company to acknowledge receipt of Mr. Phinn's letter of the 2nd Instant, and also of two letters from Mr. Osborne dated the 4th Inst. and to express the thanks of the Governor and Committee for these communications.

I am at the same time to state that, in accordance with Mr. Osborne's request, the acknowledgements of Mr. James Anderson and Mr. Jas. G. Stewart for the awards made to them will be obtained and transmitted to the Admiralty by the earliest opportunity.

22. Letter from W. G. Smith, Hudson's Bay Company, to R. M. Bromley, Admiralty, 18 July 1856.[4]

I am directed by the Governor and Committee of the Hudson's Bay Company to

[1] Ibid., f. 303.
[2] Ibid., f. 305.
[3] Ibid., f. 304.
[4] Ibid., f. 305.

acknowledge receipt of your letter of the 10th Instant transmitting Bills on Her Majesty's Paymaster General of £10,000 and £680 on account of awards made by the Lords Commissioners of the Admiralty to Dr. Rae and his companions and to Messrs. Anderson and Stewart respectively...

23. Letter from W. G. Smith, Hudson's Bay Company, to James Stewart, 21 January 1857.[1]
With reference to my letter of the 23rd May last informing you that Her Majesty's Government had been pleased to award you the sum of £280 in acknowledgment of your services on the recent Arctic Searching Expedition, I am directed to acquaint you that the amount was deposited in the hands of the Hudson's Bay Company on your behalf, and duly passed to your credit with them, but a voucher being required by the Government for this payment I am to request that you will sign and return me the accompanying receipt by the earliest opportunity.

24. Letter from W. G. Smith, Hudson's Bay Company, to Secretary of the Admiralty, 23 October 1857.[2]
Agreeably to the request contained in Mr. Bernal Osborne's letter of the 4th July 1856, I am directed by the Governor & Committee of the Hudson's Bay Company to transmit the receipts for the awards of £400 and £280 made by the Lords Commissioners of the Admiralty to Mr. James Anderson and Mr. James G. Stewart, in consideration of their services whilst employed on the Expedition sent to confirm the report of the Esquimaux as to the fate of Sir John Franklin and his party.

25. Gratuities credited to expedition members.[3]

Ananize Joseph	For good conduct on the voyage to the coast		5 " "
Assuneyunton Baptiste	Do		5 " "
Bouché Joseph	Do	5 " "	
	For finding first trace of the missing party	1 " "	
	As Mr. Stewart's servant	3 " "	9 " "
Daniel George	For good conduct on the voyage to the coast		5 " "
Fidler John	Do	5 " "	
	As Steersman	5 " "	10 " "
Fidler Henry	For good conduct on the voyage to the coast		5 " "
Johnstone Jerry	Do		5 " "
Kippling Edward	Do		5 " "
Landrie Alexander	As Mr. Lockhart's Servt.		3 " "

[1] PAA 74.1/138.
[2] HBC A.8/17, f. 340.
[3] HBC E.15/10, f. 13.

McLennan Murdoch	As Mr. Anderson's Do	3 " "	
	For good conduct on the voyage to the coast	5 " "	8 " "
McLeod Donald	Do.		5 " "
Misteagan Thomas	Do.		5 " "
Montour Ignace	Do		5 " "
Paupaunekis Paulet	Do.	5 " "	
	As Steersman	5 " "	10 " "
Reid William	For finding first traces of the missing party	1 " "	
	For good conduct on the voyage to the coast	5 " "	6 " "
			91 " "

E.E.
York Factory
August 1856

25. Letter from Thomas Collings, Admiralty, to Thomas Fraser, Secretary, Hudson's Bay Company, 21 February 1859.[1]

I forward herewith One hundred and eighty seven Arctic Medals for the Officers and Men of the Hudson's Bay Company named in the lists transmitted in your letter of the 27th ult. to the Secretary to the Admiralty who were engaged in Arctic Exploring and Searching Expeditions between the years 1819 and 1855 with the exception of those for the persons mentioned in the margin, who have received their Medals from this Department (Names in margin: Wm. Adamson, Thos. Matthews, Willm. Matthews, John Ross, James Spence).

I request that you will have the goodness to cause the same to be distributed, and in doing so that you will be pleased to bear in mind that by the Regulations on the subject the representatives in the first degree of Relationship, viz. Widows, Fathers, Mothers, Brothers, Sisters, and Children only are entitled to Medals of deceased persons.

At the expiration of a reasonable time I further request that you will be good enough to cause this Office to be furnished with a list of the persons to whom the Medals may have been issued, returning any that may remain undistributed.

P.S. Be pleased to sign the enclosed form of receipt and return it by bearer.

Relics

26. Letter from W. G. Smith, Hudson's Bay Company, to Thomas Phinn, Admiralty, 5 June 1856.[2]

With reference to my letter of the 9th January, transmitting copy of Chief Factor

[1] HBC A.8/18, f. 12.
[2] HBC A.8/17, ff. 295–6.

Anderson's report of the Expedition to investigate the traces of the party under Sir John Franklin, found by Dr. Rae, I am directed by the Governor and Committee of the Hudson's Bay Company to place at the disposal of the Lords Commissioners of the Admiralty a portion of the relics discovered by Messrs. Anderson & Stewart which have just been received at this House...

27. Letter from Thomas Phinn, Admiralty, to W. G. Smith, Hudson's Bay Company, 7 June 1856.[1]

I have received and laid before my Lords Commissioners of the Admiralty your letter of the 5th Instant, with the case which accompanied it, discovering a portion of the relics discovered by Messrs. Anderson & Stewart in their search for traces of Sir John Franklin's Expedition.

I am commanded by my Lords to acknowledge receipt of the same, and to request you to express to the Governor and Committee of the Hudson's Bay Company their thanks for the attention shewn by the Governor and Committee in the matter.

28. Letter from John Barrow, Admiralty, to Andrew Colvile, Hudson's Bay Company, 7 July 1856.[2]

Can you inform me whether the remaining articles discovered by Messrs. Anderson & Stewart have yet arrived?

I allude to the Chain Hooks, Kettles, etc. or whether the few articles sent here are all that are at present in this country.

29. Letter from W. G. Smith, Hudson's Bay Company, to John Barrow, Admiralty, 8 July 1856.[3]

I beg to acknowledge receipt of your letter of yesterday's date to the address of Mr. Colvile, and to acquaint you that the relics of Sir John Franklin's party to which you allude, have not arrived, but are looked for by return of the Hudson's Bay Company's ship from York Factory in October.

30. Letter from W. G. Smith, Hudson's Bay Company, to Thomas Phinn, Admiralty, 12 November 1856.[4]

I am directed by the Governor and Committee of the Hudson's Bay Company to place at the disposal of the Lords Commissioners of the Admiralty the remainder of the relics of Sir John Franklin's party found on Montreal Island by Messrs. Anderson and Stewart.

These relics, a list of which is enclosed herewith, have recently been received by one of the Company's ships from York Factory, Hudson's Bay.

[1] Ibid., f. 296.
[2] Ibid., ff. 303–4.
[3] Ibid., f. 304.
[4] Ibid., ff. 308–9.

List of Sundries found on Montreal Island and at the mouth of Back's River, July and August 1855.

1 Blacksmiths Cold Chisel (iron)
1 Tomahawk
1 Piece of a Gun wash rod
3 " " Rope, with the Government mark
1 " " Copper Rudder iron broken
2 " " Iron Ditto "
1 " " Copper Ring
1 " " Mahogany Board
3 " " Pine Wood, one of them having a piece of sheet copper attached to it
1 " " Oak
1 Copper Chain Hook
3 " Nails
4 Strips of Bunting
1 Handle Dinner Knife – bone
1 Iron Hinge broken
1 Brass Ring Binnacle
1 " Nail with Do.
Part of a Theodolite Stand
A small piece of Wood belonging to Capt. Back's boat

Fort Simpson (signed) Jas. Anderson C.F.
29th November 1855

31. Letter from Thomas Phinn, Admiralty, to W. G. Smith, Hudson's Bay Company, 13 November 1856.[1]

I have received and laid before my Lords Commissioners of the Admiralty your letter of the 12th Instant, placing at their disposal the remainder of the relics of Sir John Franklin's party, found on Montreal Island by Messrs. Anderson & Stewart, and I am desired to request you will communicate to the Governor and Committee of the Hudson's Bay Company their Lordships' thanks for their courtesy and attention.

32. Letter from W. G. Romaine, Admiralty, to W. G. Smith, Hudson's Bay Company, 13 February 1858.[2]

With reference to previous correspondence I am commanded by my Lords Commissioners of the Admiralty to request that the Governor and Committee of the Hudson's Bay Company will cause me to be informed if any of the Kettles and other Articles found by Mr. Anderson at the mouth of the Great Fish River, during his last Expedition in search of Sir John Franklin have ever been sent to England in their ships from Hudson's Bay, it having been stated that the various Articles found were to be forwarded to this Department.

[1] Ibid., ff. 309–10.
[2] HBC A.8/18, f. 1.

33. Letter from W. G. Smith, Hudson's Bay Company, to W. G. Romaine, Admiralty, 17 February 1858.[1]

I am directed by the Governor and Committee of the Hudson's Bay Company to acknowledge receipt of your letter of the 13th Instant, and to inform you that all the relics of Sir John Franklin's party discovered by the expedition under Messrs. Anderson & Stewart which have been transmitted to England were forwarded to the Admiralty under the dates of the 5th June and 12th November 1856, and their receipt acknowledged on the 7th June and 13th November respectively.

The Governor and Committee desire me to state that in reference to Mr. Anderson's Report of the Expedition, a copy of which was transmitted to the Admiralty on the 9th January 1856, no mention is made of any Kettles having been found among the articles but enquiry upon the subject will be instituted and the result communicatd to their Lordships.

34. Extract from letter from H. Corrie, Admiralty, to W. G. Smith, Hudson's Bay Company, 17 May 1858.[2]

... and I am at the same to request that you will send to this Office the remainder of the Articles found belonging to the lost boat of Sir John Franklin's expedition, as Kettles etc.

35. Extract from letter from W. G. Smith, Hudson's Bay Company, to H. Corrie, Admiralty, 19 May 1858.[3]

I am directed by the Governor and Committee of the Hudson's Bay Company to acknowledge the receipt of your letter of the 17th Inst. ... requesting that the remainder of the articles of the lost Boat party of Sir John Franklin's Expedition be sent to the Admiralty.

In thanking you for this communication I am instructed to state that all the articles belonging to the Expedition which have been forwarded to England have already been placed at the disposal of the Lords Commissioners of the Admiralty; and that directions have been given that such others as were brought away by Messrs Anderson & Stewart, and remain in the Indian country be transmitted to England by the earliest opportunity.

36. Letter from Thomas Fraser, Hudson's Bay Company, to W. G. Romaine, Admiralty, 14 June 1859.[4]

With reference to several letters addressed in 1856 by the Secretary of this Company to the Secretary of the Admiralty on the subject of Chief Factor Anderson's Report of the Expedition to investigate the traces of the party under Sir John Franklin, found by Dr. Rae, I am directed by the Governor and Committee of the Hudson's Bay Company to place at the disposal of the Lords Commissioners of the

[1] Ibid.
[2] Ibid., f. 2.
[3] Ibid., ff. 2–3.
[4] Ibid., f. 15.

Admiralty an additional portion of the relics discovered by Messrs. Anderson & Stewart, and which has just been forwarded to this House.

Mr. Anderson in his letter on the subject explains the cause of the delay in forwarding the relic in question. He says: 'Mr. Hargrave was so kind as to promise to take home a piece of mahogany for me, and leave it in your care. It was picked up at the mouth of Back's Great Fish River, having been used by the Esquimaux there as a snow shovel.

As nothing could be identified by it, and as it was rather clumsy to send out with the other articles in the winter, I decided on keeping it, and getting some memorial of the Expedition made of it. Should you, however, think that it would be acceptable among the other relics preserved of poor Franklin's party, you can send it to Greenwich, where I believe the other articles are preserved.'

CHAPTER XII

ENIGMAS AND LOOSE ENDS

INTRODUCTION

There are two intriguing postscripts to the story of the Anderson/Stewart expedition. One of these mysteries can be resolved fairly satisfactorily, but the other can probably never be satisfactorily confirmed or contradicted. The first of these was publicized by Captain Sherard Osborn[1] in a letter to *The Times* in February 1857; in it he enclosed a note from an anonymous correspondent in Red River. The latter claimed that he had been told by yet another anonymous informant that a messenger from James Anderson at Fort Simpson, who had brought dispatches for Sir George Simpson at Norway House the previous July (1856), reported that Indians trading to one of the posts in the Mackenzie River District relayed the information that Indians had seen recent encampments of 10–12 Whites on an island 'where Anderson and Stewart turned back'. The implication was that Anderson and Stewart had narrowly missed rescuing a party of survivors from the Franklin expedition and that, even worse, Anderson and Simpson had heard about this, but had taken no action.

Osborn's letter gave rise to a heated exchange of correspondence in *The Times* by writers including W. G. Smith, Secretary of the Hudson's Bay Company, Sir George Simpson, Alexander Isbister (always a vigorous critic of the Company) and, predictably, Dr Richard King, another perennial enemy of the Company. The latter jumped at this opportunity to denigrate the Company's efforts and to belabour the

[1] Captain Sherard Osborn (1822–75) had participated in two of the Royal Navy's expeditions in search of the Franklin expedition. In 1850–51 he had commanded the steam tender *Pioneer*, attached to Captain Horatio Austin's squadron (HMS *Resolute*, *Assistance*, *Pioneer*, and *Intrepid*) (see Osborn, *Stray Leaves*). He had taken part in the preliminary examinations of the traces of the Franklin expedition left at its winter quarters at Beechey Island in August 1850. Then, along with the rest of the squadron, *Pioneer* had wintered off Griffith Island. In the spring of 1851 Osborn had led a sledge party which searched most of the west coast of Prince of Wales Island. Osborn returned to the Arctic, again in command of *Pioneer*, as part of Captain Sir Edward Belcher's squadron in 1852 (HMS *Resolute*, *Intrepid*, *Assistance*, *Pioneer* and *North Star*). Along with *Assistance*, *Pioneer* wintered in Northumberland Sound (north-west Devon Island) in 1852–3. From there, in the spring of 1853 Osborn led a sledge party which travelled west to the northern tip of the Sabine Peninsula (Melville Island) and on its homeward route explored the north and east coasts of Bathurst Island. *Assistance* and *Pioneer* wintered again (1853–4) near Cape Osborn, Wellington Channel; there they were abandoned in the spring of 1854, their crews falling back on the depot ship, *North Star* at Beechey Island (see Belcher, *The Last of the Arctic Voyages*).

Admiralty with a proposal for yet another expedition, to be led by himself. Simpson categorically denied that Anderson had sent him any such message, while the Governor and Committee of the Hudson's Bay Company maintained that the rumours were 'utterly destitute of foundation'.[1] On reflection, one has to agree with the Governor and Committee; the idea of Indians being in the vicinity of Point Ogle at about the same time as Anderson and Stewart is quite inconceivable. One can readily accept King's assertion that Chipewyan had been known to interact amicably with the Inuit on the Thelon, but that river lies some 400–500 km farther south.

Almost certainly the source of this rumour is to be found in a letter from Mr Lawrence Clarke, in charge of the post at Fort Rae, to James Anderson, in charge of the Mackenzie River District, dated 28 February 1856.[2] He reported that two Yellowknife Indians had arrived at Fort Rae, and told him that they had found an encampment on the Thlewycho (Back River), and had seen fires at a distance. A sketch map of the arrangement of the features of the camp, and of its relation to the fires was appended. Clarke clearly was inclined to think that this must have been the camp of some survivors of the Franklin expedition, but Anderson poured cold water on this notion. In a note appended to Clarke's letter Anderson commented that he was absolutely convinced that the camp the Indians had stumbled upon was one of the ones occupied by Stewart and himself. Clarke requested permission to travel to the Back River to investigate the rumour, but there is no record in the Hudson's Bay Company's Archives of Anderson's reply. One can safely assume, from the note mentioned above, that he did not grant Clarke permission for such an expedition which he (Anderson) clearly saw as pointless.

The second intriguing postscript to the Anderson/Stewart expedition surfaced only much later, in 1890. While engaged in pioneer geological fieldwork in the area of Lake Winnipeg the geologist Joseph Burr Tyrrell,[3] fell into conversation with Joseph Boucher, who had been a member of Anderson and Stewart's expedition. Boucher reported that three men had been sent off as an independent party while the expedition was on the arctic coast, and that one of them had spotted the masts of a ship far off amongst the sea ice. On returning to the main group they had deliberately refrained from telling Anderson and Stewart since they were afraid that an attempt to investigate would dangerously prolong the expedition and jeopardize their chances of getting back to Fort Reliance. He identified the three men as Thomas Mustegan, Paulet Papanakies and Edward Kipling; almost certainly the allusion is to the party which was sent across by Halkett boat to search Maconochie Island. This group consisted of the three men listed plus Henry Fidler.

Intrigued by the story Tyrrell managed to locate the three men, and through friends arranged for them to make depositions on the subject. All the statements are reasonably consistent with each other; they reveal that it was Papanakies who

[1] See p. 265.
[2] HBC B.200/b/32, ff. 155–8.
[3] Joseph B. Tyrrell, 'A Story of a Franklin Search Expedition', *Transactions of the Canadian Institute*, 8: 1908–9, pp. 393–402. For details of Tyrrell's career see: Alex Inglis, *Northern Vagabond. The Life and Career of J. B. Tyrrell*, Toronto, 1978.

had claimed to see the masts. If one accepts that claim, it raises the intriguing question of what vessel it might have been.

In his analysis of the story Woodman[1] quite reasonably ruled out the possibility that it could have been either *Erebus* or *Terror*, since he at least was convinced that both are known to have sunk, one off the north-west coast of Adelaide Peninsula, the other off Cape Crozier, King William Island. This would suggest that Paulet had mistaken the masts of a boat for those of a ship (despite his protestations that he was familiar with ships' masts from vessels he had seen at York Factory). But whether a ship or a boat, this begs the question of whether any sort of craft associated with the Franklin expedition would still be drifting around in the sea ice in the vicinity of Point Ogle as late as August 1855. This story, too, must remain a baffling enigma, one that is unlikely ever to be satisfactorily explained.

DOCUMENTS

1. Letter from Lawrence Clarke, Fort Rae, to James Anderson, Fort Simpson, 28 February 1856.[2]

Two Yellow Knife Indians arrived here late this evening and on making inquiries of how and where they passed the winter, in requiring of them their news, to my great surprise they related the undermentioned remarkable story which, if true, may under Providence be the means of tracing the ultimate fate of some part of Sir J. Franklin's unfortunate party. I shall relate the Indian's accounts as near as possible in their own words, and afterwards give you the answers elicited by my questions.

'Two of our tribe severally named Nakesse and Tennadzie, being on a hunting excursion along the shores of the Thlewychodesse last summer, upon the top of a high rocky hill came upon an old encampment evidently inhabited by white men the previous summer. The place where the tent had been erected was plainly visible, being marked by 4 large stones forming a square used to keep the corners of the tent expanded. Large quantities of Moss had been gathered & a bed formed inside the tent with a larger quantity at the backside for making a pillow. The feet were extended towards the door or opening of the tent, evident from the fire being placed a few feet on the outside of the square. Two large square stones had been placed a short distance apart and the fire made of Green & Dry willows kindled between, the stones evidently forming the stand of a kettle. On further search directly vis a vis the encampment they found two very larger rocks close to the water's edge around which was visible the prints of people's feet (after the lapse of a year! J.A.). These rocks had been removed to their present place from a considerable distance by human agency and was, we suppose, used for tying boats thereto (Query: could these rocks be a cairn?) (Most probably the men's cooking places;

[1] Woodman, *Unravelling*, p. 279.
[2] HBC B.200/b/32, ff. 155–8.

they always used large stones for their chimneys. J.A.). After the Indians were on their way back to their camp they saw two fires at a great distance down the River marked in the chart.'

This so far is a condensed narrative of the Indians' accounts. Before I proceed to give the queries & answers, I beg leave to state that this story was spontaneous and was spoken without a pause or one question being put by me.

Question 1. Why do you think it was white men who encamped there? **Answer.** Because of a tent having been erected there, and the way the moss was laid, besides by many other infallible signs; we know no Indians use these.

Q. 2. Where and how was the fire kindled? **A.** The fire was placed a few feet from the place where the opening of the tent stood. Two flat stones was placed a few inches apart and the fire lighted between, with green and dry willows. The stones we suppose served for a stand for a kettle.

Q. 3. How many persons slept in the tent or rather on the Moss where the supposed tent stood? **A.** Two **dents** in the moss inside the tent, in the shape of two persons. There is also another place close to the Fire where one man slept.

Q. 4. Might this not have been an Esquimaux Camp? **A.** (Emphatically) No, no. We would know it immediately.

Q. 5. Then you feel persuaded that it was white men only who slept there. **A.** Yes, we are sure.

Q. 6. Did they find nothing else in the camp, no chips of wood, or any vestiges of white men? **A.** They found the shank of a deer, with particles of the meat blackened, still adhering to the bone, nothing more.

Q. 7. How far from each other are the Rocks you found by the water's edge, and from what distance have they been removed? **A.** They are a few feet apart, but have been brought to where they now stand from many yards distance.

Q. 8. Did the men you say walked about these rocks wear English shoes or Indian Mocassins? **A.** We can't say, only my brother told me he could see the foot marks very plainly.

Q. 9. You don't think then that the Rocks could have rolled to their present position? **A.** Impossible. They were removed by human agency and with much trouble.

Q. 10. After the whites – supposing them to be white men who camped there – left, what route do you think they took? **A.** We suppose that mistaking the Ennadesse (see chart) for a continuation of the Thlewycho they entered that River, and that finding themselves lost they scattered about the country in search of Deer. My brother killed many Musk Ox who had been wounded but a short time before, and others with Balls inside of them – old wounds healed. He also saw 2 fires at a great distance in that direction. (Note. On the 26th Augt. in L. Beechy a band of musk oxen were fired on, and on the (illeg.) above L. Beechy a musk ox was severely wounded but escaped. No musk oxen were seen after that date. J.A.).

Q. 11. Who do you think wounded the cattle you found with balls inside them? **A.** White Men.

Q. 12. May it not have been Mr. Anderson's people in descending the River? **A.** That cannot be. It is too far to one side; besides we are too sure by other signs, that it was not so soon as then.

Q. 13. Of what size was the Balls found in the oxen? **A.** About the same size as our trading balls.

Q. 14. The Fires your brother saw was perhaps kindled by some of your tribe? **A.** No, no. There was only 5 of our tribe who went so far, and of those 5 none went by some days' march near where we saw the fires.

Q. 15. Why did not some of the Young men go and see who kindled the fires? **A.** After seeing the last one they did not think that the fires was made as a signal, the smoke being very high, but it was so far, and the rain began to pour down for two days without stopping & extinguished all traces of its whereabouts.

Now, my Dear Sir, it is for you to judge what degree of trust may be placed in this account. For my part, I cannot help thinking the affair looks very plausible. If you will grant me permission I will start from here on the last ice with 3 or 4 Indians and follow the deer to the vicinity and by taking advantage of the plentifulness of provisions examine well the spots indicated by the Indians.

Wishing you all happiness, I remain here,

Yours faithfully
Lawrence Clarke.

Note. There cannot be a shadow of doubt that it was our fires that were seen by the Indians. The Camping place is just as we made. J.A.

2. Letter from Captain Sherard Osborn to the Editor of *The Times*, 13 February 1857.[1]

The enclosed is from an undoubted source, and of too great importance for me to feel justified in withholding it from the public press.

Mr. Anderson, as leader of the last expedition to the shores of the entrance of the Great Fish River, and as a servant of the Hudson's Bay Company, would not have merely forwarded an idle rumour to Sir George Simpson when it reflected on his employers, and added one more to the many proofs that the starving crews of Franklin's ships reached the Hudson Bay territory, and as yet have not been reached or discovered. Indeed, their fate is still wrapped in mystery, in spite of the relics brought from Beechy [sic] Island by Penny and from America by Rae.

EXTRACT OF A LETTER, DATED RED RIVER SETTLEMENT,
Hudson Bay Territory, Dec. 6, '56.

I received a letter from Roderick by the last mail, and he expresses a wish that I should write to you by the first opportunity, and state more particularly about the reports we heard last summer about some traces of whites being seen in the north.

I have just returned from ———, who was at Norway-house last July, and saw the man who brought down an express to Sir George Simpson from Mr. Anderson in Mackenzie's River (district), stating that Indians had brought over reports to one of the trading posts in that quarter that Indians had seen two or more encampments of whites on an island on some point where Anderson and Stewart turned

[1] *The Times*, 14 February 1857, p. 10; King, *The Franklin Expedition*, pp. 231–3.

back (in 1855), and that one of the encampments particularly was quite fresh, supposed to have been abandoned a day or two before the Indians saw it, and from the traces, thought there might have been about 10 or 12 men.

I could not hear of the exact locality further than that Anderson and Stewart were within a very short distance of the place where the traces were seen. I hope you have heard more particularly about the report.

3. Letter from W. G. Smith, Hudson's Bay Company, to the Editor of *The Times*, 14 February 1857.[1]

I lose no time in noticing the report given in The Times and other papers of to-day by Captain Sherard Osborn, in reference to Sir John Franklin's Expedition, and to mention that no information bearing upon this important subject has reached this house in an official or private shape, and that, in the opinion of persons acquainted with the Indian country, it is only one of those vague rumours which have been current there from time to time, upon which no reliance can be placed, and probably arises from Indians having seen one of Anderson's own encampments on or near Great Fish River, but not near its mouth, because the Indians, as far as is known to the Hudson's Bay Company, never approach within 200 or 300 miles of it, i.e. the mouth of the river.

Sir George Simpson, whose name is particularly mentioned by Captain Osborn, will be in London next week, and will no doubt personally contradict the absurd charge made in Captain Osborn's communication.

4. Letter from Captain Sherard Osborn to the Editor of *The Times*, 16 February 1857.[2]

I am as unwilling to give my time as you must be your space to a controversy with Mr. Smith as to the value of the report about Franklin's expedition, received from the Red River settlement. The public are as well able to judge as Mr. Smith, Sir George Simpson, or myself, since the value of the report depends not upon our opinions, but the fact whether or not Sir G. Simpson ever received such a communication from Mr. Anderson, the only person capable of estimating what it is worth.

I cannot accuse myself of making any charge against Sir George Simpson. I acted as I thought right towards my friends in the lost expedition, without reference to the opinions of any man.

5. Letter from Dr Richard King to the Editor of *The Times*, 16 February 1857.[3]

I beg to state that I place implicit reliance on the statement of the Indians, communicated to you by Captain Sherard Osborn, and published in your impression of Saturday, that 'they had seen two or more encampments of whites on an island on some point where Anderson and Stewart turned back in 1855, traces of about 10 or 12 men'; and that the letter of Mr. Smith, the Secretary of the Hudson's Bay Company, published in your impression of to-day, is no contradiction, inasmuch as he is

[1] *The Times*, 16 February 1857, p. 10; King, *The Franklin Expedition*, pp. 233–4.
[2] *The Times*, 17 February 1857, p. 12; King, *The Franklin Expedition*, pp. 234–5.
[3] King, *The Franklin Expedition*, pp. 235–7.

unacquainted with the fact that the Chipewyan Indians, ever since the year 1835, have hunted on the banks of the Great Fish River, and have entered into friendly relations with the Esquimaux.

Mr. Roderick M'Leod, one of the officers of the expedition in search of Sir J. Ross in 1833–35,[1] wrote to me from Great Slave Lake, on the 2nd of July, in 1836, as follows:

> It may perhaps be in favour of your enterprise, the late intimacy that has taken place between the Chipewyan and Esquimaux tribes in the course of the last summer on the Thlewee-dezza (Fish River);[2] among the latter were many inhabitants of the Thlewee-cho-dezza (Great Fish River),[3] but the majority were those that frequent Churchill annually; to prove which they produced the articles they obtained from the Company's stores in the way of trade, and readily exchanged the same with their guests by way of cementing their friendship, and I have reason to suppose it will continue uninterrupted, now that the former have become sensible of the errors of their ancestors...

I can assure you, Sir, the want of knowledge which the Governors of the Hudson Bay Company, arising from the form of their constitution, possess of their own vast territory is such that, putting aside the necessity of a further search for the Franklin Expedition, to guard the national honour, an exploring expedition is imperatively called for in reference to a renewal of their charter, which is now before Parliament.

P.S. The fact of Indians bringing the report shows that the detachment which reached the Great Fish River pushed on into the Indian country. Time will prove whether they starved, were murdered, or still exist.

6. Letter from Sir George Simpson to the Editor of *The Times*, 17 February 1857.[4]

On my arrival in London my attention was called to a letter lately addressed to The Times by Captain Sherard Osborn, in which he gives publicity to an anonymous communication from Red River settlement, wherein it is stated that information, which I kept secret, reached me at Norway-house in July last, from Mr. Anderson (of the late expedition down the Great Fish River), of traces of Sir John Franklin's party having been discovered on the Back River by Indians. The secretary of the Hudson's Bay Company, in my absence, pointed out the improbability of this story, and my only object in now troubling you on the subject is to confirm Mr. Smith's statement, by assuring you that no information of the nature alluded to by Captain Osborn, or of any kind whatever respecting the Franklin expedition, reached me from Mr. Anderson or any other person, at Norway-house or anywhere else, in the course of the past year, although Mr. Anderson had occasion to address me frequently, his last letter being dated at the latter end of July.

In order that there may be no room for misapprehension on this point I beg to

[1] Captain George Back's expedition down the Back River.

[2] The Thelon River.

[3] Back River.

[4] *The Times*, 18 February 1857.

state that no information bearing in the most remote degree on the Franklin expedition, has ever reached me, directly or indirectly, officially or privately, excepted that obtained by Dr. Rae in 1854, which was confirmed in 1855 by the expedition under Messrs. Anderson and Stewart to the mouth of the Great Fish River. The reports of those gentlemen were given to the public without one hour of unnecessary delay.

7. Letter from Captain Sherard Osborn to the Editor of *The Times*, 18 February 1857.[1]

If I may further trespass upon your good nature, pray allow me to say that I am perfectly satisfied with Sir George Simpson's statement, that he never heard of the report, information of which I sent you on Saturday last; but I can assure you that such a report was rife in Red River Settlement, and that I hold in writing expressions of astonishment on the part of a resident there, that official notice was not taken of it by Sir George Simpson. The utter igorance of Sir George Simpson fully explains the silence of the Hudson's Bay Company.

I published no anonymous document, but I did not feel justified, nor do I now consider it right, to tell the name of the writer. Directly the Red River Settlement passes, as I trust it soon will, from the irresponsible authority of the company to the care of the Canadian Legislature, I shall be happy to do so.

8. Letter from A. K. Isbister[2] to the Editor of *The Times*, 18 February 1857.[3]

With reference to the report from Hudson's Bay relative to Sir John Franklin's party, I beg to say that the communication which Captain Osborn has made public is no anonymous one, and that the extraordinary tone adopted by Mr. W. G. Smith, in torturing a simple piece of intelligence into a 'charge' against the Hudson's Bay Company, must satisfy everyone that a judicious discretion has been exercised in not giving the name of any writer whose interests could and would be made to suffer by the company.

The report is simply the re-statement of one which reached myself last autumn from Hudson's Bay, and so notorious was it that the only wonder is the Governor of the territory was ignorant of it and of others of the same nature.

9. Letter from Thomas Phinn, Admiralty, to W. G. Smith, Hudson's Bay Company, 24 February 1857.[4]

I am commanded by my Lords Commissioners of the Admiralty to transmit to you

[1] *The Times*, 19 February 1857, p. 7.

[2] Alexander Kennedy Isbister (1822–83), was born at Cumberland House (now in Saskatchewan), the son of Thomas Isbister, of Orkney descent, and Agatha, an Indian woman. Educated in the Orkneys and at Red River, he entered the Hudson's Bay Company in 1838. For most of his short career with the Company he was located in the Mackenzie District. Leaving the Company, he moved to Britain in 1842, where he obtained the degree of MA at King's College, Aberdeen. The author of two important papers on the geology of the Mackenzie District, after leaving the Company he became the champion of the rights of the Indians and an implacable critic of the Company as regards its dealings with the Indians (Barry Cooper, *Alexander Kennedy Isbister: A Respectable Critic of the Honourable Company*, Ottawa, 1988).

[3] *The Times*, 20 February 1857, p. 7.

[4] HBC A.8/17, f. 312.

herewith the copy of a letter dated the 23rd Instant from Dr. R. King relative to the reported existence of a portion of the Expedition under the command of the late Sir John Franklin, and I am to request the Governor and Committee of the Hudson's Bay Company will furnish my Lords with any information they may possess upon the subject.

10. Letter from Dr Richard King to the Lords Commissioners of the Admiralty, 23 February 1857.[1]

Upon the early decision of H.M. Govt rests the probable fate of 12 Englishmen, of 12 Servants of the Crown, despatched many years ago upon a perilous errand to an inhospitable region, – of 12 men who have been long since officially recorded as dead, but who, there is nevertheless reasonable ground for believing, are yet alive, and may be rescued from death by an immediate and vigorous effort.

These men form a portion of the long lost Expedition commanded by Sir John Franklin, and intelligence has been received from the Hudson's Bay Territory of their existence at the mouth of the Great Fish River on the continent of North America.

On the 14th Inst. Capt. Sherard Osborn forwarded to the Times Newspaper an extract of a letter addressed to him from the Red River Settlement, by a person whose name he omits to state, to the effect that an express was on its road to Sir George Simpson with the information that Indians had seen two or more encampments of white men on an Island on some point where Messrs. Anderson & Stewart (the leaders of the searching party sent by the Hudson's Bay Coy. in 1855) had turned back, and that one of these encampments was quite fresh and had probably contained 10 or 12 men.

Sir George Simpson has published a denial that the Express alluded to by Captn. Osborn ever reached him; the Secretary of the Hudson's Bay Company has stated that no information upon the subject in an official or private shape has reached the Company, and both he and Sir George Simpson urge that it is a mere Indian rumour, upon which no reliance is to be placed.

To these denials Captn. Osborn has replied that the subject was the topic of common conversation and remark at the Company's settlements; he has declined to state the name of his informants but has expressed his determination, so soon as the Red River Settlement has passed into the hands of the Canadian Legislature, that he will do so; and from a letter addressed to the Times on the 19th Inst. it appears that although the Governor of the Hudson Bay Company has been kept in the dark on the subject, Mr. Isbister had heard of the rumour many months ago! Surprise and suspicion as to the trustworthiness of Capt. Osborn's informant may be excited by the fact that his name is withheld, and similar distrust may also be felt with regard to Mr. Isbister's statement, but to those who are acquainted with the policy of the Hudson's Bay Company, an assemblage of traders whose very existence as a body is at this moment threatened with annihilation in consequence of

[1] Ibid., ff. 313–17; King, *The Franklin Expedition*, pp. 237–46.

the approaching expiration of their Charter, the circumstance that persons in their employment or subject to their influence should object to the publication of their names as having ventured without the knowledge or concurrence of the superior officers etc. to volunteer information which may lead to the journey through the Company's settlements of an Officer of the Crown at a moment when it is essential for the interests of the Company that the knowledge of the capabilities of the country should be confined to their own servants and dependants, to those, I say, who are aware of the policy pursued by the Company, a fact of this description may be a subject of comment but not of surprize.

That such a rumour is in existence therefore among the Settlements of the Hudson's Bay Company I see no reason to doubt, and as regards the degree of credibility to be attached to an Indian or Esquimaux tale of this description, I think sufficient proof has been adduced during the search for Sir John Franklin, not only that the information obtained from the natives is not to be disregarded with impunity, but that if the traces furnished by them had been promptly followed up on a former occasion, we should at this moment be in possession of more ample and satisfactory information respecting the lost Expedition than we have yet obtained.

In proof of this assertion I need only refer your Lordships to the circumstance that the first information to Dr. Rae by the Esquimaux was on the 21st April 1854. The tale then related to him was to the effect that 35 or 40 white men had perished from starvation near the mouth of a large river, at a distance of about 10 or 12 days journey. Unhappily, as he informed your Lordships in his letter of the 10th April last, he thought the information too vague to be depended upon; he made no attempt to reach or examine the spot; the opportunity was lost; and 12 months elapsed before another expedition could be fitted out. If, on the contrary, Dr. Rae had endeavoured to settle the accuracy of the rumour, and had sought and examined the locality vaguely pointed at by the Esquimaux, there can be little doubt that such definite and decided information would have been obtained as to render all further expeditions unnecessary. The truth of the rumour is therefore more than possible, and if we take into consideration the circumstance that the neighbourhood of the spot where the white men are said to have been seen is known to teem with animal life, and that where an Indian or Esquimaux can exist, an European possessed of superior tools and weapons can also find the means of subsistence, the rumour assumes a shape of probability which it would be culpable to ignore.

Under these circumstances I submit to your Lordships that there is a reasonable probability that 12 Englishmen are wandering about in an apparently hopeless attempt to escape from the frozen shores of the Polar Sea, and the question arises whether these men, dispatched by H.M. Government on the service of the Crown, are to be suffered to perish without an effort being made to restore them to their native country, or if they should be dead, to unravel their fate.

The precise spot to which an Expedition ought to be directed is known; the district to be searched is limited in extent, and two small expeditions to act jointly by sea and by land, could be fitted out at a trifling cost, a cost of which your Lordships are fully cognizant, and could make such a close and combined search as should set at rest for ever the question as to the fate of Franklin and his companions. The expedition by sea should consist of a small screw steamer to proceed through Barrow Strait and Peel Sound; the expedition by land should consist of a small party to travel in bark canoes down Great Fish River.

Experience has shown that the most extensive results in Arctic discovery have been obtained with, comparatively speaking, the most slender means. Naval expeditions, on which thousands have been profusely squandered, have proved to be repeated failures, while land parties, equipped at a cost of a few hundred pounds, have been almost uniformly successful in their objects.

In the present instance the two expeditions would support each other, would be fitted and dispatched at little cost and if officered by men of energy and resolution, inured to the climate, accustomed to command, and to the control of the Indian tribes, would be calculated, as far as human efforts can ensure success, to attain the object in view.

It is however important that these expeditions should be conducted by officers appointed by the Crown. The last land party, which returned from an exploration of Montreal Island and Great Fish River with such meagre results, was entrusted by your Lordships to the Hudson's Bay Company, by whose management the expedition was dispatched without an interpreter, without a proper map, without even the leader of the party being made acquainted with the fact that there was a particular spot on Montreal Island that Franklin or his officers would search, and where in all human probability, the survivors would deposit a record of the fate of the Expedition. On the 8th December last a proposal for the dispatch of a joint expedition in the manner alluded to, was submitted to your Lordships, and my services in conjunction with those of an Officer of distinguished merit, Lieut. Bedford Pim, R.N. were placed at your disposal, but I learn that no provision has yet been proposed in the estimates of the year for a further Arctic Search.

The subject is now pressing; another naval Expedition on the ordinary scale would probably be as fruitless as those which have hithero been dispatched on a similar errand. A joint expedition of the nature pointed out to your Lordships offers every element of success and may be sent at a fourth part of the cost; but if it be dispatched at all, the chief of the land party must leave this country before the end of the first week in March.

The facts are now before your Lordships; the honor of this Country is at stake. Are the sons of England, the faithful servants of the Crown to be abandoned to drag out a miserable existence in an inhospitable region, while there remains one spark of hope that they may be restored to their country? From my own experience as a traveller in the land where traces of our unhappy countrymen are said to have been found, I know that Europeans can there obtain the means of subsistence; the Indian tale points to the conclusion that some may yet remain alive, and while that probability exists, I cannot believe that the House of Commons or the people of England will suffer their unhappy countrymen to remain in hopeless despair, 'chewing' to use the words of Sir Francis Beaufort, 'the bitter cud of their country's want of gratitude, want of faith and want of honour'.

11. Letter from W. G. Smith, Hudson's Bay Company, to Thomas Phinn, Admiralty, 25 February 1857.[1]

I am directed by the Governor and Committeee of the Hudson's Bay Company to acknowledge receipt of your communication of 24th Instant, transmitting a copy of a letter from Dr. King to the Lords Commissioners of the Admiralty dated the 23rd Instant, relative to the reported existence of the Expedition under the command of Sir John Franklin, and requesting to be furnished with any information which they may possess on the subject.

In reply I am to state that the Governor and Committee do not possess any information on the subject in question, and believe the rumours alluded to to be utterly destitute of foundation.

I am also instructed to observe that the Governor & Committee do not consider it necessary to take any notice of the reflections cast upon the servants of the Hudson's Bay Company by the writer of the letter addressed to their Lordships.

12. Extract from: J. B. Tyrrell, 'A Story of a Franklin Search Expedition', *Transactions of the Canadian Institute*, 8 (1908–9): pp. 393–402.

... Messrs. Anderson and Stewart found no records or remains of any of the missing men of the Franklin expedition, and little or nothing was added by them to the knowledge that had previously been obtained by Dr. Rae.

Very little else was known of the history of this expedition, which is chiefly noted because it was the last party up to the present time to descend the Great Fish River to its mouth, until in the autumn of 1890 while travelling on the east shore of Lake Winnipeg, a little old French-Canadian named Boucher came to my camp. He said that he had been cook on the Anderson and Stewart expedition down the Great Fish River, and that he had an interesting story to tell if I would but listen to him.

He told of hardships and dangers that he and the other men encountered, or thought they did, but chiefly of the agony they suffered through fear that they would never again be brought back to their homes. However the most interesting part of his story was the statement that three men who were sent northward beyond Montreal Island (or Maconochie Island) to look for any signs of Sir John Franklin or his party saw one of the ships far out in the ice, but returned and reported that they had seen nothing, fearing that if they reported a ship in sight, their masters would take them to it, and they would not be able to get back to Fort Resolution that fall, and would all perish of starvation and exposure. This was doubtless the party sent on to Maconochie Island in the canvas boat on the 8th of August. The names of the three men given by him were Thomas Mustagan, Edward Kipling and Paulette Papanakies.

Thomas Mustagan was well known to me. He was the chief of the band of Ojibway Indians which had its headquarters at Norway House, near the north end of Lake Winnipeg, and though rather old at the time was a splendid type of physical manhood, besides having a good reputation as an honest, industrious man.

The others were not personally known to me, but after some enquiry Papanakies,

[1] HBC A.8/17, f. 318.

who also was an Ojibway Indian, was found to be living at Fisher River, while Kipling, a half-breed, was living at Keewatin.

Through the kindness of some friends the following statements were obtained from these three men.

EDWARD KIPLING'S STORY

Tom Mustagan, Paulet Papanakies, Henry Fidler and I were sent to the island.

In the morning we divided at an unknown island, Henry Fidler and I went to the west, and the other two to the east. We returned to the starting point in the evening and thence to the camp, where Paulet Papanakies told of having seen the ship far out in the sea. This information was not communicated to Messrs. Anderson and Stewart for the men were all tired of the expedition and were anxious to get home.

Next day we set out for Montreal Island, where Henry Fidler and I were sent on the west side. We saw the mark of the keel of a small boat that had been dragged across the island, and found the boat broken in small pieces on the eastern side. There were remains of Eskimo camps close to the broken boat. The Eskimo were supposed to be cannibals (Note. – The story of the Eskimos being cannibals proved a very real difficulty in the way of obtaining canoemen for my journey through the Barren Lands in 1893. J.B.T.) We could find no other trace than the broken boat. We returned to camp that evening and next morning began the homeward journey. As the camps were broken up that evening preparatory to moving and we (Henry Fidler and I) were longer on the search than we had expected, nothing was reported about what we had seen on the east side of the island, namely the broken boat. When we returned to Fort Garry a portion of my wages was kept back until a settlement could be made. The balance has not been paid yet. We were told that medals had been sent to us to York Factory, but we never received them.

(Sgd.) Edward Kipling.

THOMAS MUSTAGAN'S STORY
as reported by Mr. J. A. Campbell

With regard to your inquiry I have interviewed old Tom Mustagan and others respecting 'The Tale of a Ship' and have elicited the following information: – One of the Anderson and Stewart party, Paulet Papanakies, on his return from the expedition repeatedly stated that he saw a ship far out in the ice at the mouth of the Great Fish River. Those whom I have spoken to say they have no reason to doubt the man's veracity. On some one suggesting that he might have been mistaken Paulet replied, 'It was a very clear day, and I have seen the ship at York[1] too often to be deceived.'

Old Tom, who was on this expedition, gives the following account: – Paulet, Edward Kipling and I were sent off from the camp on the mainland in an inflated waterproof canoe to examine a chain of islands running far out to sea. There was open water between these islands and we proceeded from island to island, searching for remains as we went along. We found something on one of these islands, but I do not remember what it was. When we came to the last island but one, it was

[1] York Factory.

thought advisable that I, being the heaviest man in the party, should get out of the canoe, which was hardly up to our weight, and allow the other two to go on to the last islands, which lay a long distance off. I was accordingly left behind. When my companions returned from this island, which was high and rocky, they reported that they had seen nothing. The expedition turned back shortly afterwards.

After we were disbanded Paulet told me and others at Norway House that he had seen a ship from the lofty island in question, and that he had begged Kipling to say nothing about it, because, if it were known that the ship was there, an attempt would be made to reach it, their frail craft would be crushed to pieces in the moving ice, and they would surely perish.

Tom goes on to say: – I believe Paulet saw the ship. Dr. Rae had previously been told that the ship was there, and had stated that he believed such to be the case. In fact it was on account of Dr. Rae's expressed belief in the Eskimo tale that the expedition under Messrs. Anderson and Stewart was organized.

PAULET PAPANAKIES' STORY
as reported by Mr. J. B. Johnston

After saying the party left Norway House he thinks it was in the month of February, and other unimportant details, he goes on to say that a 'Husky' who was fishing at the mouth of a river, the name of which he does not know, told him (Paulet) and Thomas Mustagan that a ship had been 'ruined' and plenty people dead. They did not understand his language, but he made signs which they could readily follow, and pointed to the place where the ship was all 'broken'. Paulet and Mustagan then proceeded to see if they could find anything, one going one way and one another, and it was Paulet only who from the summit of a rocky island saw quite distinctly what he still believes to be two masts of a ship. He says had there been more sticks standing around it would have been easy to have made a mistake, but there was nothing but rocks and ice as far as he could see. And in default of any kind of wood they were obliged to use the moss on the rocks to boil their kettle. Upon my enquiring why he did not tell the chief of the expedition what he had seen, he replied in the most simple manner possible, 'Well, I was tired of the whole thing, and was thinking long to be home, and was afraid if I said anything about it, we should have to go back and see what it was, so I thought I would keep it to myself yet awhile anyhow.' And it was some time after that he related to some of his comrades what he had seen. Pieces of iron and wood, portions of a boat, were found in the vicinity of where they were then camped.

The statements of these men are given here for what they are worth. They were made in 1893, thirty-eight years after the events to which they refer. In the main points they agree with each other fairly well. That Paulet Papanakies believed that he saw a ship there can be little doubt, so that the tale cannot be set aside as simply unworthy of credence. Furthermore, it is difficult for a white man, unless he has lived a long time among natives, to appreciate the keenness of sight, and accuracy in observation common to Indians in the wilderness, and especially to such Indians as the Ojibways, who rank among the best hunters and woodsmen on this continent. Many travellers and others who know them best will bear me out in saying that inaccurate observations are almost unknown among them.

That these men were sent across to Maconochie Island is evident from Mr. Anderson's journal, and from their own accounts, it would appear that they left that island in the little portable boat, but what other island they reached is not known, and on account of our imperfect knowledge of the geography of the district it is impossible even to make a rational conjecture. From the high point said to have been reached it is not impossible that Paulet may have seen the boat which had been taken to Starvation Cove, west of Point Richardson, where so many of the party died, or that the ship which finally sank at Oot-loo-lik first drifted into Simpson Strait to a point where it would be within range of vision, and then drifted back again westward to where it was unintentionally wrecked by the Eskimos.

In any event the story is interesting, not only as adding something to the knowledge of what became of Franklin's ships, but also as furnishing some slight insight into difficulties to be overcome by travellers who venture into remote parts of northern Canada.

CHAPTER XIII

LATER SEARCHES AND ASSESSMENT

It must have been a source of real satisfaction for Anderson that his suggestions to Lady Franklin for a further search were adopted almost to the letter. In 1857 she dispatched a private expedition under the command of Captain Leopold McClintock, who had already participated in three earlier searches: as Lieutenant on board HMS *Enterprise* under Sir James Clark Ross in 1848–9, wintering at port Leopold; as Lieutenant on board HMS *Assistance*, under Captain Erasmus Ommanney in 1850–51, wintering off Griffith Island; and as commander of HMS *Intrepid* in 1852–4, wintering first at Dealy Island, and then off Cape Cockburn. Particularly on the last two expeditions he had established his reputation as the most effective and experienced sledge-traveller in the Royal Navy.

McClintock sailed from Aberdeen in the steam-yacht *Fox* on 2 July 1857.[1] After calling at Frederikshåb, Godthåb and Godhavn in West Greenland, *Fox* became solidly beset in the pack ice of Melville Bay and drifted slowly southwards for the full length of Baffin Bay and Davis Strait over the winter of 1857–8. Once his ship was freed McClintock called at Holsteinsborg and Godhavn again, then pushed north once again; this time he got across Baffin Bay with relative ease, entered Lancaster Sound, then pushed south through Prince Regent Inlet; the expedition spent its second winter in much more comfortable conditions, in snug winter quarters at Port Kennedy at the east end of Bellot Strait.

In the late winter (February–March 1859) McClintock undertook a reconnaissance sledge trip through Bellot Strait and south along the west side of Boothia Peninsula. At Cape Victoria he encountered a group of Inuit who were able to tell him, through his Greenlandic interpreter, Carl Petersen, that a group of Whites had starved to death on an island to the south-west. As proof of this story, McClintock bought from them a variety of relics including six silver spoons and forks, a silver medal inscribed to Alexander McDonald, assistant surgeon aboard *Terror*, and several buttons.[2] The Inuit also reported that a three-masted ship had been crushed by the ice west of King William Island.

[1] McClintock, Francis Leopold, *The Voyage of the 'Fox' in the Arctic Seas: A Narrative of the Discovery of the Fate of Sir John Franklin and his Companaions*, London, 1859.
[2] Ibid., pp. 233–4.

Encouraged by these finds McClintock returned to his ship and made preparations for a more extensive search in the spring. On 2 April, accompanied by Lieutenant William Hobson, he set off again, with the aim of searching the shores of King William Island. Some distance north of Cape Victoria, on the west side of Boothia Peninsula, they again met the same group of Inuit. Once again McClintock was able to purchase various items which had clearly come from *Erebus* and *Terror*. This time Petersen's questioning revealed that not one, but two ships had been seen west of King William Island; one had been crushed by the ice and had sunk in deep water; the other had been driven ashore by the ice.

At Cape Victoria, on 24 April McClintock and Hobson separated: Hobson proceeded west across James Ross Strait to Cape Felix to search the west coast of King William Island for confirmation of the Inuit reports. McClintock, meanwhile, continued south along the east coast of the island, meaning to make a complete circuit of it. On 7 May McClintock ran across an Inuit camp on the east side of King William Island and purchased from its inhabitants silver spoons and forks with the crests or initials of Franklin, Crozier, Fairholme and McDonald, and a range of other items from the missing ships. Farther south, at a single snow house on the sea ice off Point Booth on 10 May, he found 8–10 fir poles, a kayak paddle ingeniously made from two oar blades, and two large snow shovels, 1–2 m long, fashioned from what had probably been the bottom boards of a boat.

From here McClintock cut south to Point Ogle and Montreal Island. This visit, less than four years after that of Anderson and Stewart, was even less productive (in terms of throwing any light on the Franklin mystery) than theirs had been:

> ... that evening [14 May 1859] we encamped 2 miles from some small islands which lie off the north end of Montreal Island.
>
> On the morning of the 15th we made only a short march of 6 miles, as one of the men suffered severely from snow-blindness, and I was anxious to recommence night-travelling; encamped in a little bay upon the N.E. side of Montreal Island. The same evening we again set out, although it was blowing very strongly, and 'snowing for a wager', as the men expressed it, but it was only necessary for us to keep close along the shore of the island: we discovered, however, a narrow and crooked channel which led us through to the west side of the island, and, one of the men appearing seriously ill, we encamped about midnight.
>
> Whilst encamped this day, explorations were made about the N.E. quarter of the island; islets and rocks were seen to abound in all directions; eventually it proved to be a separate island upon which we had encamped. The only traces or relics of Europeans found were the following articles, discovered by Petersen, beside a native mark (one large stone set upright on the top of another), at the east side of the main – or Montreal – island: A piece of a preserved meat tin, two pieces of iron hoop, some scraps of copper, and an iron-hoop bolt. These probably are part of the plunder obtained from the boat, and were left here until a more favourable opportunity should offer, or perhaps necessity should compel the depositor to return for them.
>
> All the 16th we were unable to move, not only because Hampton was ill, but the weather was extremely bad, and snow thickly falling with temperature at zero; certainly strange weather for the middle of May! We have not had a single clear day since the 1st of the month.

On the 17th the weather, though dull, was clear, so Mr. Petersen, Thompson, and I set off with the dog-sledge to complete the examination of Montreal Island, leaving the other three men with the tent: we also hoped to find natives, but had not seen any recent traces of them since passing Point Booth. Petersen drove the dog-sledge close along shore round the island to the south, and as far up the east side as to meet our previously explored portion of it, whilst Thompson and I walked along on the land, the one close down to the beach, and the other higher up, examining the more conspicuous parts: in this order we traversed the remaining portion of the island.

Although the snow served to conceal from us any traces which might exist in hollows or sheltered situations, yet it rendered all objects intended to serve as marks proportionably conspicuous; and we may remember that it was in its winter garb that the retreating crews saw Montreal Island, precisely as we ourselves saw it. The island was almost covered with native marks, usually of one stone standing upright upon another, sometimes consisting of three stones, but very rarely of a greater number.

No trace of a cairn could be found.

In examining, with pickaxe and shovel, a collection of stones which appeared to be arranged artificially, we found a quantity of seal's blubber buried beneath; this old Esquimaux câche was near the S.E. point of the island. The interior of the island and the principal islets adjacent were also examined without success, nor was there the slightest evidence of natives having been here during the winter: it is not to be wondered at that we returned in the evening to our tent somewhat dispirited. The total absence of natives was a bitter disappointment; circles of stones, indicating the sites of their tenting places in summer, were common enough.

Montreal Island is of primary rock, chiefly grey gneiss, traversed with whitish vertical bands in a N. and S. direction (by them I often directed my route when crossing the islands). It is of considerable elevation, and extremely rugged. The low beaches and grassy hollows were covered with a foot or two of hard snow, whilst all the level, the elevated, or exposed parts were swept perfectly bare; had a cairn, or even a grave, existed (raised as it must be, the earth being frozen hard as a rock), we must at once have seen it. If any were constructed they must have been levelled by the natives; every doubtful appearance was examined with the pickaxe.

A remark made by my men struck me as being shrewd; they judged from the washed appearance of the rock upon the east side of Montreal Island that it must often be exposed to a considerable sea, such as would effectually remove everything not placed far above its reach; when looking over the smooth and frozen expanse one is apt to forget this.

Since our first landing upon King William's Island we have not met with any heavy ice; all along its eastern and southern shore, together with the estuary of this great river, is one vast unbroken sheet formed in the early part of last winter where no ice previously existed; this I fancy (from the accounts of Back and Anderson) is unusual, and may have caused the Esquimaux to vary their seal-hunting localities. Mr. Petersen suggested that they might have retired into the various inlets after the seals; and therefore I determined to cross over into Barrow's Inlet as soon as we had examined the Point Ogle Peninsula.

Upon Montreal Island I shot a hare and a brace of willow-grouse. Up to this date we had shot during our journey only one bear and a couple of ptarmigan. The first recent traces of reindeer were met with here.[1]

[1] Ibid., pp. 266–70.

On their short overland journey from Barrow Inlet to Simpson Strait McClintock and his party must have come within a few kilometres of Starvation Cove, where Schwatka, Gilder and Klutschak would later find the remains of a boat, abundant skeletal remains and relics from the Franklin expedition.[1] Heading north across the sea ice, they struck the coast of King William Island again just west of the Peffer River, and headed west, through Simpson Strait. They found the first skeleton about halfway between Gladman Point and Cape Herschel. From clothing remnants he appeared to have been a steward or officer's servant. The skeleton was later identified as that of Harry Peglar, captain of the foretop aboard *Terror*.[2]

At Cape Herschel McClintock found the massive cairn erected by Dease and Simpson in 1839 partially demolished; his men pulled the rest of it apart but found no messages. But some 19 km farther west McClintock found a cairn which Hobson, coming from the opposite direction, had left six days earlier. The most exciting part of the message left in the cairn was the news that he had found a document in a cairn near Point Victory, on the north-west coast of the island. As McClintock noted:

> ...the record paper was one of the printed forms usually supplied to discovery ships for the purpose of being enclosed in bottles and thrown overboard at sea, in order to ascertain the set of the currents, blanks being left for the date and position; any person finding one of these records is requested to forward it to the Secretary of the Admiralty, with a note of time and place; and this request is printed upon it in six different languages. Upon it was written, apparently by Lieutenant Gore, as follows:
>
> 28 of May, 1847. H.M. ships 'Erebus' and 'Terror' wintered in the ice in lat. 70°05'N., long. 98°23'W.
> Having wintered in 1846–47 at Beechey Island, in lat. 74°43'28''N., long. 91°39'15''W., after having ascended Wellington Channel to lat. 77°, and returned by the west side of Cornwallis Island.
> Sir John Franklin commanding the expedition.
> All well.
> Party consisting of 2 officers and 6 men left the ships on Monday 24th May, 1847.
> Gm. Gore, Lieut.
> Chas. F. Des Voeux, Mate.[3]

McClintock was quick to spot that there was an obvious error in this report, since the inscriptions on the graves at Beechey Island had already established that Franklin's ships had wintered there in 1845–6, not 1846–7. McClintock's description continues:

> ...round the margin of the paper upon which Lieutenant Gore in 1847 wrote those words of hope and promise, another hand had subsequently written the following words:
>
> April 25, 1848. H.M. ships 'Terror' and 'Erebus' were deserted on the 22nd of April, 5 leagues N.N.W. of this, having been beset since 12th September, 1846.

[1] Gilder, *Schwatka's Search*; Klutschak, *Overland*; Stackpole, *The Long Arctic Search*.

[2] Richard Cyriax and A. G. E. Jones, 'The Papers in the Possession of Harry Peglar, Captain of the Foretop, H.M.S. *Terror*', *Mariner's Mirror*, 40, 1954, pp. 186–95.

[3] McClintock, *The Voyage*, pp. 283–4.

> The officers and crews, consisting of 105 souls, under the command of Captain F. R. M. Crozier, landed here in lat. 69°37′42′′N., long. 98°41′W. Sir John Franklin died on the 11th June, 1847; and the total loss by deaths in the expedition has been to this date 9 officers and 15 men.
>
> F. R. M. Crozier — James Fitzjames
> Captain and Senior Officer. — Captain H.M.S. Erebus
> and start (on) to-morrow, 26th, for Back's Fish River.[1]

Encouraged by the fact that at least one message had been discovered, even if somewhat cryptic, McClintock continued westwards. On 29 May he and his party reached the western tip of King William Island, which was named Cape Crozier, and on the 30th camped on the south shore of Erebus Bay beside a ship's boat, already discovered by Hobson. It rested on a solidly-built sledge; the combined weight of boat and sledge was estimated by McClintock to be 1,400 lbs. Two skeletons, minus skulls, lay in the boat, along with a remarkable assortment of items. They included five watches, two double-barrelled guns, and some small books, including a Bible and a copy of *The Vicar of Wakefield*, a selection of assorted clothing, seven or eight pairs of assorted footwear, twine, nails, saws, files, bristles, candle-ends, sailmakers' palms, powder, bullets, shot, cartridges, lengths of slowmatch, several bayonet scabbards cut down into knife sheaths, and two rolls of sheet lead. McClintock was astounded that a shipwrecked party would have attempted to haul so much apparently useless dead weight. There was no evidence of any food, apart from some tea and about forty pounds of chocolate.

Some indication of the identity of members of the party was provided by twenty-six silver spoons and forks; Franklin's own crest appeared on eight of these while the others bore the initials or crests of Gore, Le Vesconte, Fairholme, Couch, Goodsir, Hornby and Thomas. The crest of Mr Edward Couch, mate on board HMS *Erebus*, appeared on one of the watches.

McClintock was puzzled by the fact that the sledge was pointing towards the north-east, and surmised that those who had been hauling it may have represented a splinter group who, having abandoned the retreat south, were trying to regain the ships. The skeletons were perhaps those of two men unable to go any farther, and for whom their companions had hoped to return, once they had reached the ships and recruited their strength.

Pushing on northwards along the west coast of the island, on 2 June McClintock reached Victory Point, where Hobson had found the solitary record of what had happened to the Franklin expedition. McClintock placed a copy of the all-important document in the cairn, along with records of his own and Hobson's visits. He also buried a further message beneath a large rock 3 m north of the cairn, in case the cairn were disturbed.

At the site which has since become known as Crozier's Landing (since Crozier was in command by the time the ships were abandoned) an amazing assortment of abandoned clothing and equipment lay strewn about. It suggested that, having sledged it all ashore, possibly over a considerable period, the party had selected

[1] Ibid., p. 286.

what was considered essential for the sledge trip south and had abandoned everything else here. It included four sets of boats' cooking stoves, shovels, pickaxes, iron hoops, old canvas, rope, a large single block, a small medicine chest, a Robinson dip circle and a small sextant with the name 'Frederick Hornby' engraved on it (Hornby was the mate on HMS *Terror*). The clothing alone formed a large stack 1.2 m high. Every item of clothing was searched but all the pockets were empty and there were no names marked on any of the items.

McClintock ended his search here and headed back to *Fox*'s winter quarters without further stops; Hobson, however had earlier found two more cairns and various relics between Cape Victory and Cape Felix. Both men felt sure that the Inuit had not visited the section of coast between Cape Crozier and Cape Felix since the Franklin expedition had retreated along it.

McClintock got back to his ship on 19 June and on 15 August the ice had broken up sufficiently to allow *Fox* to start for home. She reached Portsmouth on 21 September 1859, and by that evening McClintock was in London. Once the news broke, it generally seems to have been accepted in Britain that all that was likely to be learned of the fate of the Franklin expedition had been learned, and no further searches were mounted from Britain.

But this does not mean that no other searches occurred. Over the period 1864–9 the American Charles Francis Hall, working from a base in the Repulse Bay area, pursued every lead he could find among the Inuit as to what had happened to the Franklin expedition.[1] Then in 1878–80 a small party led by the American, Lieutenant Frederick Schwatka, made the first summer-time search of the coasts of King William Island and Adelaide Peninsula where they discovered abundant skeletal remains and abandoned equipment.[2] Further stories were recorded from the Inuit, or skeletal remains or relics recovered, by members of Roald Amundsen's expedition (1903–5);[3] by Knud Rasmussen during his Fifth Thule Expedition;[4] by Major L. T. Burwash in 1925–6 and again in 1929;[5] by William Gibson in 1931;[6] by Inspector Henry Larsen of the Royal Canadian Mounted Police in 1949;[7] by Paul F. Cooper in 1954;[8] by members of the Canadian Armed Forces (code-name *Operation Franklin*)[9]

[1] J. E. Nourse, ed., *Narrative of the Second Arctic Expedition Made by Charles F. Hall: his Voyage to Repulse Bay, Sledge Journeys to the Straits of Fury and Hecla and to King William's Land, and Residence among the Eskimos during the Years 1864–'69*, Washington, 1879.

[2] Gilder, *Schwatka's Search*; Klutschak, *Overland*; Stackpole, *The Long Arctic Search*.

[3] Roald Amundsen, *The Northwest Passage; being the Record of a Voyage of Exploration of the Ship 'Gjoa' 1903–1907*, London, 1908; G. Hansen, 'Toward King Hakon VII's Land', in *The North West Passage, being the Record of a Voyage of Exploration of the Ship 'Gjoa' 1903–1907*, R. Amundsen, ed., London, 1908, vol. 2, pp. 296–364.

[4] Knud Rasmussen, *Across Arctic America: Narrative of the Fifth Thule Expedition*, New York, 1927.

[5] L. T. Burwash, 'The Franklin Search', *Canadian Geographical Journal*, 1, 7, 1930, pp. 587–603; *Canada's Western Arctic*, Ottawa, 1931.

[6] W. Gibson, 'Some Further Traces of the Franklin Retreat', *Geographical Journal*, 79, 1932, pp. 402–8; 'Sir John Franklin's Last Voyage: A Brief History of the Franklin Expedition and an Outline of the Researches which Established the Facts of its Tragic End', *The Beaver*, 268, 1, 1937, pp. 44–75.

[7] Cyriax, *Sir John Franklin's Last Arctic Expedition*.

[8] Paul Fennimore Cooper, 'A Trip to King William Island in 1954', *Arctic Circular*, 8, 1, 1955, pp. 8–11.

[9] William Wonders, 'Search for Franklin', *Canadian Geographical Journal*, 76, 4, 1968, pp. 116–27; W. G. McKenzie, 'A Further Clue in the Franklin Mystery', *The Beaver*, 299, 1969, pp. 28–32.

in 1967; by a further military group from the Royal Canadian Regiment in 1973,[1] and by anthropologist Dr Owen Beattie in 1981.[2]

Even this is probably not a complete list of all the parties or expeditions which have searched for relics of the Franklin expedition, and the search will no doubt continue. That identifiable remains, skeletal and otherwise, can still be recovered was clearly demonstrated by the finds made by Barry Ranford on a small tidal island in Erebus Bay in 1992[3] and recently analysed by Keenleyside, Bertulli and Fricke.[4]

One point which has emerged from these numerous searches, finds and analyses, and especially from Woodman's analysis of the Inuit accounts recorded in Charles Hall's field notebooks,[5] is that the story of the final days of the Franklin expedition was probably much more complex than was assumed by McClintock and by many later analysts. There seems strong evidence that while the 105 survivors indeed abandoned the ships to the north-west of King William Island in April 1848, an unspecified number returned to the ships and lived on board for an unspecified time thereafter, before attempting again to escape. Moreover there may well have been more than four boats (two in Erebus Bay, one at Starvation Cove, and one on Montreal Island), involved in these attempts at escape. And, as Woodman has discussed at length in his latest book[6] Inuit oral traditions as recounted to Hall would suggest that survivors were present on Melville Peninsula considerably later than 1848.

The relevance of all this to the efforts of the Anderson/Stewart party, is that, with their problems of communication due to the lack of an interpreter, they probably tended to oversimplify what the Inuit were trying to tell them. The sites and incidents to which the Inuit were referring probably were spread over a much wider area than simply Montreal Island and the Point Ogle area, and were probably spread over a considerable time period. Even with competent interpreters Hall had considerable difficulty in unravelling the full story; how much greater was the possibility of error in Anderson's and Stewart's attempts at deciphering the mimings of their Inuit informants.

Nonetheless the efforts of Anderson and Stewart narrowed the field of search and allowed McClintock (and all subsequent searchers) to focus their attention on the area where the Franklin expedition came to grief, namely King William Island and Adelaide Peninsula on the adjacent mainland.

While none are as derisive as Dr Richard King's analyses more recent assessments of the achievements of the Anderson/Stewart expedition tend to have been somewhat dismissive. Thus Neatby wrote:

> Without an interpreter they could not find the beach where the bodies lay. Their canoes were too frail and damaged to make the crossing to King William Island. All

[1] R. T. Walsh, 'Quest in the North', *Sentinel*, 10, 5, 1974, pp. 23–4.
[2] Beattie and Savelle, 'Discovery'.
[3] Ranford, 'Bones'.
[4] Keenleyside, Bertulli and Fricke, 'The Final Days'.
[5] David C. Woodman, *Unravelling the Franklin mystery: Inuit testimony*, Montreal and Kingston, 1991.
[6] David C. Woodman, *Strangers Among Us*, Montreal and Kingston, 1996.

> they accomplished was to establish beyond doubt the identity of the river near which the men had perished.[1]

Stripping the story to the bare bones Berton wrote: 'Anderson found a few relics but no graves'.[2]

On the topic of the lack of an Inuktitut interpreter Wright was more sympathetic than most recent commentators, noting that 'none of his critics seemed to have realized the difficulty of procuring an English-speaking Eskimo or an Eskimo-speaking Indian in the heart of the North-West Territories'.[3]

One of the most recent assessments, that of Woodman (1991) is perhaps the least sympathetic and also the least accurate. For example he wrote: 'Hampered by lack of provisions and bad weather, they could not reach King William Island itself.'[4] In fact both Anderson and Stewart stressed that they were forced to turn back because beyond Point Ogle Simpson Strait was still completely ice-bound. Anderson even remarked on 6 August at the point where he cached his canoes and continued the search of Point Ogle on foot: 'Had we had open water, there would have been no difficulty in reaching Cape Herschel [on the south-west coast of King William Island], or even farther.'[5]

A comprehensive, neutral assessment of this expedition is long overdue. Firstly one can only be amazed at the speed and efficiency with which the expedition was mounted. Dr John Rae delivered the report of his findings concerning the Inuit stories about the fate of the Franklin expedition to the Admiralty on 23 October 1854. Less than a month later, during which time the Admiralty had asked the Hudson's Bay Company to mount a search, and the Governor and Committee had relayed this request to Sir George Simpson in Lachine, the latter was broadcasting letters of instructions to the officers who would be involved in various capacities, officers at posts as widely spaced as Fort Edmonton, Fort Simpson, York Factory and Fort Garry. The three best Iroquois canoemen from Sir George Simpson's own canoe crew were dispatched westwards from Lachine in late November 1854. The participants in the expedition (Chief Factor James Anderson from Fort Simpson, Chief Trader James Stewart from Fort Carlton, the three Iroquois, and voyageurs from Fort Garry and Norway House) had foregathered at the rendezvous, Fort Resolution on Great Slave Lake, by 22 June 1854. To reach that point the Iroquois had travelled some 4,200 km in winter; James Stewart some 1,125 km, and James Anderson some 500 km up the Mackenzie River. An Inuktitut interpreter, William Oman, set off from Fort Churchill on a winter trip of some 1,200 km to Fort Chipewyan to join the expedition, but went lame and had to abandon his trip.

The expedition set off by canoe from Fort Resolution on 22 June, and was back at that post by 17 September, having made a round trip of some 2,850 km, about half

[1] Leslie H. Neatby, *The Search for Franklin*, Edmonton, 1970, p. 247.

[2] Pierre Berton, *The Arctic Grail: The Quest for the North West Passage and the North Pole 1818–1909*, Toronto, 1988.

[3] Noel Wright, *Quest for Franklin*, London, 1959, p. 124.

[4] Woodman, *Unravelling*, pp. 27–8.

[5] HBC B.200/a/31. See p. 168.

of that distance being down and up the Back River, recognized by modern canoeists as one of the most challenging rivers on the continent. Apart from some 84 rapids, it flows through a series of large tundra lakes where navigation is difficult due to the low shorelines and complex lake configurations which make identification of lake inlets and outlets problematical; in addition, the chance of becoming windbound for long periods on these large lakes is high. Moreover, it was a late season, which meant that the party had to cope with ice until well down the Back River, and with the first new ice and the first snowfalls on the way back upriver. Both expedition leaders were coping with personal problems throughout the expedition: Anderson was suffering from varicose veins in his legs, while Stewart was coping with the emotional stress of a lengthy separation from a young wife, from whom he separated only some eight months after their wedding.

From the journals of both men, and even more so from their later correspondence, it is clear that there were some frictions between the two leaders. Although Anderson insisted that 'During our expedition we were and still are excellent friends, except in two or three instances when I was compelled to speak to him,'[1] these frictions cannot have been good for morale, although it is unlikely that they affected the outcome of the expedition in any real fashion.

Despite all the handicaps (and the fact that from Montreal Island northwards, the sea was obstructed with drifting or solid pack ice), the party spent two weeks on the arctic shores of Chantrey Inlet, into which the Back River debouches. From Inuit they encountered at the mouth of the river they elicited reports of the Franklin expedition practically identical to those gathered by Rae from the Pelly Bay area, with the addition of the important piece of information (reported only by Stewart) that one of the Inuit women reported that she had seen one of the Franklin people alive. They purchased an array of relics from the Inuit and on Montreal Island recovered even more artefacts derived from the missing expedition, from Inuit caches. Here too, they found a spot where, from the scatter of wood chips, they deduced that a boat had been cut up.

Given the distance the expedition had to cover, firstly to reach its base, and secondly from that base to the search area and back, along with the foul weather, the late season, and the lack of an Inuktitut interpreter (despite the Company's best efforts), these results are very creditable. Their real significance is that they greatly narrowed the search area for the subsequent expedition, that of Captain Francis Leopold McClintock which in 1859 found a trail of skeletons and abandoned equipment along the west and south shores of King William Island and, at Victory Point, the only record ever found which, however cryptic, cast some light on the final fate of the Franklin expedition.

For all that, Anderson and Stewart were undoubtedly quite frustrated by the meagreness of their findings, and especially by the lack of bodies, graves or records, after their very creditable efforts. One has to sympathize with Anderson's *cri de coeur* in his letter to Lady Franklin: 'What an opportunity Cap. Collinson lost!'.[2] What he

[1] BCA AB40 An 32.2, f. 90. See p. 169.

[2] Ibid., f. 106. See p. 178.

meant was that in 1852–3 Collinson had wintered in a well-found ship (HMS *Enterprise*) with abundant provisions and equipment at Cambridge Bay,[1] only some 375 km from Point Ogle (as against the 1,425 km from Fort Resolution to Point Ogle), and only some 275 km from the west coast of King William Island where McClintock would find such a profusion of abandoned equipment and skeletons. In the spring of 1854 Collinson had made a sledge trip northwards along the east coast of Victoria Island, and at his closest point came within about 100 km of the west coast of King William Island. One might, of course raise the objection, that Collinson had no particular reason to cross to King William Island, but he had no better reason to search where he did – and might just as easily have dispatched sledge parties in both directions. And ironically, as Collinson himself discovered, the coast he did search had already been searched by Rae in the spring of 1851.

[1] Collinson, *Journal*.

BIBLIOGRAPHY

PRINTED SOURCES

Amundsen, Roald, *The North West Passage, being the Record of a Voyage of Exploration of the Ship 'Gjoa' 1903–1907*, London, 1908.

Anderson, James, 'Letter from Chief Factor James Anderson to Sir George Simpson, F.R.G.S., Governor in Chief of Rupert Land', *Journal of the Royal Geographical Society*, vol. 26, 1856, pp. 18–25.

—— 'Chief Factor James Anderson's Back River Journal of 1855', *Canadian Field-Naturalist*, 54, 1940, pp. 63–7, 84–9, 107–9, 125–6, 134–6; 55, 1941, pp. 9–11, 21–6, 38–44.

Armstrong, Alexander, *A Personal Narrative of the Discovery of the North-west Passage; with Numerous Incidents of Travel and Adventure during Nearly Five Years' Continuous Service in the Arctic Regions while in Search of the Expedition under Sir John Franklin*, London, 1857.

Back, George, *Narrative of the Arctic Land Expedition to the Mouth of the Great Fish River, and along the Shores of the Arctic Ocean, in the Years 1833, 1834, and 1835*, London, 1836.

—— *Narrative of an Expedition in H.M.S. Terror, undertaken with a View to Geographical Discovery on the Arctic Shores, in the Years 1836–7*, London, 1838.

Balikci, A., 'Netsilik', in Damas, D., ed., *Handbook of North American Indians, Vol. 3, Arctic*, Washington, 1984, pp. 415–30.

Banfield, A. W. F., *The Mammals of Canada*, Toronto, 1974.

Barr, William, *Back from the Brink: The Road to Muskox Conservation in the Northwest Territories* (Arctic Institute of North America, Komatik Series No. 3), Calgary, 1991.

Beattie, Owen, 'A Report on Newly Discovered Human Skeletal Remains from the Last Sir John Franklin Expedition', *The Musk-Ox* 33, 1983, pp. 68–77.

Beattie, Owen and Geiger, John, *Frozen in Time: the Fate of the Franklin Expedition*, London, 1987.

Beattie, Owen and Savelle, James, 'Discovery of Human Remains from Sir John Franklin's Last Expedition', *Historical Archaeology*, 17, 1983, pp. 100–105.

Beechey, Frederick, W., *Narrative of a Voyage to the Pacific and Beering's Strait, to Co-operate with the Polar Expeditions, Performed in His Majesty's Ship Blossom, under the Command of Captain F. W. Beechey in the Years 1825, 26, 27, 28*, 2 vols, London, 1831.

—— *A Voyage of Discovery towards the North Pole, Performed in His Majesty's Ships 'Dorothea' and 'Trent', under the Command of Captain David Buchan, R.N., 1818*, London, 1843.

Belcher, Edward, *The Last of the Arctic voyages, being a Narrative of the Expedition in H.M.S. Assistance, under the Command of Captain Sir Edward Belcher, C.B., in Search of Sir John Franklin, during the Years 1852–53–54*, 2 vols, London, 1855.

Bellot, Joseph-René, *Memoirs of Lieutenant Joseph René Bellot ... with his Journal of a Voyage in the Polar Seas, in Search of Sir John Franklin*, 2 vols, London, 1855.

Berton, P., *The Arctic Grail: The Quest for the North West Passage and the North Pole 1818–1909*, Toronto, 1988.

Boosfield, H., 'Bernard Rogan Ross', in *Dictionary of Canadian Biography*, X (1871–80), 1972, p. 629.

Briggs, Jean L., *Never in Anger: Portrait of an Eskimo Family*, Cambridge, 1970.

Brown, Jennifer and Van Kirk, Sylvia, 'George Barnston', *Dictionary of Canadian Biography*, XI (1881–90), Toronto, 1982, pp. 52–3.

Burgess, Joanne, 'John Stewart', *Dictionary of Canadian Biography*, VIII (1851–60), Toronto, 1985, pp. 837–9.

Burwash, L. T., 'The Franklin Search', *Canadian Geographical Journal*, 1, 7, 1930, pp. 587–603.

—— *Canada's Western Arctic*, Ottawa, 1931.

Collinson, Richard, *Journal of H.M.S. Enterprise, on the Expedition in Search of Sir John Franklin's Ships by Behring Strait, 1850–55*, London, 1889.

Cooper, Barry, *Alexander Kennedy Isbister: a Respectable Critic of the Honourable Company*, Ottawa, 1988.

Cooper, Paul F., 'A Trip to King William Island in 1954', *Arctic Circular*, 8, 1, 1955, pp. 8–11.

Cyriax, Richard J., *Sir John Franklin's Last Expedition: a Chapter in the History of the Royal Navy*, London, 1939.

—— 'Recently Discovered Traces of the Franklin Expedition', *Geographical Journal*, 117, 1951, pp. 211–14.

—— 'The Voyage of H.M.S. *North Star*, 1849–50', *Mariner's Mirror*, 50, 1964, pp. 307–18.

Cyriax, Richard and Jones, A. G. E., 'The Papers in the Possession of Harry Peglar, Captain of the Foretop, H.M.S. *Terror*', *Mariner's Mirror*, 40, 1954, pp. 186–95.

de Bray, Emile F., *A Frenchman in Search of Franklin: De Bray's Arctic Journal 1852–1854*, W. Barr, trans. and ed., Toronto, 1992.

Fairholme vs Fairholme's Trustees, Record, Proof and Documentary Evidence. In Declarator, Court and Reckoning etc., George K. E. Fairholme, Esq. against the Trustees of Adam Fairholme, Esq., Scottish Court of Session, First Division (Scottish Record Office, 1856).

Fleming, R. Harvey, ed., *Minutes of Council, Northern Department of Rupert Land, 1821–31*, London, 1940 (Hudson's Bay Record Society, III, p. 427).

Franklin, John, *Narrative of a Journey to the Shores of the Polar Sea, in the Years 1819, 20, 21 and 22*, London, 1823.

—— *Narrative of a Second Expedition to the Shores of the Polar Sea in the Years 1825, 1826, and 1827, including an Account of the Progress of a Detachment to the Eastward by John Richardson*, London, 1828.

Galaburri, Richard, 'The Rat Lodge Revisited', *Arctic*, 44(3), 1991, pp. 257–8.

Galbraith, John, *The Little Emperor: Governor Simpson of the Hudson's Bay Company*, Toronto, 1976.

—— 'Sir George Simpson', *Dictionary of Canadian Biography*, VIII (1851–60), Toronto, 1985, pp. 812–18.

Gibson, W., 'Some Further Traces of the Franklin Retreat', *Geographical Journal*, 79, 1932, pp. 402–8.

—— 'Sir John Franklin's Last Voyage: a Brief History of the Franklin Expedition and an Outline of the Researches which Established the Facts of its Tragic End', *The Beaver*, 268, 1, 1937, pp. 44–75.

Gilder, W. H., *Schwatka's Search: Sledging in the Arctic in Quest of the Franklin Records*, New York, 1881.

Gilpin, James D., 'Outline of the Voyage of H.M.S. Enterprize and Investigator to Barrow Strait in Search of Sir John Franklin', *Nautical Magazine*, 19, 1850, pp. 8–9, 82–90, 160–70, 230.

Godfrey, W. E., *The Birds of Canada*, Ottawa, 1966.

Goossen, N. J., 'William MacTavish', *Dictionary of Canadian Biography*, IX (1861–70), Toronto, 1976, pp. 529–32.

Great Britain. Parliament, *Return to an Address of the Honourable The House of Commons, dated 7 February 1851; – for, 'Copy or Extracts from any Correspondence or Proceedings of the Board of Admiralty…', 'Copies of any Instructions from the Admiralty to any Officers in Her Majesty's Service, Engaged in Arctic Expeditions…' and 'Copy or Extracts from any Correspondence or Communications from the Government of the United States … in Relation to any Search to be Made on the Part of the United States…'*, London (House of Commons, Sessional Papers, Accounts and Papers, 1851, 33, 97).

—— *Further Papers Relative to the Recent Arctic Expeditions in Search of Sir John Franklin, and the Crews of H.M.S. 'Erebus' and 'Terror'*, London, 1855.

Further Papers Relative to the Recent Arctic Expeditions in Search of Sir J. Franklin and the Crews of Her Majesty's Ships 'Erebus' and 'Terror' etc., presented to the House of Commons, London, 1856.

Hamelin, Jean, 'Alexandre-Antonin Taché', *Dictionary of Canadian Biography*, XII (1891–1900), Toronto, 1990.

Hanbury, David T., *Sport and Travel in the Northland of Canada*, London, 1904.

Hansen, G., 'Towards King Haakon VII's Land', in *The North West Passage, being the Record of a Voyage of Exploration of the Ship Gjoa 1903–1907*, R. Amundsen, ed., London, 1908, II, pp. 296–364.

Hooper, William H., *Ten Months among the Tents of the Tuski, with Incidents of an Arctic Boat Expedition in Search of Sir John Franklin, as far as the Mackenzie River and Cape Bathurst*, London, 1853.

Household Words, 'The Lost Arctic Voyagers', *Household Words*, 245 (2 December 1854), pp. 361–5.

—— 'The Lost Arctic Voyagers', *Household Words*, 246 (9 December 1854), pp. 385–93.

—— 'The Lost Arctic Voyagers', 248 (23 December 1854), pp. 433–7.

Houston, Stuart, ed., *To the Arctic by Canoe, 1819–1821: the Journal and Paintings of Robert Hood, Midshipman with Franklin*, Montreal and London, 1974.

—— ed., *Arctic Ordeal: The Journal of John Richardson, Surgeon-Naturalist with Franklin, 1820–1822*, Kingston and Montreal, 1984.

—— ed., *Arctic Artist: The Journal and Paintings of George Back, Midshipman with Franklin, 1819–1822*, Montreal and Kingston, 1994.

—— 'John Rae (1813–1893)', in *Lobsticks and Stone Cairns*, Richard C. Davis, ed., Calgary, 1996, pp. 87–9.

Inglis, Alex, *Northern Vagabond: The Life and Career of J. B. Tyrrell*, Toronto, 1978.

Jones, A. G. E., 'Captain Robert Martin: a Peterhead Whaling Master in the 19th century', *Scottish Geographical Magazine*, 85, 1969, pp. 196–202.

Kane, Elisha Kent, *The U.S. Grinnell Expedition in Search of Sir John Franklin: a Personal Narrative*, London, 1854.

Keenleyside, A., Bertulli, M. and Fricke, H., 'The Final Days of the Franklin Expedition: New Skeletal Evidence', *Arctic* 50(1), 1997, pp. 36–46.

Kennedy, William, *A Short Narrative of the Second Voyage of the Prince Albert in Search of Sir John Franklin*, London, 1853.

King, Richard, *Narrative of a Journey to the Shores of the Arctic Ocean in 1833, 1834 and 1835, under the Command of Captain Back, R.N.*, 2 vols, London, 1836.

—— *The Franklin Expedition from First to Last*, London, 1855.

—— *The Franklin Expedition: To His Grace the Duke of Newcastle, Secretary of State for the Colonies. A Letter of Appeal*, London, 1860.

Klutschak, Heinrich, W., *Overland to Starvation Cove: With the Inuit in Search of Franklin, 1878–1880*, W. Barr, trans., and ed., Toronto, 1987.

MacInnes, Joe, *The Breadalbane Adventure*, Montreal and Toronto, 1982.

Mackinnon, C. Stuart, 'James Anderson', *Dictionary of Canadian Biography*, IX (1861–70), Toronto, 1976, p. 5.

—— 'James Green Stewart', *Dictionary of Canadian Biography*, XI (1881–90), Toronto, 1982, pp. 854–5.

McClintock, Francis Leopold, *The Voyage of the 'Fox' in the Arctic Seas: A Narrative of the Discovery of the Fate of Sir John Franklin and his Companions*, London, 1859.

McCormick, Robert, *Voyages of Discovery in the Arctic and Antarctic Seas, and round the World, Being Personal Narratives of Attempts to reach the North and South Poles; and of an Open-Boat Expedition up the Wellington Channel in the Year 1852, under the Command of R. McCormick, R.N., F.R.C.S., in H.M.B. 'Forlorn Hope', in Search of Sir John Franklin*, 2 vols, London, 1884.

McDougall, George F., *The Eventful Voyage of H.M. Discovery Ship 'Resolute' to the Arctic Regions in Search of Sir John Franklin and the Missing Crews of H.M. Discovery Ships 'Erebus' and 'Terror', 1852, 1853, 1854*, London, 1857.

McKenzie, W. G., 'A Further Clue in the Franklin Mystery', *The Beaver*, 299, 1969, pp. 28–32.

Mickle, S., 'The Hudson Bay Expedition in Search of Sir John Franklin', *Women's Canadian Historical Society of Toronto Transactions*, 20, 1919–20, pp. 11–45.

Neatby, L. H., *The Search for Franklin*, Edmonton, 1970.

—— 'François Beaulieu', *Dictionary of Canadian Biography*, X (1871–80), Toronto, 1972, p. 38.

Nourse, J. E., ed., *Narrative of the Second Arctic Expedition made by Charles F. Hall: his*

Voyage to Repulse Bay, Sledge Journeys to the Straits of Fury and Hecla and to King William's Land, and Residence among the Eskimos during the years 1864–'69, Washington, 1879.

Osborn, Sherard, *Stray Leaves from an Arctic Journal; or, Eighteen Months in the Polar Regions, in Search of Sir John Franklin's Expedition, in the Years 1850–51*, London, 1852.

—— ed., *The Discovery of the North-west Passage by H.M.S. 'Investigator', Capt. R. M'Clure, 1850, 1851, 1852, 1853, 1854 ... from the Logs and Journals of Capt. Robert M. M'Clure*, London, 1856.

Parry, William Edward, *Journal of a Voyage for the Discovery of a North-West Passage from the Atlantic to the Pacific, Performed in the Years 1819–20, in His Majesty's Ships Hecla and Griper*, London, 1821.

—— *Journal of a Second Voyage for the Discovery of a North-West Passage from the Atlantic to the Pacific, Performed in the Years 1821–1822–1823, in His Majesty's Ships Fury and Hecla, under the Orders of Captain William Edward Parry*, London, 1824.

—— *Journal of a Third Voyage for the Discovery of a North-West Passage from the Atlantic to the Pacific Performed in the Years 1824–25, in His Majesty's Ships Hecla and Fury, under the Orders of Captain William Edward Parry*, London, 1826.

Pullen, H. F., ed., *The Pullen Expedition in Search of Sir John Franklin*, Toronto, 1979.

Rae, John, *Narrative of an Expedition to the Shores of the Arctic Sea in 1846 and 1847*, London, 1850.

—— 'Journey from Great Bear Lake to Wollaston Land', *Journal of the Royal Geographical Society*, 22, 1852, pp. 73–82.

—— 'Recent Explorations along the South and East Coast of Victoria Land', *Journal of the Royal Geographical Society*, 22, 1852, pp. 82–96.

Ranford, B., 'Bones of contention', *Equinox*, 74, 1994, pp. 64–87.

Rasmussen, Knud, *Across Arctic America: Narrative of the Fifth Thule Expedition*, New York, 1927.

Rich, Edwin E. and Johnson, A. M., eds, *John Rae's Correspondence with the Hudson's Bay Company on Arctic Exploration 1844–1855*, London, Hudson's Bay Record Society, 1953.

Richards, Robert L., *Dr John Rae*, Whitby, 1985, pp. 576–8.

—— 'John Rae', *Dictionary of Canadian Biography*, XII (1891–1900), Toronto, 1990.

Richardson, John, *Arctic Searching Expedition: a Journal of a Boat-Voyage through Rupert's Land and the Arctic sea, in Search of the Discovery Ships under Command of Sir John Franklin*, 2 vols, London, 1851.

Roberts, K. G. and Shackleton, P., *The Canoe: A History of the Craft from Panama to the Arctic*, Toronto, 1983.

Ross, James C., *A Voyage of Discovery and Research in the Southern and Antarctic Regions during the Years 1839–43*, London, 1847.

Ross, John, *Narrative of a Second Voyage in Search of a North-west Passage, and of a Residence in the Arctic Regions during the Years 1829, 1830, 1831, 1832, 1833*, London, 1835.

Ross, M. J., *Polar Pioneers: John Ross and James Clark Ross*, Montreal and Kingston, 1995.

Scott, W. B. and Crossman, E. J., *Freshwater Fishes of Canada*, Ottawa, 1973.
Simpson, Thomas, *Narrative of the Discoveries on the North Coast of America, Effected by the Officers of the Hudson's Bay Company during the Years 1836–39*, 2 vols, London, 1843.
Seemann, Berthold, *Narrative of the Voyage of H.M.S. Herald during the Years 1845–51, under the Command of Captain Henry Kellett, R.N., C.B., being a Circumnavigation of the Globe, and Three Cruises to the Arctic Regions in Search of Sir John Franklin*, 2 vols, London, 1853.
Smith, D. Murray, *Arctic Expeditions from British and Foreign Shores from the Earliest Times to the Expedition of 1875–76*, Edinburgh, 1877.
Snow, William Parker, *Voyage of the Prince Albert in Search of Sir John Franklin: a Narrative of Every-Day Life in the Arctic Seas*, London, 1851.
Stackpole, E. A., ed., *The Long Arctic Search: the Narrative of Lieutenant Frederick Schwatka, U.S.A., 1878–1880, Seeking the Records of the Lost Franklin Expedition*, Mystic, CN, 1965.
Sutherland, Peter C., *Journal of a Voyage in Baffin's Bay and Barrow Straits, in the Years 1850–1851, Performed by H.M. Ships, 'Lady Franklin' and 'Sophia', under the Command of Mr. William Penny, in Search of the Missing Crews of H.M. Ships Erebus and Terror*, 2 vols, London, 1852.
Tyrrell, Joseph Burr, 'A Story of a Franklin Search Expedition', *Transactions of the Canadian Institute*, 8, 1908–9, pp. 393–402.
Van Kirk, Sylvia, 'John Ballenden', *Dictionary of Canadian Biography*, vol. VIII (1851–60), Toronto, 1985, pp. 59–60.
Wallace, Hugh, *The Navy, the Company and Richard King: British Exploration in the Canadian Arctic, 1829–1860*, Montreal, 1980.
Walsh, R. T., 'Quest in the North', *Sentinel*, 10,5, 1974, pp. 23–4.
Wilson, Clifford, *Campbell of the Yukon*, Toronto, 1970.
—— 'Footnotes to the Franklin Search, I: Halkett's Air Boat', *The Beaver*, 285, 1, 1955, pp. 46–8.
Wilson, Malcolm, 'Sir John Ross's Last Expedition, in Search of Sir John Franklin', *The Musk-Ox*, 13, 1973, pp. 3–11.
Wonders, William, 'Search for Franklin', *Canadian Geographical Journal*, 76, 4, 1968, pp. 116–27.
Woodman, David C., *Unravelling the Franklin mystery: Inuit testimony*, Montreal and Kingston, 1991.
—— *Strangers among Us*, Montreal and Kingston, 1996.
Woodward, F. J., *Portrait of Jane: A Life of Lady Franklin*, London, 1951.
Wright, Noel, *Quest for Franklin*, London, 1959.

INDEX

The names of Capes, Forts, Lakes, Points and Ships are grouped under those headings. The Roman numerals immediately following the names of geographical features and individuals refer to the map or maps where these sites may be located, or to relevant illustrations.

Aberdeen, 6, 8, 249
Abhaldessy, *see* Lockhart River
Abitibi, 43 n.2
accounts, method of handling, 50, 53, 56, 182; to be sent to Montreal by winter express, 170
Acres Lake (**XII**), 147 n.3
Adelaide Peninsula (**XI**), 82 n.4, 139, 175, 178, 179, 180, 236, 254, 255
Admiralty, British, 15, 17, 18, 19, 35; mounts Franklin searches, 6–11; asks Hudson's Bay Company to mount a search, xii, 25, 27, 31–2, 33, 190, 200, 256; Lords Commissioners of, 19, 27, 31–2, 33, 35, 49; reluctant to mount a further search, 25, 62 n.2, 200; seeks Rae's advice, 25; seeks Back's advice, 26; declines King's offer to lead another expedition down the Back River, 210–11; requests that the Hudson's Bay Company submits claims, 220; pays the bills submitted by the Company, 222–3; rewards Anderson and Stewart for their services, 224–8; forwards Arctic Medals to Hudson's Bay Company for distribution 229; artefacts from Franklin expedition forwarded to by Hudson's Bay Company, 229–33; list of artefacts forwarded to, 231; asks the Hudson's Bay Company for any information *re* King's allegations concerning reports of a possible camp of Franklin survivors on Back River, 241–2
Agatha, 241 n.2
agrets, 84, 84 n.3, 88, 88 n.2, 121, 121 n.5
Ah-hel-dessy, *see* Lockhart River
Ahtudesy, *see* Lockhart River
Akuaq, 129 n.7
Alaska, 11, 82 n.3
Albany, 44 n.3
alcohol, not allowed by the Company in Mackenzie District, 29; greatly needed on the expedition, 130, 130 n.5; indulged in at Point Pechell, 136 n.9; recommended by Anderson for future expeditions, 166 n.1
alder (*Alnus crispa, A. rugosa*), 122, 140
American Fur Company, 58
ammunition, 22, 49, 51, 54, 80; to be fetched from Fort Chipewyan, 84
Amundsen, Roald, 254
Amundsen Gulf, 10
Anarin, Joseph, *see* Anarize, Joseph
Anarin Lake(**VII**), 113 n.12
Anariz, Joseph, *see* Anarize, Joseph
Anarize, Joseph, 59 n.4, 108, 121, 172; signs contract, 58, 59; wages received, 223
Anderson, Alexander, 170
Anderson, Eliza Charlotte, 37
Anderson, Georgina L., 170
Anderson, James (a) (**III**), xii, 29 n.1, 35, 41, 42, 45, 48, 50, 51, 56, 57, 59, 61 n.2, 69, 71, 78, 81, 83, 84, 88, 93, 95, 96, 98, 99, 100, 101, 102, 105, 107, 121, 162, 172, 235, 256; biographical sketch, 37–8; earlier published versions of journal of, xii; as possible expedition leader, 26, 30, 35; handles Hannah Bay murders, 37; promoted Chief Trader, 37; promoted Chief Factor, 37, 98; health damaged by expedition, 38; letter of appreciation for, 38; retires, 38; death, 38; qualifications to lead expedition, 42, 60; appointed co-leader of expedition, 42, 46, 55, 56, 83, 190, 192, 200; varicose veins, 42, 82, 87, 168, 178, 180, 182, 198, 257; instructions to from Sir George Simpson, 46–51; to start for Fort Resolution immediately, 47, 85; promised promotion to Chief Factor, 49, 50; receives letter from Lady Franklin, 61 n.4; receives letter from Sophia

Cracroft, 63–6; travels upriver from Fort Simpson to Fort Resolution, 71; delays his start till after breakup on the Mackenzie, 71, 81, 83, 85, 88, 91; delayed by ice on Great Slave Lake, 71; reprimands Stewart for large number of men taken on trip from Fort Chipewyan to Fort Resolution and back, 74 n.1, 92; taken aback by appointment to expedition, 81; to resume control of Mackenzie District after the expedition, 82, 91, 99, 168–9, 179; feels Collinson missed an excellent opportunity to find the Franklin expedition, 82, 82 n.4, 96, 257–8; convinced wintering on the coast impractical, 87, 89, 96, 103; considers Stewart's promotion premature, 87; choice of route to Back River, 89, 100; reveals plans for searching the coast, 86, 89; considers Back River route unpromising and dangerous, 90, 96; thinks Franklin's ships may have been crushed in winter quarters, 90, 95, 101, 168; starts from Fort Simpson, 93; supports Rae *re* reports of cannibalism, 95; pessimistic appraisal of expedition's chances, 96, 100, 101; makes preparations for a second year of searching, 96–7; orders boats built for a second year of searching, 96–7, 100, 103, 159 n.1; detained by ice at Big Island, 99–101; reaches Fort Resolution, 102, 183; weighs relative merits of routes to Back River, 92, 100, 103; exploration, 105; maps produced by, 105, 169, 182, 182 n.2; collects plants, 109; refutes Stewart's statement that one Inuit woman had seen a Franklin survivor alive, 130 n.2, 167, 212–13, 218; has row with Stewart, 139 n.5, 211; returns to Fort Resolution, 148; recalls Lockhart to Fort Reliance, 160, 186; leaves Fort Resolution for Fort Simpson to resume his duties, 161; returns to Fort Simpson 161, 162, 181, 182; preliminary reports for Simpson, 166, 167–70; more detailed reports for Simpson, 166, 170–77, 181–2; sends report to Lady Franklin, 166, 177–9; fair copy of journal, 166, 166 n.1, 169, 182; recommendations for future expedition 90, 96, 166 n.1, 168, 178; reports to Eden Colvile, 166, 179–81; suffers from lack of sleep on Back River, 169, 179; found Stewart useless on Back River, 169, 179, 211, 215; accuses Stewart of sleeping on way down the river, 169, 169 n.1, 179, 211; accuses Stewart of lack of initiative, 169; claims Stewart's observations were inaccurate, 169, 180; commends Lockhart very highly, 170, 180; would like to have Lockhart as an officer in Mackenzie River District, 170; requests elastic stocking and knee cap, 170; requests that any newspaper cuttings of his report be sent to relatives, 170; not interested in another expedition, 178; wants to take furlough, 179, 182; relieved to get back to Great Slave Lake, 180, 181; children at Red River doing well, 181; recommends Bernard Ross and Hardisty for promotion, 181; specifies that Ignace Montour should take dispatches from Red River to Lachine, 181–2, 213, 214, 215; sends Franklin expedition artefacts to Norway House, 182; report published in newspapers, 198; copies of newspaper accounts sent to relatives, 198; criticizes Stewart's competence on the Back River, 211; criticizes Stewart for delaying dispatches by picking up his wife, 212, 213, 214; finds fault with newspaper accounts based on interviews with Stewart, 167, 212, 218; criticizes Stewart for awarding Laferté same wages as other men, 213, 214, 215; requests of Swanston that Stewart's and his wife's expenses and Laferté's wages be charged to Stewart's account, 213, 214, 215; finds Stewart's gun and sends it after him, 213; sells Stewart's frock coat, 214; lodges official complaint against Stewart for delaying official dispatches, 214; at Simpson's recommendation, withdraws official complaint against Stewart, 212; award of £400 received from British Government for services, 219, 224–8; forwards to London a piece of mahogany used by the Inuit as a snow shovel, 233; allegedly reported to Simpson that Indians had found camp recently occupied by Whites on the Back River, 234, 238, 240; rejects the possibility of the said camp having been occupied by Franklin expedition survivors, 235; convinced this was a camp occupied by Stewart and himself, 235, 238; his proposals for a further expedition put into practice, 249; distance travelled to reach Fort Resolution, 256

Anderson, James (b) (cousin), 55 n.1, 81

Anderson, James (uncle), 170
Anderson, Margaret, 38, 88, 181; gives birth to a son, 87
Anderson, Robert, 37
Anderson, William, 44, 55; attempts to provide Inuktitut interpreter, 69, 78–80
Anderson and Stewart, confirm Inuit stories relayed by Rae, 106, 184, 186, 190, 194, 195, 201; recover artefacts from Franklin expedition, 106; find no skeletal remains; 106; find no documents, 106; start from Fort Resolution, 107, 183, 187, 200, 256; daily swim, 116 n.7; frictions between, 211, 257; allegedly had heard story of Indians finding a camp occupied by Whites, but had not acted upon it, 234; narrowed field of search for later searchers, 255, 257
Anderson Falls (**XII**), 149, 155
Anderson River, 38, 71
Anderson's Hill, 140
Antarctica, 2, 4
Arctic Blue Books, 167
Arctic coast, 66; suggestions of wintering at, 49, 60; impracticability of wintering at, 87, 89, 96, 103, 131, 131 n.1, 131 n.5, 136 n.9, 166 n.1, 168, 175, 178, 180, 185
Arctic Council, rules out possibility of Franklin heading for Back River, 206–7
Arctic expeditions, provisions shipped north for, 29, 29 n.1, 71, 75, 77, 83, 99; unused provisions to be assumed by the Hudson's Bay Company, 99, 221; bill for provisions for submitted to the Admiralty, 221
Arctic Medal, 220; 187 copies forwarded to Hudson's Bay Company for distribution, 229
Armadale, Laird of, 11 n.1
Arrowsmith, Aaron, charts of, 21, 35
Artillery Lake (**VI, XII**), 26, 27, 89, 92, 103, 105, 108, 108 n.5, 118 n.1, 144, 144 n.4, 145 n.1, 147 n.3, 149, 151, 155, 156, 159, 170, 171, 175, 178, 183, 185; still ice-bound, 105, 111, 171; climate much worse than Great Slave Lake, 147 n.8
Assaminton, Baptiste, *see* Assanayunton, Jean Baptiste
Assanayneton, Jean Baptiste, *see* Assanayunton, Jean Baptiste
Assanayunton, Jean Baptiste, 59 n.4, 108, 121, 172; signs contract, 58, 59; builds canoes, 69; repairs canoes, 139 n.1; eyesight possibly failing, 168; had made round trip from Montreal to Norway House, prior to joining the expedition, 180; wages received, 223
Assiniboia, Governor of, 43 n.3
Assinienton, Jean-Baptiste, *see* Assanayunton, Jean-Baptiste
Assistance Harbour, 8
Athabasca, *see* Fort Chipewyan
Athabasca boat, 148, 148 n. 4, 161, 161 n.1, 162
Athabasca District, 30 n.1, 37, 45, 45 n.1, 45 n.2, 48, 51 n.3, 55 n.4, 57
Athabasca River (**IV**), 67, 92 n.3, 161, 163 n.3, 163 n.5
Athabasca Tar Sands, 163 n.5
aurora borealis, 143, 144, 144 n.3, 147, 162; seen for first time, 139; exceptionally active, 147
Austin, Captain Horatio, 6, 8, 234 n.1
Aylmer, Baron (Whitworth-Aylmer, Matthew), Governor-General of Canada, 116 n.1
Aylmer Lake (**VI, VIII**), 26, 90, 105, 108, 111, 115, 116 n.1, 118 n.1, 121, 143, 169, 176, 178, 181, 182, 183, 185; breakup on, 85, 108 n.5, 118; still icebound, 105, 116–18, 171, 183; ice on margins, 143

Back, George, xi, 20, 22, 43, 61, 103, 108 n.5, 116, 116 n.6, 119 n.2, 119 n.3, 119 n.4, 122 n.2, 147 n.1, 174; voyage in HMS *Terror*, 2, 4; advice sought by Admiralty, 26; outline of Back River expedition, 26–7; recommendations to Admiralty, 27, 31, 32, 35; narrative of Back River expedition, 31, 36, 81, 85, 90, 92, 100, 119, 172, 190; recommends canoes over boats, 31; recommends depot at Fort Reliance, 31; attempts to clarify rumours about his men killing Inuit near Point Ogle, 65 n.2; route to Artillery Lake, 89; description of Hoarfrost River, 100; ascends HoarfrostRiver, 105; speed of travel, 106; description of route accurate, 109, 119, 122, 123, 193; estimates of heights of hills exaggerated, 109; rejects Mountain Portage route, 111; reaches Sandhill Bay, 118 n.1; almost loses boat, 121, 121 n.6; camps, 122, 122 n.3, 125, 126, 138; map inaccurate, 126, 144 n.1, 183; map on too small a scale to be useful, 126, 128, 143, 144, 169, 172, 180; builds boats, 145
Back Lake (**VI, VII**), 105, 115, 115 n.4
Back River (**VI, VIII, IX, X**), xii, 12, 13, 15, 17,

19, 20, 22, 23, 26, 27, 30, 31, 38, 43, 44, 46, 52, 55, 57, 62, 62 n.1, 65 n.2, 75, 78, 83 n.3, 86, 89, 100, 103, 105, 111, 111 n.1, 137 n.9, 194, 200, 207, 235, 236, 240; recommended by Rae as the expedition's route, 28; Hanbury on, xi; difficulties of, 60, 86, 96, 100, 106, 168, 172, 190, 193, 257; expedition reaches headwaters of, 118, 118 n.5, 159 n.1; worst possible route, in Anderson's view, 166 n.1; Franklin survivors heading for, 253
Backhouse, John, 137 n.7
Back's River, *see* Back River
Baffin Bay, 1, 5, 6, 7, 249
Baffin Island, 6
Baillie, George, 123 n.6
Baillie's River, 123, 123 n.6, 141
Ballenden, John, 44, 45, 51, 52, 57, 59; biographical sketch, 44 n.4; instructed to engage voyageurs, 44; receives instructions from Simpson, 52–4, 58–9
Baltic Sea, 25
Banks Island, 9, 10
Banksian pine, *see* jackpine
Barclay, Archibald, Secretary, Hudson's Bay Company, 21, 27, 31, 35; solicits Rae's advice, 29
Bark Mountains, *see* Birch Mountains
Barnston, George, 59, 77, 92; biographical sketch 44 n.3; as naturalist, 44 n.3; receives instructions from Simpson, 54–6; sends letter of introduction with Norway House voyageurs, 76–7
Barnston Lake (**VII**), 112 n.3, 113 n.7
Barnston River (**VII**), 105, 112 n.3, 113 n.8
Barren Lands, 111
Barrière, 164
Barrow, John, 133 n.4, 230
Barrow Inlet (**IX**), 106, 252
Barrow Strait, 5, 8, 62, 204
Bathurst Inlet (**IX**), 123, 123 n.4
Bathurst Island, 9, 234 n.1
Batture Rapids, 98
Batty Bay, 8
Beads, Jacob, recommended by Rae for expedition, 30, 52, 53
bear (*Ursus* spp), 93; signs, 124
Beattie, Dr Owen, 255, confirms reports of cannibalism, 18
Beaufort, Sir Francis, Hydrographer to the Admiralty, 25, 28, 62, 62 n.3, 131 n.1
Beaufort Sea, 10
Beaulieu, Francois, 74, 151 n.4, 162; biographical sketch, 74 n.5; accompanied Alexander Mackenzie, 74 n.5; on Franklin's second expedition, 74 n.5; supported three wives, 74 n.5; holds salt monopoly, 74 n.5
Beaulieu, Jacques, 74 n.5
Beaulieu, John, 74
Beaulieu, 'King', 69, 78, 82; member of Lockhart's party, 74 n.5, 149, 151, 153 n.1, 153 n.2; wages, 84; sent back to Mountain Portage to read message, 153; cutting wood, 154; wife unable to make nets, 155; cutting shingles, 156; hunting caribou, 158; making meat rack, 158; reconnoitres south end of Pike's Portage, 158; making meat scaffold, 158; salting tongues and smoking meat, 159; hunting, 159; left at Fort Reliance to continue trading for meat, 160, 182
beaver (*Castor canadensis*), 93
Beaver Lodge, 144 n.4, 145, 171
Beddome, Mr, 79
Bedeau, 94
Beechey, Captain Frederick William, 4, 122 n.13
Beechey Island, 9, 10, 62 n.2, 204; traces of Franklin expedition found at, xi, 7, 234 n.1; fresh appearance of all the artefacts left by the Franklin expedition, 201; confirmed as wintering site of *Erebus* and *Terror*, 252
Beechey Lake (**IX**), 122, 123, 141, 142, 176; Back found icebound, 123 n.1
Beghula River, *see* Anderson River
Beirnes Lake (**VII**), 113 n.12
Belcher, Captain Sir Edward, 8, 16, 62 n.2, 206, 234 n.1; court martial, 25
Bell, John, 45, 59, 78, 84; biographical sketch, 45 n.1; receives instructions from Simpson, 57; reports Stewart's arrival at Fort Chipewyan, 73; reports difficulty in supporting Stewart's party, 73
Bellot, Enseigne-de-vaisseau Joseph René, 8
Bellot Strait, 8, 12, 14, 62, 62 n.1, 86, 90, 90 n.5, 95, 168, 178, 249
Bering Strait, 1, 4, 5, 6, 9, 10, 11, 17, 31 n.4
berries, ripe, 145
Berton, Pierre, 256
Bertulli, M., 255
Big Island (**VI**), 71, 81, 83, 85, 93, 99–101, 101 n.4, 162, 180; men sent here from Fort Resolution to fish, 75, 84, 92, 94, 98, 98 n.2; fishery relatively poor, 181, 182
Big River (**IV**), 67, 72

birch (*Betula papyrifera*), 93, 101, 112, 145; dwarf (*B. glandulosa*), 118 n.4
Birch Mountains, 92, 92, n 4
birchbark, 69, 75, 78, 84, 91; difficulties of obtaining, 92; of poor quality, 107 n.7, 121, 168
birch wood, 97
Bird, Captain Edward, 5
Bissett, Alexander, 43 n.1
Bissett, James, 53 n.3, 60 n.4, 61 n.2; biographical sketch, 43 n.1; travels from Lachine to Red River, 54; letter of introduction for, 58
blackflies (Simuliidae), 155, 156
Blackfly Point, name suggested by Lockhart, 152
Blanky, Thomas, 4
blueberries (*Vaccinium uliginosum*), 157
boats, York, 42, 55 n.4, 60 n.2, 164 n.1; Anderson's orders for second season, 96–7, 100, 103; damaged, 164
Bolieu, King, *see* Beaulieu, King
bomb vessels, 2
Boothia, Gulf of, 1, 168
Boothia, Isthmus of, 1, 14, 62, 90, 168, 178
Boothia Peninsula (**IX**), 2, 12, 13, 14, 19, 30 n.2, 61, 62, 86, 90, 96, 101, 249, 250
Bouché, Joseph, *see* Boucher, Joseph
Boucher, Joseph, 108, 121, 173; finds first Franklin artefacts on Montreal Island, 133, 133 n.4; receives reward for finding first traces, 134, 219, 228; finds horses at Portage la Loche, 163; wages received, 224; relays to Tyrrell rumour that a ship had been seen off arctic coast, 235, 245
boutes, *see* bow-men
Bouvier, steersman, 97
bow-men, 85
breakup, on Aylmer Lake, 85; Clinton-Colden Lake, 85; Great Slave Lake, 86; Liard River, 88, 91, 91 n.5; Mackenzie River, 91–93; late, 105
Bridport Inlet, 9
Britain, 42
British Columbia, Northern 40
British Government, 29, 46, 49, 50, 52, 55; undertakes to cover expenses of expedition, 27, 32, 35, 52, 54, 219; undertakes to reward exceptional service on expedition, 27, 32, 35, 49, 52, 54, 198, 219; asks Hudson's Bay Company to mount a search, 46, 52, 53, 54, 56, 57, 58, 59; believes expedition achieved everything humanly possible, 196; does not plan to mount any further expeditions, 196, 198, 199; will pay the reward for solving the fate of the Franklin expedition to Rae, 198; makes awards to Anderson and Stewart, 219
British Museum, 44 n.3, 67 n.1
Brockville, 45 n.4
Bromley, R.M., Accountant General, Admiralty, 227–8
Brough (member of boat's crew), 97
Bruce County (Ontario), 45 n.1
Buchan, Captain David, 3
Buffalo Lake, *see* Peter Pond Lake
Buffalo Narrows, 67
Buffalo River (**IV**), 74, 102, 102 n.2, 103, 213
Bullen, Sir Charles, 125 n.2
Bullen's River (**IX**), 125, 140
Burr Lake (**XII**), 147 n.3
Burwash, Major L.T., 254
Bury, 16
Bush Portage, 164
Bushnan Cove, 201
butterflies (Lepidoptera), 98

caches, meat, 119 n.5, 123; Inuit, 124, 133, 174, 251; recovered, 138 n.4, 139, 140, 140 n.3, 141, 142
Cahoutchella Point, *see* Kahochella Peninsula
Calcutta, 37
Cambridge Bay (**IX**), 8, 11, 82 n.1, 82 n.4, 86, 258
Camden Bay, 11, 83 n.2
Campbell, Robert, 37, 40–41, 60 n.3, 83, 88, 91, 96, 151, 170; given temporary charge of Mackenzie River District, 50, 55, 82, 85, 99; as back-up for co-leader of the expedition, 50; reaches Fort Resolution from Portage la Loche, 157; sends boats and canoes from Fort Simpson to Fort Reliance, 159 n.1; leaves Fort Simpson for Fort Liard, 181, 182; afflicted by curse of Glencoe, 181
Campbell Lake (**VII**), 105, 113 n.12
Campement des Anglais, 72
camps, Indian, 113, 121; Inuit, 124, 124 n.1, 126, 128, 135, 135 n.3, 137, 174, 175, 183, 185
Canada, 12, 60; East, 58; Lower, 39, 40; Upper, 37; West, 45 n.4
Canada jay, *see* whisky jack
Canadian Arctic Archipelago, 4, 10
Canadian Armed Forces, 254
Canadian Shield, 70

Canadians, 12
candling, 95, 95 n.4
cannibalism among Franklin survivors, 15, 20, 22, 95, 95 n.8, 201, 203, 207; Rae criticized for relaying reports of, 15, 16, 17–8, 25, 61, 65, 90 n.3, 92; Dickens comments upon, 18, 65, 92–3; reports confirmed, 18
canoes, recommended over York boats, 28, 47; north, 28, 31, 47, 91, 91 n.6; recommended by Back, 31; built for expedition, 42, 45; to be built at Fort Chipewyan and Fort Resolution, 47, 48, 51, 57, 84; three recommended for expedition, 47; going out of use in interior, 60; built at Fort Resolution, 67, 69, 74 n.1, 75, 77, 78, 81, 85; materials for, 69, 75, 78, 84; built at Fort Chipewyan, 69, 77, 88, 182; of excellent design, but birchbark poor, 170; Stewart's experiment of lining with canvas, 76, 92, 168 n.2; limitations of, 83, 168; Chipewyan model unsuitable, 84; three used on expedition initially, 85; used by Anderson in travelling to Fort Resolution, 88; made at Fort Liard, 88, 91; length of, 28, 31, 47, 91, 91 n.6; damaged, 71, 93, 94, 101, 102, 116, 116 n.5, 129, 135, 137, 138, 138 n.3, 139, 139 n.1, 141, 142, 171, 172, 174, 178, 183; ordered by Simpson from Fort Simpson, 97, 147, 147 n.6; too heavy, 102, 107 n.7, 115, 138, 141, 142, 168, 170; almost crushed by ice, 106, 116; loads of, 107 n.7; leaky, 108, 141, 152,175; shipping water, 109, 142, 144, 145, 147; worst one left behind, 122, 173; repairs to, 122, 129, 139; broken up for firewood, 142; best possible craft for Back River, 166 n.1; loads of, 170
'canot du nord', *see* canoe, north
CAPES:
Beaufort, *see* Point Beaufort
Bowden, 204
Britannia (**XI**), 82, 86, 89
Bunny, 204
Cockburn, 9, 10, 249
Colville, 14
Crozier, 236, 253
Felix, 197, 250; identified by Osborn as location of *Erebus* and *Terror*, 204
Fullerton, 15
Horn, 1
John Herschel (**IX**), 2, 86; cairn at, 86, 86 n.3, 89, 135 n.5, 168, 252, 256
of Good Hope, 11
Osborn, 9, 234 n.1
Parry, 10
Porter, 207
Prince Alfred, 10
Riley, 7
Spencer, 7
Victoria, 249, 250
Walker, 4, 204
Capitaine, 154
Capot, 128 n.3
Capot Blanc, 119 n.9
Capot Rouge, son hired to hunt by Lockhart, 154
Cariboo Lake, *see* Reindeer Lake
caribou (*Rangifer tarandus groenlandicus*), 13, 87, 112 n.10, 113, 118 n.4, 122, 123, 124, 125, 126, 128, 129, 130, 134, 135, 135 n.11, 137, 139, 140, 141, 142, 142, 144, 153, 176; migration 87 n.1, 89, 137, 139, 140, 151, 157–8, 175; tracks, 116, 133, 137, 139, 175, 251; rotting carcases, 121, 126; calves, 122, 129; carcases at Inuit camp, 125; tongues, 125 n.3; calves at Inuit camp, 125; killed on Montreal Island, 134; killed on Point Ogle, 136; killed on Maconochie Island, 136; drowned, 141, 141 n.2; migration reaches Fort reliance, 157–8; intensive hunting of from Fort Reliance, 158; skins, 173; numerous on Adelaide Peninsula, 175
Caribou Island, 109 n.3, 147
caribou lichen (*Cladonia* spp.), 112, 112 n.9, 113
Carlton House, *see* Fort Carlton
Cascades (**IX**), 123, 128, 141; (Clearwater River) 163
Cassette Portage, 74, 78
Cassette Rapids, 74 n.3, 162 n.7
Castor and Pollux Bay, 2
Castor and Pollux River, 12, 14, 207
Cator, Captain Bertie, 6
Central America, 5
Chantrey Inlet (**IX, XI**), 11 n.1, 86, 130 n.6 breakup, 86; ice conditions on, 105, 131 131 n.7, 134, 134 n.5, 134 n.7, 135, 135 n.10, 137, 173–4, 177, 184, 190, 257; west shore searched, 134–5, 174, 178, 184
char, Arctic (*Salvelinus alpinus*), 129 n.8
Charlton Island, 11 n.1
charts, recommended by Rae, 31; provided for the expedition, 35–6, 49, 85
Chesterfield Inlet, 12–13, 125; Inuit from, 124 183
Chilkat Indians, 41

China, 4
Chipewyans, 70; guide Oman, 78, 79, 80; interacting amicably with Inuit on the Thelon River, 235, 240
chronometers, provided for the expedition, 36
Chukchi Sea, 6, 122 n.13
Churchill (**IV**), 15, 28, 30, 43, 44, 48, 52, 55 n.3, 56, 56 n.3, 59, 60, 69, 70, 76, 77, 78, 79, 90, 256
Churchill River (**IV**), 45, 73 n.1, 161, 164, 213, 215
circles, sorted, 124 n.1
Clarke, Lawrence, 94, 157; relays report by Yellowknife Indians that they had found a camp recently occupied by Whites on the Back River, 235, 236–8; believes this camp had been occupied by survivors of the Franklin expedition, 235; requests permission to investigate the matter himself, 235, 238
Clarke, Mrs, 157
Clarke's Lake (**V, VII**), 112 n.6
Clear Lake, 72
Clearwater River (**IV**), 67, 73 n.2, 161, 163, 163 n.6
Clinton-Colden Lake (**VI, VIII**), 26, 103, 105, 111, 118 n.1, 143, 144, 176, 183; breakup on, 85, 108 n.5; still ice-bound, 105, 171; Narrows of, 143; difficulties of navigation on, 143, 144, 185
Clinton-Colden Straits, 119; searching for, 143, 144, 144 n.1
clothing, recommended by Anderson for future expeditions, 166 n.1
cloudberries (*Rubus chamaemorus*), 139, 139 n.10
Clover Bar, 42
Collings, Thomas, Admiralty, 229
Collinson, Captain Richard, 10–11, 17, 25, 26, 35, 86, 90, 96; proposed search for, 27, 28, 30, 31 n.4, 36, 37; emerges from Arctic, 37; summary of expedition, 82 n.1; missed opportunity, 82 n.4, 90, 96, 178, 257–8; dispatches from, 82, 82 n.3, 86; searches east coast of Victoria Island, 82 n.1, 258; discovers this area already searched by Rae, 258
Collinson, Captain T.B., 26
Collinson Peninsula, 8, 11 n.1
Columbia District, 44 n.3, 44 n.4
Colvile, Andrew, 32 n.1, 230; death of, 198
Colvile, Eden, 84, 100
Colvile, Mrs 88
Committee Bay, 11 n.1, 14, 22
compasses, 169; provided for expedition, 36; Anderson plans to use, 83; found useless, 128 n.4
Cooper, Paul F., 254
Copenhagen, Battle of, 3
Copper Indians, *see* Yellowknife Indians
Coppermine River, 2, 3, 5, 8, 11 n.1, 62, 62 n.1, 87 n.2, 97 n.2, 204
Cornwallis Island, 4, 8, 252
Coronation Gulf (**IX**), 11, 28 n.2
Corrie, H., Admiralty, 232
Couch, Edward, silverware and watch recovered, 253
courts martial, 16, 25
Couteaux Jaunes, *see* Yellowknife Indians
Couteaux Jaunes River, 93
Covent Garden, 15
Cracroft, Sophia, 63, 92, 95; argues that evidence brought back by Rae is inconclusive, 63; asks Anderson to bury any corpses he finds, 63; asks Anderson to bring back any records, 64; casts doubts on Ouligbuck's reliability, 64
cranberries (*Vaccinium vitis-idaea*), as an antiscorbutic, 11 n.1
Cree, 70
Cresswell's Tower, 204
Crimea, 25
Crimean War, xii, 25, 196
crow berry (*Empetrum nigrum*), 112, 112 n.8, 141
crows (*Corvus brachyrhynchos*), 95, 95 n.2, 121, 124, 141
Crozier, Captain Francis Rawdon Moira, previous career, 3–4; probable knowledge of Inuktitut, 4; silverware recovered, 14, 19, 20, 22, 250; knowledge of stores at Fury Beach, 86 n.5; signs final message from Franklin expedition, 253
Crozier's Landing, vast amount of equipment and clothing abandoned at, 253–4
Cumberland House (**IV**), 42, 45 n.1, 45 n.2, 53, 69, 76 n.2, 148 n.3, 164, 241 n.2
Cumberland House District, 45 n.2, 55 n.4, 76
Cut Rocks, 147
cypres, *see* jackpine

dags, *see* daggers
daggers, 125, 125 n.7, 173
dandelions (*Taraxacum* spp.), 124

Daniel, George, 66, 108, 121, 172; wages received, 224
Davis Strait, 8, 249
Deal, 15
Dealy Island, 9, 10, 249
Dease, Peter, 1, 11 n.1, 12, 14, 32 n.1, 86 n.3, 252; locates King's cache on Montreal Island, 197
Dease Strait (**IX**), 11
deer, *see* caribou
deer passes, 87, 87 n.3, 89, 90, 131, 131 n.4, 141
De Haven, Captain E. J., 7
Demren, Francis, 66
Deschambault, Mr, *see* Deschambeault, George
Deschambeault, Colonel Louis, 45 n.2
Deschambeault, George, 45, 59, 67, 72; biographical sketch, 45 n.2; receives instructions from Simpson, 57
Desjarlais, guides Lockhart over Pike's Portage, 156
Desmarais, Francis, 66
Des Marais Islands, 101, 101 n.4
Des Voeux, Charles, signs message left on King William Island, 252
Detroit, 58
Detroit Clear Lake, 164
Devon Island, xi, 4, 6, 7, 9, 234 n.1
diabase, 109 n.7, 111 n.6
Dibble, Colonel, 58
Dickens, Charles, comments on reports of cannibalism among Franklin expedition survivors, 18, 65
divers, 113, 118
Dogrib Indians, 103
dogs, 67, 72, 73, 92; exhausted by Stewart's winter travels, 67, 74, 75; charged to the expedition, 76
Dolphin and Union Strait, 5, 11, 11 n.1
dragon flies (*Somatora franklini*), 98
driftwood, 108
Duck Isle Portage, 163
ducks, 93, 109, 112, 153, 162, 163, 164; Esquimaux, *see* eiders, king
Dundas Peninsula, 8
Dunnett, Captain, 5

eagles, 109, 164; bald (*Haliaeetus leucocephalus*) 109 n. 9; golden (*Aquila chrysaëtos*), 109 n.9
East Arm (**VI**), 27, 81 n.4, 81 n.5, 105, 109 n.7, 149; landscape of, 112, 112 n.1
Eastmain, 55 n.1, 81
Ecours, 94
Edinburgh, 44 n.3, 44 n.4
Edmonton, 42
Edward's Butte, 141 n.4
Ee,so,ey, 81
Egg Lake, 44, 48, 53, 61
Eglinton Island, 9
eiders, King (*Somateria spectabilis*), 126 n.8, 129, 131, 133, 134, 135
Ellice, Edward, 13
Ellice's River (**IX**), 86, 89
Elliott, Captain, 131 n.9
Elliott's Bay (**XI**), 131, 131 n.7, 136 n.9, 173, 174, 178, 184
Embarras River, 92, 163
Emma Bay, 5
English Channel, 15, 18
English River, *see* Churchill River
English River District, 45 n.2, 76
Erebus Bay (Beechey Island), 7, 9
Erebus Bay (King William Island) 18, 253, 255
ermine (*Mustela erminea*), 128, 139 n.2, 139
Escape Rapid (**X**), 128, 139
Esquimalt, 43 n.1
Esquimaux, *see* Inuit *or* Inuktitut
Euphrates River, 4

Fairholme, Adam, 194 n.1
Fairholme, George K.E., 194 n.1
Fairholme, Lieutenant James, 194 n.1; silverware recovered, 250, 253
Fairholme case, 133 n.4, 167, 194, 194 n.1, 212
falcons, Peregrine (*Falco peregrinus*), 128 n.12
Falls of Anthony, Officer Commanding US Troops at, 58
Felix Harbour, 1
Fiddler, Harry, *see* Fidler, Henry
Fidler, Henry, 66, 108, 121, 172, 235; recommended by Rae for expedition, 30, 52, 53; one of the group who searched Maconochie Island, 136 n.7, 246; wages received, 224; along with Edward Kipling found boat broken up on Montreal Island, 246
Fidler, John, 66, 108, 121, 134, 143, 144, 172; recommended by Rae for expedition, 28, 30, 52, 53; rewarded as steersman, 142, 228; reconnoitring route, 145; wages received, 224
Fifth Thule Expedition, 254
Finlayson, James, 72
fireplaces, found by Pike, 119 n.9

First Cascade, 124, 124 n.2
Fish River, *see* Back River
Fisher River, 246
fishing, 98, 101, 108, 109, 112, 113, 115, 116, 121, 122, 151, 152, 154, 157
fishing eagle, *see* osprey
Fitzjames, Commander James, 4; signs final message from Franklin expedition, 253
Flour Point, 165
Fond-du-Lac (Lake Athabasca) (**IV**), 70, 76 n.2, 77, 78 n.4, 79, 81 n.4, 91; guide arrives at, 81
Fond-du-Lac (Great Slave Lake), 78, 103, 145, 180; hunters to be hired, 82, 83, 84
Forsyth, Captain Charles, 6; brings first news of the Franklin expedition to Britain, 8
FORTS:
Alexander, 45 n.4
Anderson, 38
Carlton (**IV**), 42, 44, 51, 53, 54, 55, 57, 59 n.2, 61, 67, 217, 256; Stewart departs from, 67, 72, 72 n.2
Chipewyan (**IV**), 28, 30, 30 n.1, 32 n.1, 37, 42, 45, 45 n.1, 48, 49, 51, 51 n.3, 53, 54, 55, 56, 56 n.3, 57, 59, 60, 67, 69, 70, 72, 73, 75; 76 n.1, 76 n.2, 77, 78, 79, 80, 81 n.4, 85, 88, 90, 97, 161, 164 n.2, 180, 182, 256; arrival of voyageurs from Norway House, 77, 77 n.2; supplies to be fetched from, 84; Stewart and Lockhart return to, 163, 163 n.1
Churchill, *see* Churchill
Confidence, 5, 8, 11 n.1, 45 n.1, 86 n.4, 87 n.2
Edmonton (**IV**), 45, 45 n.4, 54 n.1, 57, 256
Enterprise, 3
Frances, 45 n.4
Good Hope, 41, 45 n.1
Garry (**IV**), 11 n.1, 43 n.2, 44, 51 n.2, 52, 53, 84 n.6, 161, 166, 167, 256; Lower, 66, 165; Upper, 43 n.3, 44 n.4, 58, 66, 165
Langley, 44 n.3
Liard, 40, 45 n.1, 50 n.2, 55 n.5, 88, 91, 151, 181, 182
McPherson, *see* Peel River Post
Nez Percés, 44 n.3
Norman, 67 n.1
Providence, 98
Rae (**VI**), 41, 90, 97, 162, 235, 236
Reliance (**I, IV, VI, XII**), 27, 31, 43, 61 n.4, 63 n.2, 69, 74 n.5, 90, 97, 97 n.6, 103, 108 n.5, 145 n.4, 147, 147 n.3, 149, 153, 153 n.5, 166, 167, 171, 176, 178; winter quarters built at, 75, 119 n.8, 147; Anderson leaves, 147; Stewart and Lockhart leave, 148 n.1; Lockhart arrives at, 149; Back's buildings in ruins, 149, 153; Lockhart reuses fireplaces and chimneys, 149, 153–4, 180; Lockhart rebuilds, 154, 156, 180–85; Back's description of site, 153 n.5; Back's description of original establishment, 153 n.5; ruinous state of Back's establishment, 154 n.1; Lockhart returns from trip to Fort Resolution, 157, 182
Resolution (**IV, VI**), 26, 28, 30 n.3, 40, 41, 42, 43, 43 n.2, 44, 45, 48, 51, 57, 67, 67 n.1, 71, 73, 75, 78, 81, 82 n.2, 83, 85, 88, 90, 92, 95, 96, 97, 98 n.2, 99, 101, 102, 106, 149, 166, 211, 256, 258; selected as expedition's starting point, 47, 50, 59, 190, 193; Stewart arrives on his first visit, 74; expedition departs, 107, 190, 193; Anderson, Stewart and Lockhart return to, 148, 194; Lockhart leaves, 149; boats from Fort Simpson arrive, 159
Selkirk, 37, 40, 41
Simpson (**IV, VI**), 11 n.1, 37, 38, 40, 41, 42, 45 n.1, 48, 50, 55 n.4, 63 n.2, 67 n.1, 71, 77, 83, 84, 88, 93, 95, 96, 103, 147 n.6, 161, 161 n.2, 162, 256
Smith, 74 n.3, 162 n.6
Vancouver, 44 n.4
William, 40
Yukon, 38, 41, 43 n.2
Foxe Basin, 11 n.1
Foxe Channel, 2
foxes (*Alopex lagopus*), 126, 139
Frances Lake, 40
Frances Lake Post, 37, 40, 41, 99
Frances River, 40
Franklin, Jane, Lady, 6, 8, 63, 63 n.2, 92, 95; offers opinion as to fate of the Franklin expedition 61; criticizes Rae for relaying Inuit report of cannibalism, 61; requests that Anderson recover locks of Franklin's hair, 62; urges Anderson to recover Franklin's private letters and papers, 62, 95; offers reward for letters and papers, 62; suggests area which should be searched, 62; predicts failure of the expedition, 200; Anderson sends report to, 166; less than satisfied with expedition's results, 196, 198; plans a further search expedition, 198, 200; dispatches McClintock's expedition in *Fox*, 249

Franklin, Sir John, xi, 1, 46; career of, 3; first overland expedition of, 3, 32 n.1; second overland expedition of, 3, 32 n.1; as Governor of Tasmania, 3; silverware and silver plate recovered, 15, 19, 21, 22, 250, 253; boat allegedly found with name on it, 189; death of, 253

Franklin's last expedition, outline of, 1–5, 192; orders, 4; earlier searches for, xi, 5–11; survivors seen by Inuit 13–14, 19–20, 21–2; corpses found by Inuit, 13–14, 19–20, 21–2; artefacts recovered, 15, 19–21, 22, 23–4; ammunition and gunpowder in possession of survivors, 22; records to be searched for and recovered, 32, 47, 49; corpses of to be buried, 49; artefacts from found on Montreal Island, 133, 133 n.4, 174, 188; no traces found on west shore of Chantrey Inlet, 135, 174; bodies probably buried in sand, 136, 136 n.9, 168, 175, 179, 180, 185, 189, 191, 193, 195; records probably destroyed by weather or animals, 136 n.9, 168, 175, 179, 180; records possibly thrown away by Inuit, 168, 179; artefacts forwarded to Norway House, 182; artefacts preserved at Greenwich Hospital, 195; assumed tried to ascend Back River, 184, 185; Stewart believed no members surviving, 195; mystery of what happened to skeletons in light of preservative effect of arctic climate, 201; artefacts handed over by Hudson's Bay Company to Admiralty, 223, 229–33; list of artefacts forwarded to Admiralty, 231; some members may have returned to ships, 255; may have travelled with more than four boats, 255

Franklin Lake (**IX, XI**), 27, 129, 138, 173, 177, 183, 200; still ice-covered, 129

Fraser, Thomas, Secretary, Hudson's Bay Company, 229

Frederikshåb, 249

freeze-up, early, 105

freezing, 142, 143, 144, 162, 163, 176

French Lake (**XII**), 147 n.3

Fricke, H., 255

Frog Portage, 161, 164, 164 n.3

Frozen Strait, 3

Fury Beach, 4, 90 n.4; stores left at by Parry, 86 n.5, 90, 96, 101 n.1, 204

game list, 176

Garry, Nicholas, 125 n.6

Garry Lake (**IX, X**), 27, 125, 125 n.6, 126, 140, 173; icebound, 105, 126, 139, 140, 141, 169, 173, 183

Gateshead Island, 11, 82 n.5

geese (Anserinae), 93, 109, 111, 112, 126; Ross's (*Chen rossii*), 67 n.1; Canada (*Branta canadensis*, 109, 109 n.6, 115, 116, 118, 122, 123, 124, 128, 129, 130, 134, 140, 141, 143, 145, 163; White-fronted (*Anser albifrons*), 119, 119 n.1, 122, 123, 125, 135, 141, 142, 145, 147; moulting birds run down and killed, 122, 122 n.11, 123, 124, 128, 129, 176; laughing, *see* white-fronted; have completed moult, 138, 138 n.6, 139, 139 n.3; migrating south, 140, 141, 142, 143, 145; Snow (*Chen caerulescens*), 143, 144; have all gone south, 162

geology, of East Arm area, 109, 109 n.7, 111; of Aylmer Lake area, 116; of Sussex Lake area, 118; of Muskox Lake area, 119; of Malley's Rapids area, 122; of east side of Chantrey Inlet, 131; of west side of Chantrey Inlet, 135

Geral'da, Ostrov, 6

Gibraltar Point, 111 n.5

Gibson, William, 254

Gilder, Colonel W. H., 106, 107 n.1, 252

Gjoa Haven, 129 n.7

Glacier Creek (**XII**), 147 n.3

Gladman Point, 252

Glencoe, curse of, 181, 181 n.3

Glen Garry, 13

gneiss, 119, 124, 251

Godhavn, 249

Godthåb, 249

Goodsir, Harry, silverware recovered, 253

gooseberries (*Ribes oxyacanthoides*), 95

Gore, Lieutenant Graham, 4; silverware recovered, 19, 253; message left by on King William Island, 252

Graham and Simpson, 32 n.1

Graham, Sir James, First Lord of the Admiralty, 27, 31, 35; rejects offers by King, Osborn, Pim and McCormick to lead expedition, 205

Grand Batture, 162

Grand Detour, 74

Grand Diable Rapid, 164

Grand Marais, 72

Grand Noir, 94

Grand Rapid (**IV**), 161, 165, 211, 213, 214

granite, 111, 115, 116, 118, 131, 135, 153 n.5
Grant Point, 108 n.2
grasshoppers, 95
graves, none found on Montreal Island, 134, 174, 177
Gray, member of boat's crew, 97
Great Bear (constellation), 139
Great Bear Lake (**IV**), 5, 8, 11, 11 n.1, 28 n.2, 45 n.1, 86 n.4, 87 n.2
Great Fish River, *see* Back River
Great Slave Lake (**IV, VI, VIII, XII**), 26, 27, 28, 31, 42, 43, 47, 48, 50, 51, 52, 53, 54, 55, 56, 57, 59, 71, 74, 75, 78, 79, 81, 83, 85, 88, 90, 92, 97, 101 n.4, 101 n.5, 105, 108 n.5, 113, 118, 145, 145 n.3, 147 n.3, 149, 161, 164 n.2, 166, 171, 176, 181, 185, 191, 256; breakup on, 75, 82, 83, 85, 88; ice still solid, 78, 94, 102, 103, 109; ice conditions, 71, 86, 88–9, 101, 109–12; climate much better than Artillery Lake, 147 n.8
Great Slave Lake Post, 45 n.2
Greenhithe, 1
Green Island, 93
Green Island Rapids, 94 n.1
Green Lake (**IV**), 67, 72
Greenland, 3, 5, 6, 249
Greenwich Hospital, 195, 233
Grey, boatbuilder at Fort Simpson, 96–7
Griffin, Captain Samuel, 7
Griffith Island, 8, 234 n.1, 249
grindstone, 154
Grinnell Peninsula, 9
Grosse Roche Portage, 163
Gros Ventre's, 164
ground squirrel, Arctic (*Spermophilus parryii*), 113 n.15
grouse, white, *see* ptarmigan
gulls (Laridae), 102, 109, 112, 113, 118, 124, 129, 131; black-headed, 129; Bonaparte's (*Larus philadelphia*), 129 n.6, 133, 141
gum (for canoes), 69, 75, 84, 84 n.1, preparation of, 84, 84 n.1
gumming, 93, 93 n.6, 115 n.5, 137, 139, 141, 143, 147, 152
gunpowder, 22, 80, 84; to be fetched from Fort Chipewyan, 84
Gwich'in, 83, 83 n.2

Hackland, James, 80
hail, 98, 125, 141, 175, 191
Halkett, Lieutenant Peter, 28 n.2
Halkett boat, 28, 86, 106, 107 n.7, 135 n.5, 136, 171, 174, 175, 178, 235, 245; details of, 28 n.2; previously used by Rae, 28 n.1, 48; supplied to the expedition, 48, 56; used for examining offshore islands, 134 n.2, 137
Hall, Charles Francis, 254, 255
Hall of Clestrain, 11 n.1
Hamilton, W. A. B., Secretary to the Admiralty, 27, 35; requests assistance of Hudson's Bay Company, 31
Hampton, Robert, 250
Hanbury, David, encounters Inuit on Back River, xi
Hannah Bay, murders at, 37
Hanningayarmiut, 124 n.11
Hansard, 199
Hardisty, William Lucas, 41, writes letter of appreciation to Anderson, 38; forwards dispatches from Collinson, 82; recommended by Anderson for promotion, 87, 181
Hardisty's Lake (**VII**), 113 n.12
hares, Arctic (*Lepus arcticus*), 134, 251; Varying (*Lepus americanus*), 93
Hargrave, James, 43 n.3, 44 n.4
harpoons, 91
Harrison, D. A., 29 n.1, 170; to send off returns from Fort Liard, 82
Harrison Lake, 113 n.1
Harry Lake (**XII**), 147 n.3
Hawaii, *see* Sandwich Islands.
Hawk Rapids (**IX**), 124, 124 n.6, 140–41
hawks (Accipitidrae), 124 n.6, 128, 128 n.12; Rough-legged (*Buteo lagopus*), 128 n.12
Hay River (**IV**), 101 n.5, 102
'Head of the line', 94, 94 n.1
heather, Arctic (*Cassiope tetragona*), 115, 115 n.3. 131 n.2, 135 n.4, 137, 143
heights, estimated, 109, 111, 115, 153 n.5
Herald Island, *see* Geral'da, Ostrov
Her Majesty's Government, *see* British Government
Hoarfrost River (**VI**), 26; difficulties of, 100, 103, 105, 147, 171, 183; water depth in lake off, 158
Hobson, Lieutenant William, 63 n.4, 90 n.5, 250, 253; conducts independent search of west and south coasts of King William Island, 250; leaves message reporting his finds, for McClintock, 252; finds only message so far recovered, as to the fate of Franklin expedi-

tion, 252–3, 257; finds cairns between Point Victory and Cape Felix, 254
Holsteinsborg, 249
Hongkong, 10, 11
Honolulu, 32 n.1, 43 n.1
Hook's Island, 74
Hooper, Lieutenant William, 6
Hopkins, Edward, M., 58; forwards Anderson's report and newspaper cuttings to London, 186–7; relays Stewart's report to the *Montreal Herald*, 189–92
horizons, artificial, provided for expedition, 36
Hornby, Reverend Edward, 16, 17
Hornby, Frederick, 16; sextant recovered, 254
horses, at Portage la Loche, 161
Household Words, 18, 65, 92
House of Commons, 196
Hudson Bay (**IV**), 11 n.1, 12, 44 n.3, 45 n.4, 55 n.1, 69, 121 n.7
Hudson's Bay, *see* Hudson Bay
Hudson Strait, 2, 11 n.1
Hudson's Bay Company, 1, 11 n.1, 12, 15, 16, 17, 18, 19, 20, 21, 26, 28, 37, 46, 52, 53, 54, 56, 57, 58, 66, 67 n.1, 70, 74 n.5; undertakes search, xii, 25, 59, 61, 192; finances Sir John Ross's search, 6; sponsors Rae's search, 8; Committee of, 16, 28, 31, 32 n.1, 33, 40, 198; Governor of, 28, 31, 33, 40, 198; North American headquarters of, 43 n.1; Archives, 105, 212; expedition casts positive light on, 196, 198; submits bills to the Admiralty, 219, 221–223; suggests to Admiralty that Anderson and Stewart merit awards for their services, 224–5; receives Arctic Medals from Admiralty for distribution, 229; forwards artefacts from Franklin expedition to Admiralty, 229–33; list of artefacts forwarded to Admiralty, 231; denies rumour about Indians finding a camp recently occupied by Whites, 235
Hudson's Bay House, 212
Hull, 4
Hunt, Mr, 164

ice conditions, on Mackenzie River, 71; on Great Slave Lake, 71, 78, 81, 83, 101–2, 109–12, 171; on Aylmer Lake, 116–17, 143, 172, 183; on Garry Lake, 126, 173; on Chantrey Inlet, 131–2, 131 n.7, 134, 134 n.5, 134 n.7, 135, 135 n.10, 137, 173–4, 177, 184, 190, 257; on Saskatchewan River, 164, 165
ice-masters, 4
Icy River (**VIII**), 119
Igloolik, 3
Ile-à-la-Crosse, (**IV**), 45, 45 n.2, 55 n.4, 57, 67, 69, 72, 74, 76 n.2, 148 n.3, 163, 215; Stewart and Lockhart return to, 164
Ile aux Bouleaux, 98
Ile aux Morts, 162
Ile du Raquet, 162
Iles des Marais, 101 n.4
inconnu (*Stenodus leucichthys*), 152
Indian, hunts, 94; tents, 145; canoes, 145
injuries, 93, 115, 116, 118, 119, 119 n.7, 143
instruments, recommended by Rae, 30–31; provided for the expedition, 35–6, 49, 61, 85; Anderson plans not to use, 82, 85
Inuit, met by Hanbury on Back River, xi; report seeing Franklin expedition survivors, xii, 13–14, 17, 19–20, 21–2, 32, 46, 130 n.2, 189, 190, 257; articles from Franklin expedition in possession of, xii; 13–14; 19–21, 22, 23–4, 46, 130, 138, 173, 177, 184, 191, 200, 257; find corpses from Franklin expedition, 14–5, 17, 19–20, 22, 46; report cannibalism among Franklin survivors, 15, 17, 20, 22; reports of cannibalism confirmed, 18; trade goods for, 30, 47, 48, 54, 55, 56; to be hired to point out the sites where the Franklin expedition members died, 30, 47; of Liverpool Bay area, 38; may have attacked Franklin survivors, 65, 65 n.2; possibly killed by members of Back's party, 65, 65 n.2; cairns, 123–4, 124 n.1, 128, 129, 133, 133 n.4, 141; camps, 124, 124 n.1, 126, 128, 135, 135 n.3, 137, 174, 251; caches, 124, 133, 174, 251; first group encountered, near McKinley River, 124, 172, 183; canoes, 124; gifts made to, 125, 138, 140, 173, 175, 183, 185; tents, 124, 125, 129, 130, 140, 173, 183, 184; appearance, 125, 140; contact with Churchill Inuit, 125, 140, 173; group encountered below Franklin Lake, 129–30, 173, 177, 183–4, 200; fishing sites, 129 n.7; knew of death of the Franklin expedition members, 130, 130 n.2, 138, 173, 177, 183, 193; one woman reported having seen a survivor alive, 183–4, 185, 186, 188, 191, 193, 194, 212–13, 257; reported a wrecked boat, 184; possessed no books or documents, 130, 130 n.4, 138, 138 n.1, 173, 175, 177, 180, 185, 191; tools, 131; Franklin

Lake group encountered again, 138, 175, 185; artefacts purchased from, 138, 175, 257; had made caches of fish, 138, 175; near Pelly Lake, 140; group near McKinley River encountered again, 140, 175; tattoos, 140; recover paddle, 140 n.5; honesty, 140 n.5; deny killing Franklin expedition members, 185; had no European clothing or weapons, 201 may not have cut up boat on Montreal Island, 196, 202; may have taken advantage of weakness of Franklin expedition members, 202–3; interacting amicably with Chipewyans on Thelon River, 235, 240; tell McClintock of Whites starving to death and of ships being crushed in the ice, 249, 250; oral traditions suggest presence of Franklin survivors on Melville Peninsula, 255

nuktitut (language), Franklin survivors unable to understand, 14, 21; McLellan's and Mustegan's command of 106, 124 n.10, 125, 177; Washington's vocabulary, 177

nuktitut interpreter, 76; recommended by Rae, 26, 69; fails to appear 69, 76, 77, 85, 91, 102, 103; attempts to provide, 67, 70, 78–9; general difficulties in obtaining, 71, 81; to travel across country from Churchill to Fort Chipewyan, 78; lack of, 106, 107 n.7, 200, 257; fails to overtake expedition, 109; lack of regretted, 124 n.10, 125, 130, 171, 173, 177, 180; limitations of lack of, 255

nvernessshire, 13

ron, for sled runners, 97

roquois canoemen, 12, 43, 44, 48, 53, 59, 60, 61, 67, 69, 84, 106, 121, 213, 214, 256; travel from Lachine to Red River, 54; contract signed by, 58; rate of pay, 58–9, 60, 219; meet with Anderson's approval, 85; superb skills of, 128, 168, 172, 183, 190; left to repair canoes at Point Pechell, 135, 174, 178; travel with Stewart to Lachine, 186; reach Lachine, 186, 198; complete round trip from Lachine to Point Pechell and back in 13 months, 191; distance travelled to reach Fort Resolution, 256

sbister, Alexander, 234; biographical sketch, 241; claims that the rumour of an Indian report of a camp recently occupied on the Back River was widespread, 241, 242; suggests that anyone who openly reported such a story might expect retribution from the hudson's Bay Company, 241

Isbister, James, recommended by Rae for expedition, 52

Isbister, Thomas, 241 n.2

Island Portage, 164

Isle aux Freines, 73

Isles de Pierre, 74

Itimnaarjuk, 129 n.7

Itkelek, reports seeing Whites on the Back River, xi

jackfish (*Esox lucius*), 152

jackpine (*Pinus banksiana*), 112, 112 n.2

Jambe de Bois, 78; wife refused to make nets, 154–5; hunting caribou, 158; reconnoitres south end of Pike's Portage, 158; making meat scaffold, 158; hunting, 159

James Bay, 11 n.1, 37

James Ross Strait, 250

Jervois River (**IX**), 124

Jobin, Anbroise, 66, 151; fishing, 154; cutting wood, 154, 156; working in sawpit, 158, 159; making meat rack, 158; tanning and mending nets, 158; wages received, 224

Johnson, Jeremiah, 66, 108, 121, 134, 172; recommended by Rae for expedition, 30, 53; wages received, 224

Johnston, James, *see* Johnson, Jeremiah

Johnstone, Jerry, *see* Johnson, Jeremiah

Jones, Major, takes testimony *re* Inuit allegedly killed near Point Ogle, 65 n.2

Jones Sound, 6

Journal of the Royal Geographical Society, 167

Kahochella Peninsula (**VI**), 109 n.7, 111, 111 n.5, 183

Kahoochellah, *see* Kahochella Peninsula

kayaks, 129 n.5, 130, 140, 173

Keenleyside, A., 255

Keewatin (Lake of the Woods), 246

Keith Island, 109 n.4

Kellett, Captain Henry, 5, 6, 9, 10; court martial, 25

Kelvington, 44 n.6

Kennedy, Charles, recommended by Rae for expedition, 30, 52, 53

Kennedy, William, 8, 12, suggested leader for further search expedition, 198

Kensal Green Chapel, 63 n.3

Kent Peninsula (**IX**), 3

kettles, 125, 128 n.3, 129 n.1; copper and tin,

130, 133, 173, 184; stone, 140; design recommended by Anderson, 166 n.1

King, Dr Richard, 26, 234; hears rumours of members of Back's party killing Inuit near Point Ogle, 65 n.2; criticizes Admiralty and British Government for entrusting search to Hudson's Bay Company, 197, 205, 210; criticizes Anderson and Stewart for failing to take an Inuktitut interpreter, 197, 205, 208, 244; criticizes Anderson and Stewart for not erecting a granite monument, 197, 205; left cache on Montreal Island, 197, 208; argues that Franklin would have left message in his cache on Montreal Island, 197, 209; criticizes Anderson and Stewart for not checking his cache, 197, 205, 208, 210, 244; exhorts Admiralty to commission him to lead another expedition down Back River to check his cache, 197, 199, 208; complains that Anderson and Stewart did not extend their search to King William Island, 199, 210; well acquainted with Montreal Island and area, 205; offer to lead expedition declined, 205; proposes expedition down Back River in 1845, 206; proposes to lead expedition in search of the Franklin expedition via the Back River in 1847, 206; predicts Franklin's ships are off west coast of Somerset Island, 206; repeats his offer in 1850; predicts location of Northwest Passage, 206; criticizes Rae for not immediately proceeding to Chantrey Inlet to check the Inuit's story, 207; repeatedly emphasizes that his predictions had been right, 208; argues that Anderson and Stewart were inadequately equipped by the Hudson's Bay Company, 210; fully believes the Indian report of a camp recently occupied by Whites on the Back River, 239; accuses Hudson's Bay Company of ignorance of its own territories, 240; pushes yet again for another search expedition to be led by himself, 234–5, 240, 242–4; suggests that no employee of the Company would openly relay the information about possible survivors of the Franklin expedition having been recently reported on the Back River, 242–3; argues that the Indian report is probably reliable, citing the demonstrated reliability of the stories told by the Inuit to Rae concerning the Franklin expedition, 242; proposes a rescue expedition by screw steamer, and a canoe party by the Back River, 244; had earlier proposed such a joint expedition, in conjunction with Lieutenant Pim, 244

King's College, Aberdeen, 241 n.2

King's Posts, 44 n.3

King William Island (**IX**), 1, 2, 11, 11 n.1, 17, 18, 61, 63, 63 n.4, 82 n.4; 86, 86 n.3, 89, 90 n.5, 95, 96, 178, 197, 204, 236, 250, 252, 254, 255, 256, 257, 258; Franklin survivors seen on, 14, 19, 46; ice prevented expedition from reaching, 168

King William's Land, *see* King William Island

Kipling, Edward, 66, 108, 121, 128, 134, 143, 172, 246; wages received, 224; one of group who searched Maconoche Island, 246; member of the independent group, one of whom had seen ship off arctic coast, 235, 245; gives statement concerning incident of ship seen in the ice off arctic coast, 246; along with Henry Fidler found a boat broken up on Montreal Island, 246; did not relay this information to Anderson or Stewart, 246

Kipling Lake (**XII**), 147 n.3

Kippling, George, 92, 151; surveys ruins of Fort Reliance, 153; cutting shingles, 54; reconnoitres Lockhart River with Lockhart, 155; sick, 156; hunting caribou, 158; reconnoitres south end of Pike's Portage, 158; tanning & repairing nets, 158; laying floor, 159; wages received, 224

Kirkwall, 11 n.1

Kittson, H. 58

Klutschak, Heinrich, W. 106, 107, 107 n.1, 25

Kotzebue Sound, 5, 6

La Bonne Mountain, 163

La Bonne Portage, 73, 163

Labrador, 81

Labrador tea (*Ledum groenlandicum*), 112, 11 n.7, 131 n.2

Lac a la Crosse, 72

Lac à Primeau, 164

Lac Cruche, 72

Lac des Bois, 164

Lac la Biche, 164 n.2

Lac la Loche, *see* La Loche Lake

Lac qu'il doit permit, 72

Lachine, 12, 28, 32 n.1, 37, 42, 43 n.2, 44, 4

50, 52, 53, 54, 56, 57, 58, 59, 76, 98, 161, 211, 256
Lachine Canal, 43 n.1
La Cimetière, 164
Laferté, Alfred, 6, 108, 213; invalided from expedition, 119 n.7, 172; reaches Fort Reliance, 156; hunting caribou, 158; adjusts saw, 158; sawing, 158; laying floor, 159; sent with message to recall Lockhart to Fort Reliance, 160; dispute between Anderson and Stewart over wages, 213–15; wages, 224
La Grande Rivière, *see* Big River
LAKES:
Athabasca (**IV**), 42, 51 n.3, 70, 76, 77, 78, 79, 81 n.4, 92 n.3
Beechy, *see* Beechey Lake
Huron District, 44 n.4
Marmonance, 73
MacDougall (**X**), 126–8, 128 n.2, 139, 140, 169, 176
Superior, 12, 109
Superior District, 40
Walmsley (**VI**), 26
Winnipeg (**IV**), 12, 161, 165 n.1, 211, 235, 245
La Loche Lake, 67, 73
La Loche River, 67, 73
Lamalaice, steersman, 159
Lancaster Sound, 6, 7, 8, 249
Land Arctic Searching Expedition, 50, 53; proposed composition of, 47, 48, 55; to winter on the coast if necessary, 49; composition of, 119, 171, 190; members, 108, 121, 172; casts positive light on Hudson's Bay Company, 196, 198; leaves many questions unanswered, 201; total bill for, 219; wages to members, 219, 223–4; bonus of £5 for good conduct paid to all members, 219, 228–9; other gratuities paid to members, 227–8 all members received Arctic Medal, 220; £3000 voted for financial year 1855–6, 220; assessment of, 256–8; distance travelled by, 256
Landrie, Alexander, 66, 151, 153, cutting wood, 154, 156; working in saw-pit, 158, 159; making a meat rack, 158; making meat scaffold, 158; sent after Stewart with his gun, 213; wages, 224; receives bonus as Stewart's servant, 219, 228
Lapierre's House, 41
larch (*Larix laricina*), 145
Laret-esch, arrives at Fort Reliance with his band, 156
Larsen, Inspector Henry, 254
lava flows, 109 n.7
Lefroy, Lieutenant John Henry, 11 n.1
Le Mesurier River, 136, 174
lemmings, Brown (*Lemmus sibiricus*), 128 n.6; Collared (*Dictostonyx torquatus*), 128 n.6;
Le Noir, B., 93
Leopold Harbour, *see* Port Leopold
Les Isles aux Morts, 102
Les Isles Brulés, 102
letter-clip, brass, 130, 173, 177, 187, 191; given to Franklin by his wife, 130 n.3
Le Vesconte, Henry, silverware recovered, 253
Lewes River, 40, 41
Liard River (**IV, VI**), 40, 71; still icebound, 88; breakup on 88, 91, 91 n.5
Libby, H.W., 58
limestone, 135
Little Athabasca River, *see* Clearwater River
Little Shaginnah, 164
Liverpool, 27, 35, 43 n.3
Lochbroom Parish, 32
Lockhart, James (**XIII**), biographical sketch, 43 n.2, 44 n.6, 76, 97, 103, 164, 213, 214; reaches Fort Carlton, 67; leaves Fort Carlton, 72; travels from Fort Chipewyan to Fort Resolution, 69, 78; feeling low, 73; to build winter quarters at Fort Reliance, 74 n.5, 119 n.8, 149; to collect provisions (fish and caribou) for a possible wintering, 75; to transport boats, provisions and clothing to head of Back River, 75, 97, 100, 144 n.5, 147 n.4, 159 n.1; sand cliff named after, 141, 149; misses Anderson and Stewart, 147 n.3, 151, 159 n.2; recalled to Fort Reliance, 147 n.4, 151; returns to Fort Resolution, 148, 151; leaves Fort Resolution, 149, 151; reaches Fort Reliance, 149; reuses Back's fireplaces and chimneys, 149, 153–4; reconnoitres route to Artillery Lake via Lockhart River, 149, 155; reconnoitres Pike's Portage, 149, 156; fetches supplies from Fort Resolution, 149; starts north with boats for head of Back River, 151, 159, 176, 186; praised by Anderson, 151; finds messages from Anderson and Stewart, 152; rebuilds Fort Reliance, 154, 156, 185; buys canoe from Laret-esch, 156; starts for Fort Resolution to meet boats with supplies, 157; reaches Fort Resolution, 157; starts back for Fort Reliance, 157; gets back to Fort Reliance, 157; hunting caribou,

158, 159; hires Indians to go to head of Back River, 158; reconnoitres south end of Pike's Portage, 158; smoking meat and salting tongues, 159; recalled to Fort Reliance by Anderson, 160, 176, 186; ordered to close up Fort Reliance and return to Fort Resolution, 160; starts from Fort Reliance for Fort Resolution, 160; accompanies Stewart with dispatches to Fort Garry, 161, 162, 177; shoots eagle, 164; highly commended by Anderson, 170, 180; Anderson would like to have him as an officer in Mackenzie River District, 170; wages received, 223
Lockhart River (**XII**), 108 n.5, 145 n.1, 153 n.5, 171 n.1, 176, 185; totally impassable, 145 n.1, 149, 155; Lockhart reconnoitres, 149, 155; spectacular falls and rapids on, 155, 155 n.2
London, 1, 11 n.1, 32 n.1, 254
Londonderry, 67 n.1
Long Rapids, 142
loons (Gaviidae), 113, 131, 134; Arctic (*Gavia arctica*), 113 n.5, 135; Common (*G. immer*), 113 n.5; Red-throated (*G. stellata*), 113 n.5; Yellow-billed (*Gavia adamsii*), 113 n.5
Lop Stick Point, 94
Lord Mayor Bay, 96
Lord Mayor's Harbour, *see* Lord Mayor Bay
lupins (*Lupinus arcticus*), 124

Macbean, G.A., silverware recovered, 19, 20
McBeath, Adam, 40
McClintock, Francis Leopold, xii, 9, 255; sledge journeys of, 8, 249; solves the Franklin mystery, 63 n.4, 90 n.5; finds only document which throws light on fate of Franklin expedition, 63 n.4, 90 n.5, 197; publishes charitable assessment of Anderson's and Stewart's expedition, 197, 210; criticizes Hudson's Bay Company for not supplying Inuktitut interpreter, 197, 210; finds ptarmigan bones left by Parry on Melville Island, 201; takes part in James Ross's expedition, 1848–9, 249; takes part in Austin's expedition in *Assistance*, 1850–51, 249; commands *Intrepid*, 1852–4, 249; commands search expedition in *Fox*, 249–54, 257; sails from Aberdeen, 249; undertakes reconnaissance sledge trip, 249; buys silverware and buttons from Inuit at Cape Victoria, 249; mounts major sledge trip, 250–54; searches east coast of King William Island, 250; buys more silverware from Inuit on east coast of King William Island, 250; finds various wooden artefacts from missing ships near Point Booth, 250; visits Montreal Island, 250–51; crosses Simpson Strait to King William Island, 252; finds first skeleton, 252; finds no messages at Cape Herschel, 252; finds cairn with message from Hobson, 252; finds boat and sledge at Erebus Bay, 253; finds two skeletons in boat, 253; finds vast amount of equipment and clothing abandoned at Crozier's Landing, 253–4; returns to London, 254
M'Clure, Captain Robert, 10, 14, 16, 22, 36, 62 n.2, 204, 206; court martial, 25
M'Clure Strait, 10
McCormick, Robert, offer to lead expedition rejected, 205
McDermot, Sarah, 43 n.3
McDonald, Alexander, 4; silverware recovered, 19, 20, 249, 250
McDonald, John, recommended by Rae for expedition, 30, 52, 53
MacDougall, Lt Colonel, 126 n.6
McFarlane, Roderick, 37
McFarlane's Lake (**VII**), 113 n.1
McGill University, 44 n.3
Mackenzie, Alexander, 74 n.5
McKenzie, I., 164
Mackenzie, Roderick, 38
Mackenzie Delta, 5, 6, 10, 11 n.1, 71, 87 n.2, 121 n.7
Mackenzie Lake (**VII**), 105, 113 n.12
Mackenzie Mountains, 45 n.1
Mackenzie River (**IV, VI**), 3, 5, 27, 29 n.1, 30, 32, 37, 41, 42, 59 n.3, 71, 73 n.1, 91 n.5, 97, 98, 161, 162 n.2, 204; Upper, 81, 82, 85; breakup on, 91, 93; ice drifting on, 93–5, 98, 100
Mackenzie River District, 11 n.1, 29, 29 n.1, 30, 37, 38, 40, 41, 43 n.2, 45 n.1, 45 n.2, 47, 48, 50, 52, 55, 55 n.4, 56, 61, 67 n.1, 89 n.1, 91, 97 n.3, 99, 161, 215, 235, 241 n.2; left in Robert Campbell's care, 50, 99; geology of, 241 n.2
McKenzie's River Brigade, *see* Portage la Loche Brigade
McKay, James, 53, 92, 129 n.3; alleged to have killed Inuit near Point Ogle, 65 n.2
McKay, William, 44, 44 n.6, 48, 51, 53, 61 n.2; to

be second-in-command of expedition if necessary, 50
McKay, Mary, 45 n.4
McKay's Peak (**IX**), 129, 138, 138 n.3
McKay's Peak Rapid, 138
Mackinlay River, *see* McKinley's River
McKinley, Rear-Admiral, 124 n.7
McKinley's River (**IX, X**), 124, 140, 173
Mackinnon, C. Stuart, 42
McLellan, Murdoch, 44, 45, 55, 56, 69, 77, 108, 121, 173, 182 n.5; recommended by Rae, 30, 52; knowledge of Inuktitut, 106, 124 n.10, 125, 130, 173 n.1; receives bonus as Anderson's personal servant, 219, 229; wages, 224
McLellan Lake (**VII**), 113 n.12
McLennan, Murdoch, *see* McLellan, Murdoch
McLeod, Alexander, R., 44 n.4, 147 n.1; builds Fort Reliance, 153 n.5; informs Richard King of recent amicable interaction between Inuit and Chipewyans on the Thelon River, 240
McLeod, Donald, 66, 108, 121, 134, 172; wages, 224
McLeod, Sarah, 44 n.4
McLeod Bay, 153 n.5
Maconochie, Captain, 136 n.3
Maconochie Island (**XI**), 106, 136, 136, 170, 178, 245; searched, 136, 175, 178, 235, 248
MacTavish, Dugald, 43 n.3
MacTavish, John George, 43 n.3
MacTavish, William, 43, 55, 59, 69, 77, 78, 79, 80; biographical sketch, 43 n.3; appointed Governor of Assiniboia, 43 n.3; appointed Governor of Rupert's Land, 43 n.3; receives instructions from Simpson, 56
'Made Beaver', 93, 93 n.5, 94, 142, 154, 154 n.3
Magnetic Pole, 62, 86, 90, 168, 204
Maitland, Phelps & Co, 35
'mal de raquettes', 70
Malley, William, 122 n.2
Malley's Rapids, 122
maps, drafted by Anderson, 104, 105
Margaret Lake (**VII**), 105, 113 n.13
marmots (*Marmota* spp.), 113, 113 n.15, 115, 118
marsh marigold, 101, 101 n.2
Marshal's Point, 165
marten (*Martes americana*), 94
Martin, Captain Robert, 5
mathematical instruments, provided for expedition, 36
Maufelly, 119 n.2
Meadowbank, Lord, 128 n.10
Meadowbank River (**IX**), 128
Melville Bay, 5, 6, 249
Melville Island, 1, 8, 9, 201, 234 n.1
Melville Peninsula, 11 n.1, 28 n.2, 255
Mercy Bay, 9, 10, 62 n.2
Methy Portage, *see* Portage la Loche
Mexico, 6
Michipicoten, 40, 44 n.3
Miles, clerk at Fort Simpson, 71, 82, 91, 170; sends provisions for Anderson's party, 94
Miles Lake (**V, VII**), 112 n.11
Mills, *see* Miles
Mindota, 58
Mingan, 38
Minnesota Territory, 58, 61
Minnesotian, 167, 187–8
Misteagun Lake (**VII**), 113 n.12
Misteagun, Thomas, *see* Mustegan Thomas
Mistegan, Thomas, *see* Mustegan, Thomas
Mistigan, Thomas, *see* Mustegan, Thomas
moccasins, requisitioned from Fort Chipewyan, 84
Monson Lake, 164
Montagnais, 163, 164
Montour, Ignace, 59 n.4, 108, 121, 144, 168, 172; signs contract, 58, 59; praised by Anderson; had made two round trips from Montreal to Red River prior to joining the expedition, 180; selected by Anderson to relay dispatches from Red River to Lachine, 181–2, 213, 214, 215; Simpson considers unreliable because of an alcohol problem, 216; wages, 224
Montreal, 26, 31 n.1, 43 n.1, 44 n.3, 59, 91 n.2, 131 n.8, 161, 166, 167, 180, 191, 212
Montreal Department, 50, 53, 56
Montreal Herald, 167, 189–92
Montreal Gazette, 167, 192–4
Montreal Island (**XI**), 2, 20, 22, 86, 89, 105, 131 n.7, 137, 137 n.3, 179, 190, 193, 205, 207, 245, 255; searched, 131–4, 174, 177, 180, 184, 188, 204; location where boat had been cut up, 106, 133 n.4, 174, 177, 184, 186, 190, 193, 208, 257; Franklin artefacts found on, 133, 133 n.4, 170, 177, 184, 208, 257; ice solid beyond, 135, 168, 178, 257; King leaves cache on, 197; natural location for Franklin expedition members to wait for breakup on the Back River, 203; visited and

searched by McClintock, 250–51; some metal articles from Franklin expedition found, 250; geology of, 251; appeared to have been inundated by the sea, 251
Montresor River (**IX**), 129
Montresor, Lt General Thomas, 129 n.2
Moore, Captain Thomas, 5
Moose Factory, 11 n.1, 37
Moose Lake, 53
moostigues, *see* sandflies
Morrin's, 73
Morrison, William, obtains statement *re* Inuit allegedly killed near Point Ogle, 65 n.2
mosquitoes (*Aedes* spp.), 69, 78, 95, 101–2, 107, 112 n.12, 113, 115, 116 n.7; 118 n.4, 119
Mount Erebus, 2
Mount Meadowbank (**IX**), 128, 129
Mount Royal Cemetery, 32 n.1
Mountain Portage (**V, VII**), 92, 104, 105, 108, 108 n.5, 111, 147, 153, 155, 171, 177, 183; anticipated advantages of, 100, 103; Indian assesses as difficult, 108 n.5; ascent of, 112–15, 171; location of start of, 112; landscape of, 112, 115, 181
Mountain Rapids (Slave River), 162 n.7; (Churchill River) 164
mouse, beaver, 128
Muddy Lake, 164
Mull, 45 n.1
Munro (potential Inuktitut interpreter), 69; fails to arrive at Churchill, 79, 80; finally appears, 80; wanted for another expedition, 80
musk oxen (*Ovibos moschatus*), 13, 119 n.2, 119 n.9, 121, 122, 123, 128, 128 n.7, 139, 141, 141 n.6, 142, 176; distribution of, 121 n.7, 124; near-extermination of, 121 n.7; recovery of, 121 n.7; bull killed, 122; bull escaped, 142; meat dubious, 122, 122 n.6; calves, 123; hides used for Inuit boot soles, 124; hides used for tents, 130; Utkuhikalingmiut reliant upon, 130 n.1; skins, 173; shot by Yellowknives and found to contain musket balls, 237
Muskox Lake (**VIII**), 119, 142, 143, 172, 176, 183
Muskox Rapid (**VIII**), 119, 142, 156, 172, 183
Mustegan, Thomas, 14, 44, 45, 55, 56, 69, 108, 121, 142, 143, 172, 182 n.5; recommended by Rae, 28, 30, 52, 77; knowledge of Inuktitut, 106, 124 n.10, 125, 130, 173 n.1; lame, 134; one of the group who searched Maconochie Island, 246; reconnoitring route, 143, 145; very reliable guide, 148; familiar with Churchill (English) River, 213, 215; wages, 224; member of group, one of whom had seen ship off arctic coast, 235, 245; chief of Ojibway band at Norway House, 245; reported that after expedition Paulet Papanakies repeatedly affirmed he had seen a ship far out in the ice, 246; believed Papanakies's story, 247
Mustegon, Thomas, *see* Mustegan, Thomas

Nahoway, 45 n.4
Nakessie, 236
Napier, Joseph, accuses Government for inadequate support for expedition, 196, 199; Lady Franklin's chief supporter in House of Commons, 199 n.1; presses Government for loan of a ship and captain to mount a further search, 199 n.1; suggests some members of Franklin expedition might still be alive, 199
Napoleonic Wars, 3
Nardarl-yousa, guides Oman, 79–80; delivers Oman's letter of introduction, 81
Nar darl yousa, *see* Nardarl-yousa
Nar,darl,yousa, *see* Nardarl-yousa
Narrows, Garry Lake, 125, 139
Navy Board Inlet, 6
Neatby, Leslie H., 255
Nelson Head, 10
net sinks, *see* net weights
net weights, 151
nets, making, 154–5
Netsilingmiut, 124 n.11
Newcastle, Duke of, 210
newspaper accounts of expedition, 167, 187–94
New York, 7, 33 n.3, 35
Nipigon, 37, 109 n.7
Nipigon Bay, 109, 111
Northern Department, 43 n.3, 59; Council of, 32 n.1, 42, 50, 215
Northern Express, 72
Northumberland Sound, 9, 234 n.1
North West Company, merger with the Hudson's Bay Company, 32 n.1, 44 n.3, 45 n.1
Northwest Passage, xi, 1, 2, 26; location of predicted by Richard King, 206
North Pole, 3

North Saskatchewan River (**IV**), 42, 59 n.2, 72 n.2
North Shore, 38, 44 n.3
North Slope, 11
North Somerset, *see* Somerset Island
Norway House (**IV**), 12, 30, 32 n.1, 40, 41, 42, 43, 43 n.3, 44, 44 n.3, 45 n.4, 48, 51, 52, 53, 54, 55, 56 n.3, 57, 67 n.1, 69, 76, 76 n.2, 77, 79, 81, 86 n.1, 91, 92 n.1, 142, 149, 161, 165, 180, 182, 198, 211, 213, 214, 217, 238, 245, 256
Norway House District, 45 n.4
noyé, *see* Rapids of the Drowned
Nut Lake, 44 n.6

Ogle, Vice-Admiral Sir Charles, 136 n.1
Ojibways, 245, 246
Okhotsk, 32 n.1
Oman, Margaret, 70
Oman, William, selected as Inuktitut interpreter, 70, 76 n.2, 78; biographical sketch, 70; abilities as an interpreter, 70, 99; sets off from Churchill for Fort Chipewyan, 70, 78, 79–80; forced to abandon journey, 70, 80, 256; letter of introduction, 70, 79; returns to Churchill, 70, 80–81; wages, 79, 224; Chipewyan guides for, 79; probable route, 91 n.1
Oman, William, senior, 70
Ommanney, Captain Erasmus, 6, 249; finds traces of Franklin expedition, 7
Oot-ko-hi-ca-lik (Inuktitut name for Back River) 20, 22, 46
Oot-koo-i-hi-ca-lik, *see* Oot-ko-hi-ca-lik
Operation Franklin, 254
Orkney Islands, 11 n.1, 44 n.4, 45 n.4, 70, 241 n.2
Orkneymen, Stewart hires two at Fort Chipewyan, 76
Osborn, Captain Sherard, 6, 9; biographical sketch, 234 n.1; finds traces of Franklin expedition, 7; interprets evidence, 196; believes Franklin survivors had refashioned boat on Montreal Island for ascent of Back River, 196, 202; suggests further traces should be sought up the Back River, 197, 202; accurately predicts where *Erebus* and *Terror* were abandoned, 197; argues that Franklin would have left records in a cairn, 201; argues that Inuit not responsible for cutting up boat on Montreal Island, 202; believes Inuit may have taken advantage of weakness of Franklin expedition members, 202–3; does not completely rule out possibility of some survivors, 203; urges British government to continue the search, 203; experience of sledging, 203–4; accurately predicts location of *Erebus* and *Terror*, 204; offer to lead expedition declined, 205; writes to *The Times* concerning rumour of Indians finding a camp recently occupied by Whites on Back River, 238–9, 240, 242; alleges that Anderson had informed Simpson about the rumour, 238, 240, 242; claims that the rumour was widespread in Red River Settlement, 241, 242
Osborne, B. Admiralty, 227
Osmer, Charles, 4
osprey (*Pandion haliaetus*), 109, 109 n.5, 109 n.9
Otter Rapids, 164
Ouligbuck, William, 45, 48, 51, 53, 55, 57, 60, 70, 99; as Rae's interpreter, 14, 56; recommended by Rae for further expedition 26, 28–9, 30, 30 n.2, 52, 69; sought after to be Inuktitut interpreter, 43, 44, 48, 59, 69; ordered to proceed from Churchill to Fort Chipewyan, 56, 69; reliability as interpreter questioned, 65, 65 n.1; runs away from Rae and party, 64, 64 n.2; makes himself unavailable, 69, 70; fails to appear at Churchill, 78, 79, 80; wanted for another expedition, 80
Outram Lake, 105, 115 n.7
Outram River, 115, 115 n.7, 171–2
Oxford House, 42, 45 n.1, 45 n.2

Pacific Ocean, 5, 9, 25, 74 n.5
paddles, 84, 107 n.7, 128 n.3, 140 n.5, 171; Inuit, 124, 138
Pambrun, P. C., 41
Panama, 5
Panpoumakis, Paulet, *see* Papanakies, Paulet
Papanakies, Paulet, 69, 77, 108, 121, 172, 182 n.5; carries letter of introduction from Barnston, 76 76 n.2, 77; leads Norway House voyageurs, 77, 77 n.2; one of the group who searched Maconochie Island, 136 n.7; 246; rewarded as steersman, 142, 228; wages, 224; saw masts of ship in the ice off arctic coast, 235–6, 245, 246, 247; decided not to tell Anderson or Stewart of this discovery, 245, 246, 247; was not mistaken since he

had frequently seen ships at York Factory, 246; Ojibway, 246
Parry, Captain William Edward, 61, 63; first expedition, 1; second expedition, 3; third expedition, 4, 86 n.5, 90 n.4; examines relics from Franklin expedition, 7; ptarmigan bones left by on Melville Island found by McClintock, 201
Parry Channel, 1, 9
Parry Falls (**XII**), 149, 155, 155 n.2
partridge, White, *see* ptarmigan
Pas Mission, 164
Paulet's Rapids, 142 n.2
Paupaunekis, Paulet, *see* Papanakies, Paulet
Peace River (**IV**), 74, 75
Peace-Athabasca Delta, 92 n.4
Pechell, Sir J.B., 135 n.1
Peddie, John S, silverware recovered, 19, 21
Peel River, 45 n.1
Peel River Post, 10, 41, 43 n.2, 45 n.1, 83 n.2
Peel Sound, 62, 63, 204
Peers, Augustus, 41
Peffer River, 252
Pegler, Harry, 252
Pelican Lake, 164
Pelican Rapids, 162 n.7
Pelly, Sir John Henry, 125 n.5
Pelly Banks Post, 40, 41
Pelly Bay, xii, 11 n.1, 13, 14, 17, 19, 21, 22, 23, 46, 56 n.2, 64 n.4, 106, 207
Pelly Lake (**IX, X**), xi, 27, 125, 125 n.5, 140, 141, 169, 173, 183
Pelly Point, 8
Pelly River, 40, 60
Pembina, 52, 58
Pembina River, 73
pemmican, 42, 54, 72 n.2, 81, 83, 87, 88, 97, 113, 116; caches of, 31, 90, 97, 122, 123, 124 n.2, 125, 126, 128, 129, 130, 138; stolen by wolves, 137, 137 n.10; rotten, 159, 182; brought back, 176
Penny, Captain William, 6, 8; finds traces of Franklin expedition, 7; experience of sledging, 204
permafrost, 101, 101 n.3
Peter Pond Lake (**IV**), 67, 72
Peterhead, 5, 204
Peters, Mrs, 157
Petersen, Carl, 249, 250, 251; finds Inuit cache of metal objects on Montreal Island, 250
Pethei Peninsula (**VI**), 109, 109 n.7
Pethenent, *see* Pethei Peninsula
Phillips, Commander Charles, 8
Phinn, Thomas, Secretary, Admiralty, 210, 225–31, 241, 245
Piché's Band, 163
Piche's House, 73
Pierre au Calumet, 73, 163
pike, Northern (*Esox lucius*), 115, 115 n.2, 152
Pike, Warburton, 119 n.9
Pike's Portage (**XII**), 108 n.5, 145–6, 145 n.4, 151, 156 n.1, 171 n.1, 176 n.2; variants of, 147 n.3; Lockhart reconnoitres, 149, 156
Pim, Lieutenant Bedford, 9, 10, offer to lead expedition declined, 220, 244
Pine Lake, 164
Pine Portage, 163
pines (*Pinus* spp.), 144
pipe, 94, 94 n.2
Pipe Stone Point, 147, 152, 157; water depth off, 158
plants, in flower, 109
plovers (Charadriidae), 118, 129, 134, 135
POINTS:
Aigle, *see* Point Ogle
au Foin, 94
Backhouse (**XI**), 137, 175, 176
Barrow, 6, 10, 11
Beaufort (**XI**), 130, 137 n., 173, 174, 175, 183, 190, 193
Booth, wooden artefacts from missing ships found by McClintock, 250
de la Guiche, 14
de St. restaux, *see* Point Saresto
Herschell, *see* Cape John Herschel
Keith, 109
Ogle (**IX, XI**), 20, 22, 27, 65 n.2, 82, 86, 89, 90, 106, 136 n.9, 142, 193, 205, 207, 208, 235, 236, 255, 256, 258; searched on foot, 135–6, 135 n.5, 174, 178, 190; landscape of, 135–6, 175; island at high tide, 136, 136 n.2; possible fate of bodies at, 136, 136 n.9, 168, 175, 179, 180, 185; Franklin artefacts found on, 136, 175, 178, 184, 200; visited by McClintock, 250
Pechell (**XI**), 106, 135, 136, 174, 178
Richardson (**XI**), 135 n.5, 136, 175, 178, 248
Saresto, 162 n.3
Turnagain, 3
Victory, 252, McClintock leaves messages in cairn at, 253, 257
Pointe au Gravoir, 72

Pointe au Saline, 73
Pointe aux Trembles, 163
Pointe des Roches, 101–2, 103, 108, 162
poles, 107 n.7, 142, 171
poling, 142 n.3
poplars (*Populus* spp.), 93; first seen on return, 147
Porcupine River, 41
Portage la Loche (**IV**), 40, 67, 69, 73 n.1, 76 n.2, 86, 89 n.1, 97, 97 n.3, 161, 161 n.2, 163, 163 n.7; Brigade, 54, 55 n.4, 86 n.1, 99, 149, 151, 151 n.2, 198, 211, 215
Port Bowen, 4, 6
Port Leopold, 5, 6, 8, 10, 204, 249
Port Kennedy, 249
Portsmouth, 254
potato crop, 162
Prairie Portage, 74
Presque Islands, 102
Prince of Wales Island, 4, 8, 204, 234 n.1
Prince of Wales Strait, 10
Prince Patrick Island, 9
Prince Regent Inlet, 1, 4, 5, 96, 168, 178, 249
Princess Royal Islands, 10
Prise, 164
Provideniya, 5
Pruden's Lake (**VII**), 113 n.1
Prudhoe Bay, 3
ptarmigan (*Lagopus* spp.), 13, 111, 113 n.14, 115; Rock (*L. mutus*), 111 n.3, 135, 135 n.9, 139, 143, 251; Willow (*L. lagopus*), 111 n.3, 121, 153, 153 n.3
Pullen, William J.S., 6, 9, 71; describes tar springs, 163 n.5

Quebec City, 38, 45 n.2, 51
Queen Maud Gulf, 11
Quoich River, 13

rabbits, *see* hares
Rabbit Point, 111, 183
Rabbitskin River, 93
Rae, Dr John, 48, 49, 52, 53, 54, 55, 56, 59, 61, 63, 86 n.4, 90, 96, 97, 101, 192; biographical sketch of, 11 n.1; earlier searches for Franklin, 5, 8, 11, 11 n.1, 27, 32 n.1, 41, 86 n.4, 87 n.2, 90, 220; finds artefacts from Franklin expedition, xii, 15, 19–21, 22, 23–4; hears Inuit accounts of Franklin expedition, 13–15, 19–20, 21–2, 32, 32 n.1, 46, 82, 207; reputation as a walker, 11 n.1; first arctic expedition, 11 n.1; as a surveyor, 11 n.1; plan of last expedition, 12; details of last expedition 13–15; game bag during wintering at Repulse Bay, 13; recovers silverware from Franklin expedition, 15, 19–21, 22–3, 46, 189; returns to England, 15; reports his findings to *The Times*, 15, 18–19; reports his findings to the Admiralty, 15, 17, 19–21, 46, 256; reports his findings to the Hudson's Bay Company, 15–16, 17, 19, 21–4, 31; criticized by Victorian England, 16, 92, 189; responds to criticisms, 16–17; exonerated, 18; hands over Franklin artefacts to Hudson's Bay Company, 20, 21; tenders advice on a further search, 26, 28–31, 35; declines command of new expedition, 26, 60, 200; recommends depot at east end of Great Slave Lake, 28; uses Halkett boat, 28 n.2; recommendations *re* sails, 30; recommends hiring Back River Inuit, 30; recommendations *re* voyageurs, 30, 52, 53; recommendations *re* instruments and charts, 30–31; held up as a model to follow, 49; confirms Ouligbuck's reliability, unless likely to stand to gain, 65 n.1; refuses to winter on arctic coast, 87, 89; comfortable wintering at Repulse Bay, 87, 89; recommendations *re* canoe length, 28, 31, 47 91 n.6; supported by Anderson *re* reports of cannibalisms, 95; did not expect to find any traces of Franklin expedition, 207
Rae, John (senior), 11 n.1
Rae, Margaret, 11 n.1
Rae Falls (**VII**), 113, 113 n.9
Rainy Lake, 12, 45 n.4
Rainy Lake District, 45 n.4
Ranford, Barry, 255
Rapid Portage, 74
Rapid River, 55; Portage, 164
Rapide a Pierre, 73
Rapide Couche, 164
Rapids of the Drowned, 162 n.8
Rasmussen, Knud, 254
Rat Indians, *see* Gwich'in
Rat Lodge, 103, 108, 108 n.5, 144, 144 n.4
ravens (*Corvus corax*), 95 n.2, 121 n.2, 124 n.9
Red River (Athabasca basin), 73
Red River Colony, *see* Red River Settlement
Red River District, 45 n.2; Lower, 44 n.4

Red River Rebellion, 43 n.3
Red River Settlement, 28, 29 n.1, 30, 43, 44 n.3, 44 n.4, 46, 48, 51, 52, 53, 53 n.2, 54, 55, 57, 58, 59, 60, 61, 66, 67, 75, 87, 92, 142, 161, 165, 180, 191, 211, 214, 241 n.2; rumour of an Indian report of a camp recently occupied by Whites on the Back River allegedly widespread there, 241
Regina Inlet, 62
Reid, James, 4
Reid, Will, 108, 121, 173; finds first Franklin artefacts on Montreal Island, 133, 133 n.4; receives reward for finding first traces, 134, 219, 228; wages, 224
Reindeer Lake, 91, 164 n.2
reindeer moss, *see* caribou lichen
Rendall, John, made snowshoes for Franklin expedition, 133 n.4
Repulse Bay, xii, 11 n.1, 13, 14, 16, 17, 18, 19, 23, 26, 28, 28 n.2, 30 n.2, 56 n. 2, 106; Rae's comfortable wintering at, 87, 89–90; a paradise compared to Chantrey Inlet, 168, 180, 193–4
Return Reef, 3
Reward, offered by Anderson for finding first traces of Franklin expedition, 134; offered men for good performance, 142; for steersmen, 142
rheumatism, men complaining of, 138
Richards, Commander George, court martial, 25; experience of sledging, 204
Richardson, Sir John, 5, 28 n.2, 61, 63, 87, 89, 136 n.6; examines relics from Franklin expedition, 7; searches for Franklin, 11 n.1, 32 n.1, 45 n.1, 220; accurately predicts where Franklin ships were wrecked, 63, 63 n.4; boats of, 97, 97 n.2; acknowledges receipt of a copy of Anderson's journal, 199; makes deposition in connection with Fairholme case, 199; submits extracts of Anderson's journal for publication by the Royal Geographical Society, 199–200; holds out little hope of records of Franklin's expedition being recovered, 200
Riel, Louis, 43 n.3
Riel Rebellion, 72 n.2
Rivière au Barrier, 73
Rivière Embaras, *see* Embarras River
Rivière la Loche, *see* La Loche River
Roberts, Edward Boyd, sold snowshoes to Stanley, 133 n.4
robin, American (*Turdus migratorius*), 113, 113 n.4
Robinson, Lieutenant Frederick, finds few stores left at Fury Beach, 204
Rock Rapids, 128, 139, 139 n.6
rock weathering, 111, 115, 119
Rocky Island, 108, 151
Rocky Point, 108 n.2, 152, 157, 171, 183
Roe's Welcome Sound, 13, 15
Romaine, W. G., Admiralty, 231–2
Ross, Bernard Rogan, 67, 69, 74, 78, 81, 82, 84, 89, 91, 92, 100, 103, 162; biographical sketch, 67 n.1 as possible expedition leader, 26, 30; as naturalist, 67 n.1; to accompany Stewart if Anderson doe not arrive in time, 75; recommended by Anderson for promotion, 87, 181; reaches Fort Resolution from Portage la Loche, 157; instructs Stewart in taking observations, 169
Ross, Christina, 67 n.1
Ross, Donald, 43 n.3, 67 n.1
Ross, Captain James Clark, 2, 14, 61, 63; explores north coast of King William Island, 1; Antarctic voyage of, 2, 4; commands Franklin search expedition, 5, 249
Ross, Captain Sir John, second expedition of, 1, 26, 27, 86 n.5, 204; mounts Franklin search expedition, 6, 8; uses stores from Fury Beach depot, 204
Ross, John (HBC), 52
Ross, John (Royal Artillery), testifies *re* Inuit allegedly killed near Point Ogle, 65 n.2
Ross Ice Shelf, 2
Ross Lake (**VII**), 105, 113 n.12
Ross Sea, 2
Rowand, Dr John, 180
Rowley, Graham, 65 n.2
Roy, François, 72
Royal Artillery, 26
Royal Canadian Mounted Police, 254
Royal Canadian Regiment, 255
Royal Geographical Society, publishes extracts from Anderson's journal, 200
Royal Navy, Crimean War operations, 25
Royal Scottish College of Surgeons, 11 n.1
Royal Society Museum, 44 n.3, 67 n.1
Rupert's Land, Governor of, 43 n.3, 84
Russell Island, 4
Russia, 32 n.1
Russian American Company, 32 n.1

Sabine, Colonel, 63; examines relics from Franklin expedition, 7
Sabine Peninsula, 234 n.1
Sageh, 155
sailing, 125, 137, 138, 139, 140, 144, 145, 147, 151, 162, 163, 164; directions in Garry Lake, 126, 126 n.1
salt, source of for Mackenzie basin, 74 n.4
Salt Plains, 74 n.4, 74 n.5
Salt River, 69, 74, 74 n.5, 78, 162
sand bars, 140, 141
Sand Cliffs (**IX**), 123, 141
sandflies, 112, 113, 115, 139
Sandhill Bay (**VIII**), 116, 118, 118 n.1, 172
Sandhill Rapid, 128
sandstone, 116, 118
sandstorm, 125
Sandwich Islands, 1, 10
Sandwick, 70
Sandy Hill Bay, *see* Sandhill Bay
Sandy Portage Lake, 112 n.4
San Francisco, 37
Saskatchewan, 44 n.6, 241 n.2
Saskatchewan River (**IV**), 161, 164 n.3
Saskatchewan River District, 42, 45 n.2, 45 n.4, 52, 53, 54 n.1, 57 n.3
Saskatoon, 59 n.2, 72 n.2
Saugeen, 45 n.1
Saulteaux, refuse to go beyond Fort Reliance, 159
Sault Ste Marie, 11 n.1, 12, 43 n.3, 44 n.4
Sault St Mary's, *see* Sault Ste Marie
Saunders, Captain James, 6
Sauvé, Norbert, steersman, 97; delivers boats from Fort Simpson to Fort
Reliance, 159 n.1
scenery, attractive, 123, 123 n.3
Schwatka, Lieutenant Frederick, 106, 107 n.1, 252, 254
Scotland, 44 n.4, 45 n.1
Scottish Court of Session, 167, 194, 194 n.1, 212
Scottish Highlands, 32 n.1
Scottish Record Office, 194 n.1
scurvy, Rae's treatment of, 11 n.1
sea ice, 106; prevents expedition from getting beyond Maconochie Island, 106
seals, 13, 131, 133, 137; Ringed (*Phoca hispida*), 131 n.3; Bearded (*Erignathus barbatus*), 131 n.3
searches, of Montreal Island, 133–4, 136 n.9; of west shore of Chantrey
Inlet, 134–6, 134 n.6, 135 n.2, 136 n.9
Selkirk, Lord, 45 n.2
Sept Iles, 45 n.1
sextants, provided for expedition, 36
Sevastopol, 25
Sheerness, 25
Shepherd, John, Deputy Governor of the Hudson's Bay Company, 32, 35, 225;
Simpson reports progress to, 59
SHIPS:
- *Advance* (USS), 7, 8
- *Assistance* (HMS), 6–7, 9, 16, 25, 234 n.1, 249
- *Bellerophon* (HMS), 3
- *Blossom* (HMS), 4, 122 n.13
- *Breadalbane* (HMS), 62 n.2
- *Dorothea* (HMS), 3
- *Enterprise* (HMS), HMS, 5, 10–11, 17, 25, 249, 258; proposed search for, 27, 28, 31 n.4, 59, 61 n.1, 82 n.5, 83, 83 n.2
- *Enterprise* (whaleman), 5
- *Erebus* (HMS), xi, 1, 4, 11, 20, 21, 32, 194 n.1; description of, 2; previous polar service of, 2; modifications to, 2–3; steam machinery of, 2–3; provisioning of, 3; name carved on chip of wood on Montreal Island, 133, 133 n.2, 133 n.4; abandoned off King William Island, 252
- *Felix*, 6, 7, 8
- *Fox*, xii, 90 n.5, 249, 254
- *Fury* (HMS), 4, 86, 86 n.5, 90, 90 n.4, 96, 101
- *Griper* (HMS), 1
- *Hecla* (HMS), 1, 4, 86 n.5
- *Herald* (HMS), 5, 6
- *Intrepid* (HMS), 6–7, 9, 10, 16, 25, 234 n.1, 249
- *Investigator* (HMS), 5, 10–11, 16, 25, 62 n.2, 204
- *Lady Franklin*, 6, 7, 8
- *Mary*, 6
- *North Star* (HMS), 6, 9, 10, 16, 234 n.1
- *Phoenix* (HMS), 9, 16
- *Pioneer* (HMS), 6–7, 9, 16, 25, 23, 24 n.1
- *Plover* (HMS), 5, 6
- *Polyphemus* (HMS), 3
- *Prince Albert*, 6, 8–9
- *Prince of Wales* (HBC), 11 n.1, 15, 18
- *Prince of Wales* (whaleman), 5
- *Rescue* (USS), 7, 8
- *Resolute* (HMS), 6–7, 9, 10, 16, 25, 234 n.1
- *Sophia*, 6, 7, 8
- *Talbot* (HMS), 9, 16
- *Terror* (HMS), xi, 1, 11, 16, 20, 21, 32, 86 n.5,

249, 252; description of, 2; previous polar service of, 2, 4; modifications to, 2–3; steam machinery of, 2–3; provisioning of, 3; name carved in piece of wood found on Montreal Island, 133, 178, 184, 187, 191, 193, 208; boat allegedly found with ship's name on it, 188; abandoned off King William Island, 252
Trent (HMS), 3
Victory, 1, 86 n.5, 204
Shoal Islands, 165
signal fires, 145 n.1
silverware, from Franklin expedition, recovered by Rae, 15, 19–21, 22–3, 46, 189; bought from Inuit by McClintock, 249, 250; found in boat at Erebus Bay, 253
Simpkinson, Lady Mary, 63, 63 n.3
Simpson, Frances, 32 n.1
Simpson, Geddes Mackenzie, 32 n.1
Simpson, James W., 58
Simpson, Mary, 32 n.1
Simpson, Sir George (**II**), 11 n.1, 12, 27, 28, 29, 29 n.1, 31, 32, 37, 38, 41, 42, 43 n.3, 44, 44 n.3, 45, 53, 54, 56, 57, 58, 63, 65 n.2, 67, 70, 71, 73, 75, 80, 81, 82, 83, 85, 87, 88, 98, 99, 102, 161, 234; biographical sketch, 32 n.1; travels of, 32 n.1; correspondence, 32 n.1; character, 32,n.1; liaisons, 32 n.1; as supporter of arctic exploration, 32 n.1; knighthood, 32 n.1; asked to mount a search expedition, 32–5; asked to select expedition leaders, 35; receives Company's instructions, 37, 190, 256; employs James Stewart, 40; selects Anderson and Stewart to lead expedition, 46, 190, 192; dispatches instructions to Anderson and Stewart, 42, 46–50, 190; private letter to Anderson, 50; private letter to Stewart, 51–2; private letter to John Ballenden, 52; reports progress to London, 59; suggests that Rae be persuaded to take command of expedition, 60; recommends training Inuit as interpreters, 81; sends further supplies for expedition, 99; satisfied with results of expedition, 196; congratulates Anderson on outcome of expedition, 198; presses Anderson's and Stewart's claim for special remuneration, 198; sends Anderson elastic stocking and knee cap, with instructions for fitting, 198–9; sides with Stewart with regard to Anderson's criticisms, 212, 215–16; advises Anderson to withdraw his official complaint about Stewart, 212, 216; threatens Stewart with expulsion from the Company for drunken brawling, 212, 216–18; efficiency of his information-gathering network, 212; considers Montour unreliable, in view of his problems with alcohol, 216; reminds Anderson he has no authority to make recommendations about Stewart's expenses, 216; submits bills to Hudson's Bay House for total of £6642.18.6, 221; denies that Anderson reported a rumour about Indians finding a camp recently occupied by Whites, 235, 240
Simpson, Thomas, 1, 11 n.1, 12, 14, 32 n.1, 86 n.3, 89, 206, 252; locates King's cache on Montreal Island, 197
Simpson Group, 109
Simpson Peninsula, 11 n.1
Simpson Strait (**IX, XI**), 2, 248, 252, 256
Sinclair, George, steersman with Back, 85, 92, 128 n.1; alleged to have killed Inuit near Point Ogle, 65 n.2
Sinclair, Samuel, recommended by Rae for expedition, 30, 52
Sinclair, William, 45, 45 n.4, 54, 591; receives instructions from Simpson, 57
Sinclair, William, Senior, 45 n.4
Sinclair's Falls (**X**), 128, 139
Sitka, 10, 32 n.1
slate, 122, 123
Slave Lake, *see* Great Slave Lake
Slave River (**IV**), 67, 69, 74 n.2, 74 n.3, 74 n.4, 75, 78, 83, 102, 102 n.2, 103, 157, 161; breakup on, 75, 84; Rapids, 162 n.6
sledges, proposed for transporting canoes, 75, 75 n.2; 75, 92
Smith, Malcolm, testifies *re* Inuit allegedly killed near Point Ogle, 65 n.2
Smith, William. G., Secretary, Hudson's Bay Company, 35, 186, 212, 218, 224–32, 234, 241; writes to *The Times* and to the Admiralty to the effect that no information had reached Hudson's Bay House concerning Indian rumour of a camp recently occupied by Whites on the Back River, 239, 245; suggests that the camp in question was probably one of Anderson's and Stewart's, 239
Smith Sound, 6
Smithsonian Institution, 44 n.3, 67 n.1
snow, falling, 105, 137, 142, 144, 162, 163, 165, 175, 191; banks, 113, 122; patches, 115,

129; snow on ground, 136, 137 n.1, 145, 170, 191, 194; storm, 164
snowbirds, *see* Snow buntings
Snow buntings (*Plectrophenax nivalis*), 135
snowshoes, 11 n.1, 70; wooden parts found on Montreal Island, 133 n.4, 178, 184, 186, 188, 191, 193; manufacture of, 133 n.4
Somerset Island, 5, 8, 61, 86 n.5, 90 n.4, 204, 206
Southampton Island, 2
speed of travel, 106, 139, 256–7; compared to Back, 106, 131 n.1, 143 n.2
Spence Bay, 1
Spencer, clerk at Fort Carlton, 51, 54, 57
Spencer's River, 93, 162
Spider Islands, 165
spruce (*Picea* spp.), 101, 112, 144 n.2; roots, *see* wattap; description of at treeline, 115; last seen, 115; first seen on return, 143
St Andrew's, Church, 42; Parish, 42
St Boniface, 45 n.2
St Lawrence, Gulf of, 38, 44 n.3
St Magnus Cathedral, 11 n.1
St Paul (Minnesota), 29, 43 n.3, 54, 161, 167, 212; Postmaster at, 58;
Agent, American Fur Company at, 58
St Petersburg, 32 n.1
Stanley, Stephen, surgeon aboard *Erebus*, 133; name carved on piece of wood on Montreal Island, 133, 133 n.4, 174, 178, 184, 187, 188, 191, 193
Starvation Cove, 106, 248, 252, 255
Stewart, James Green, xii, 45, 46, 50, 53, 57, 59, 61 n.2, 81, 82, 83, 85, 89, 92, 103, 128, 256; biographical sketch, 38–42; Rae's assessment of, 41; Simpson's assessment of, 41; activities in the Yukon, 40–41; appointed co-leader of the expedition, 42, 51, 55, 56, 83, 190, 192, 194; promoted Chief Factor, 42; given charge of Cumberland House District, 42, 212, 217; death, 42; qualifications to lead expedition 42, 60; overwhelmed by tundra and arctic coast, 42, 129 n.1, 131 n.1, 131 n.5, 136 n.9, 189; misses his wife, 42, 125 n.4; 128 n.8, 136 n.9, 143 n.4, 145 n.4, 147 n.2, 164, 211, 257; instructions to from Sir George Simpson, 46–50, 51–2; to start for Fort Chipewyan on receipt of instructions, 48, 51; promised promotion to Chief Trader, 49, 52; to take sole command of expedition if necessary, 50; travels from Fort Carlton to Fort Chipewyan, 67, 72–3, 187; makes trip from Fort Chipewyan to Fort Resolution and back, 67, 73–5; travels from Fort Chipewyan to Fort Resolution, 69; waits for Anderson at Fort Resolution, 71; receives letters from his wife, 72; reaches Fort Chipewyan, 73, 73 n.3, 182; reprimanded by Anderson for large number of men taken on trip from Fort Chipewyan to Fort Resolution and back, 74 n.1, 92; reaches Fort Resolution on first visit, 74; reaches Fort Chipewyan for second time, 75; reports on trip to Fort Resolution to Simpson, 75–6; provisional plan for expedition, 75; plans to haul canoes on sledges initially, 75; hires two Orkneymen at Fort Chipewyan, 76; lines canoes with canvas, 76, 92; cogitates on possibilities of wintering, 76; rather pessimistic as to success of expedition, 76; writes to William MacTavish, 77; starts from Fort Chipewyan for Fort Resolution, 78, 183, 187; Anderson dubious about his foresight and prudence, 85; Anderson considers promotion premature, 87; compares progress with that of Back, 106; takes observations to determine party's position, 109 n.1, 111, 112, 113, 126 n.7; reports that one Inuit woman had seen a Franklin survivor alive, 130 n.2, 138 n.3, 167, 183–4, 186, 188, 191, 194, 212–13; row with Anderson, 139 n.5; sand cliff named after, 141; never wants to see Back River again, 147 n.2; waits at Fort Reliance for Lockhart, 147, 186; returns to Fort Resolution, 148, 186; starts from Fort Reliance for Fort Resolution, 160; starts from Fort Resolution with dispatches for Lachine, 148 n.3, 161, 162, 177; deviates to Norway House to pick up his wife, 151, 211; reaches Fort Garry, 165, 186; leaves Fort Garry, bound for Montreal, 161; leaves St Paul, 188; reunited with his wife, 165; delivers report to Simpson, 166, 182–6; keeps journal, 166; slept much of the way down the river, according to Anderson, 169, 179, 211; lacking in initiative, according to Anderson, 169; packs up Fort Reliance, 176; postulates that Franklin survivors tried to ascend Back River, 184, 185; regrets that the Company had not undertaken the search five years earlier, 185; reaches Montreal, 186, 191, 192, 194, 198; deposition in

Fairholme case, 167, 194–5; believes there are no survivors from Franklin expedition, 195; severely criticized by Anderson for delaying dispatches by picking up his wife, 212, 213, 214; threatened with expulsion from the Company for drunken brawling, 212, 217–18; criticized by Anderson for paying Laferté same wages as other men 213, 214, 215; official complaint lodged against by Anderson for delaying official dispatches, 214–15; supported by Simpson in the matter of delaying dispatches, 216; official complaint withdrawn at Simpson's suggestion, 212; award of £280 received from British Government for services, 219, 224–8; distance travelled to reach Fort Resolution, 256
Stewart, John, 38, 51, 192
Stewart, Margaret (née Mowat), 42, 165; references to in Stewart's journal, 125 n.4; 128 n.8, 136 n.9, 143 n.4, 145 n.5, 147 n.2, 165; reaches her father's house, 165
Stewart's Rapids, 141 n.5
Stone Fort, *see* Fort Garry, Lower
Stoney Island, 148
Stornoway, 65 n.2
Straits of Magellan, 10
strawberries (*Fragaria glauca*), 95
Stromness, 11 n.1, 44 n.4, 70
Sturgeon Weir River, 161, 164
Sulphur Point, 102 n.1
Sulphur Springs, 102, 162
surgeon, recommended by Anderson, 166 n.1
Sussex, H.R.H Duke of, 118 n.3
Sussex Lake (**VIII**), 27, 118, 118 n.3, 143, 147 n.4, 176, 183
Svalbard, 3
swamp berries, *see* cloudberries
Swaney, member of boat's crew, 97
Swanston, John, 212, 214, 216

Taché, Bishop Alexandre-Antonin, 164, 164 n.2
talc, 118
Taltheilei Narrows (**VI**), 109, 109 n.8, 152, 171
Tal-thel-la, *see* Taltheilei Narrows
Tal-thel-leh Straits, *see* Taltheilei Narrows
tamarack, *see* larch
tar springs, 163; exploited by Hudson's Bay Company, 163 n.5
Tasmania, 3
Tavistock Hotel, 15
Taylor, Parick, alleged to have killed Inuit near Point Ogle, 65 n.2
teal, Green-winged (*Anas carolinensis*), 121, 121 n.3
telegraph, electric, 37
telescopes, provided for expedition, 36
temperature, 93, 111, 116; warmer below Beechey Lake, 123, 123 n.4
Tennadzie, 236
terns, Arctic (*Sterna paradisea*), 95, 95 n.1
Thai-Koh Antetti, *see* Sandy Portage Lake
Thames, River, 1, 5, 123 n.7
Thelon River, xi, 235, 240
The Pas (**IV**), 164 n.6
thermometers, provided for the expedition, 36
The Times (London), 15, 26, 35; publishes Rae's reports, 16, 17, 18–19; publishes Anderson's report, 167; publishes exchange concerning rumour of Indians finding a camp recently occupied by Whites, 234
The Times (St Paul), 167, 188–9
They-gee-yeh-too-ey, *see* Miles Lake
Thin Lake, 164
Thleeychodese, *see* Back River
Thlewee-cho-dezza, *see* Back River
Thle-wee-choh, *see* Back River
Thlewycho, *see* Back River
Thlewee-dezza, *see* Thelon River
Thlewychodese, *see* Back River
Thompson, Alexander, 251
thunder, 95, 98, 102, 191
Thunder Bay, 109 n.7
Timbré, 103
tobacco, to be fetched from Fort Chipewyan, 84
tongs, blacksmith's, 130
Toronto, 11 n.1, 67 n.1
Touchwood Hills, 45, 53
Toura Lake (**XII**), 147 n.3
trade goods, for Inuit, 30, 44, 48, 54, 55, 56, 88
Trafalgar, Battle of, 3
trap rocks, 109, 111
treeline, 87 n.2, 115; about two days' travel from Great Slave Lake, 85
trees, starting to bud, 78; leafing out, 93, 101; becoming scarcer, 113; getting larger, 145
trout, Lake (*Salvelinus namaycush*), 115, 115 n.1, 152, 154; caught by hand, 116
trout, salmon, *see* trout, lake
Trout Fall, 164
Trout Lake, 164
Trout River, 94 n.1

Tyrrell, Joseph Burr, interviews Joseph Boucher, concerning rumour of ship seen off arctic coast, 235, 245; collects depositions from men involved, 246–7; believes that Papanakies did see the masts of a ship, 247–8

Ullapool, 32 n.1
Union Bay, 7
United States, 29; Navy, 7; Mail, 52; Post Office, 61
University of Edinburgh, 11 n.1
Utkuhikalingmiut, 129 n.7, 130 n.1
Uvaliarlit, 124 n.11

Venus, seen for first time, 139
Victoria, HRH Princess, 130 n.6
Victoria, B.C., 43 n.1
Victoria Channel, *see* Victoria Strait
Victoria Headland (**XI**), 106, 130, 131, 137 n.7
Victoria Island (**IX**), 5, 8, 10, 11, 11 n.1, 12, 82 n.1, 86, 86 n.4, 90, 96, 204, 257
Victoria Land, *see* Victoria Island
Victoria Sound, *see* Victoria Strait
Victoria Strait, 12, 62, 63, 86, 90, 96, 101, 168, 184, 192, 193, 204
Victory Point, 1, 2
vocabularies, Inuktitut, provided for expedition, 36
vole, Northern redbacked (*Clethrionomys rutilus*), 128 n.6
voyageurs, 67, 256; recommended by Rae, 52, 53; contract signed by, 66; wages, 66; from Norway House reach Fort Chipewyan, 77; from Red River, 92; high spirits of, 112, 116, 118, 171, 176; paid off at Red River, 161, 186

Wager Bay, 2
wages, 142, 223–4
Wainwright Inlet, 6
Waldron River, 112 n.3
Walker Bay, 10, 82 n.1
Walla Walla, *see* Fort Nex Percés
Warren, Captain Samuel, 123 n.8
Warren's River (**IX**), 123, 123 n.8, 124
Washington, Captain John, 225
water depth, 112; off Pipe Stone Point, 158; off Hoarfrost River, 158
water levels, 122, 123, 123 n.7, 124, 138, 139, 140, 141, 142, 162, 168, 176, 183
Watson Lake, 40
wattap, 84, 84 n.2, 93 n.6
wavies, grey, *see* geese, white-fronted
weather, unusually cold, 105
Wedderburn and Company, 32 n.1
Wedderburn, Andrew, 32 n.1
Wellington Channel 4, 7, 8, 9, 204, 234 n.1, 252
Westminster, 123 n.7
Whalefish Islands, 5
whalers, use stores from Fury Beach depot, 204
Whirlpool Canyon, 40
whisky jack (*Perisoreus canadensis*), 112 n.12, 147, 147 n.5
Whitby (England), 4
Whitby (Ontario), 43 n.2
whitefish, Lake (*Coregonus clupeaformis*), 152, 152 n.3, 153 n.3, 154
White Mud Portage, 163
White Sea, 25
Williamson, Gilbert, 97
Willow Island, 142
willows (*Salix arctica*), 101, 119, 129, 131, 140, 236; no longer available, 143
Winnipeg River District, 45 n.1
Winter Harbour, 1, 8
Winter Island, 3
Wolstenholme Sound, 6
Wolfe Rapids, *see* Wolf's Rapids
Wolf's Rapids, 128, 138
wolf skins, 173
Wolf Straits, 157
wolverines (*Gulo gulo*), 97
wolves (*Canis lupus*), 118, 119, 128, 128 n.7, 131, 141; lying in wait for swimming caribou, 123, 138, 141; stealing pemmican, 137, 137 n.10; raid cache, 141; swimming across East Arm, 157; may have eaten bodies of Franklin expedition members, 193, 195
Woodman, David C., analyses Hall's notebooks, 255; rules out possibility that Papanakies could have seen either *Erebus* or *Terror*, 236; inaccurate assessment of the Anderson/Stewart expedition, 256
Woolwich Barracks, 65 n.2
Wright, Noel, recognizes difficulty of finding an Inuktitut interpreter, 256

Yellowknife Chief, 75
Yellowknife Indians, 74 n.5, 78, 85, 85 n.1, 103,

107 n.7, 108, 119, 171, 172, 183; hunt for the expedition from Fort Reliance and on Back River, 75, 85, 90, 92; inhospitable, 152; hired to hunt by Lockhart, 154, 56; waiting at Fort Reliance, 157; hired by Lockhart to go to head of Back River, 158; reach Fort Reliance from Back River, 158; report finding camp on the Back River recently occupied by Whites, 235, 236–8; report seeing fires in distance on Back River, 237–8; report killing muskoxen and finding musket balls in them, 237

York Factory (**IV**), xii, 11 n.1, 12, 13, 15, 16, 19, 28, 32 n.1, 43 n.2, 43 n.3, 44, 44 n.4, 48, 51, 53, 54, 56, 56 n.3, 57, 67 n.1, 69, 76 n.2, 77, 79, 80, 91, 236

York Factory District, 44 n.3

Youcon, *see* Yukon

Yukon, 37, 40, 60 n.3, 86

Yukon River, *see* Lewes River